RELIABILITY AND RISK ANALYSIS

Methods and Nuclear Power Applications

RELIABILITY AND RISK ANALYSIS

Methods and Nuclear Power Applications

NORMAN J. McCORMICK

Department of Nuclear Engineering
University of Washington
Seattle, Washington

ACADEMIC PRESS

A Subsidiary of Harcourt Brace Jovanovich, Publishers

New York London Toronto Sydney San Francisco 1981

ACADEMIC PRESS, INC.
111 Fifth Avenue, New York, New York 10003

United Kingdom Edition published by
ACADEMIC PRESS, INC. (LONDON) LTD.
24/28 Oval Road, London NW1 7DX

Library of Congress Cataloging in Publication Data

McCormick, N. J. (Norman J.)
Reliability and risk analysis.

Includes bibliographical references and index.
1. Nuclear facilities--Reliability. 2. Nuclear facilities--Accidents. 3. Risk--Statistical methods. I. Title.
TK9153.M33 621.48'32 81-2758
ISBN 0-12-482360-2 AACR2

PRINTED IN THE UNITED STATES OF AMERICA

81 82 83 84 9 8 7 6 5 4 3 2 1

Contents

PART III
OTHER RISK ASSESSMENTS

Preface

The analysis of risk to society from nuclear power and other technologies will be important in the 1980s. In order to quantify the probability per unit time that an undesirable event may occur, the fundamentals of reliability engineering are needed to quantify the expected frequency with which these events might be initiated. For this reason the concepts of reliability and risk analysis are linked together in this book, although reliability analysis is important in its own right for the study of power plant availability.

A practical itinerary is provided for senior-level or first-year graduate students who want a survey of some fundamentals of, and principal results from, reliability and risk studies primarily for nuclear power applications. The book is oriented toward both engineering students and professionals.

In Part I, basic concepts in reliability engineering are considered. Manipulations with probability distributions are briefly covered, including those most frequently encountered in reliability and risk studies. An introduction to data manipulation concepts is given, as well as a discussion of failure data sources. An analysis of the reliability and availability of simple systems, comprised of linked components, is presented. Methods to assess risk (fault tree analysis and event tree analysis) are discussed, followed by an overview of computer programs used in fault tree analysis.

Part II begins with an introduction to the definitions and procedures used to conduct risk analyses. Next, societal risks from nuclear power are considered, including those risks from light water reactors, liquid metal fast breeder reactors, and high temperature gas reactors, as well as the transportation and disposal of nuclear wastes.

In Part III other risk assessments are discussed, including some risk comparisons of nuclear versus nonnuclear systems. An introduction to risk–benefit analyses gives an example of using risk studies to allocate money to improve safety systems. Finally, the subject of the public's perception and acceptance of risk is addressed.

A prior knowledge of probability theory would be helpful for the material in Part I; likewise, a previous introduction to the engineered safety features of a nuclear reactor makes portions of Part II easier to understand. For those without this background, introductory material is provided in Chapter 2 and the appendixes.

The text has envolved from lecture notes distributed to University of Washington students over the past several years, and it is to them that the book is dedicated. Many of their comments (and especially those of Robert Colley) have resulted in improvements to the text; undoubtedly other changes could be made. To continue refining and polishing a manuscript for too long is like trying to prevent a grown child from leaving home: Once you have done the best you can with the time available, it is best to let your creation go.

I would like to thank Robert Erdmann for encouraging me to undertake this project. A particular acknowledgment and note of gratitude is due to LaDonna Kennedy, who transferred several drafts of illegible script into readable form.

July 1980

CHAPTER 1

Introduction

1-1 WHY STUDY RISK ANALYSIS?

In the latter half of the 1970s, the subject of risk analysis began to take on increased importance. We live in an era when the life expectancy of the average individual is increasing, in part because of improved, large-scale technologies such as nuclear power. Yet many people have a nagging concern that improvements in safety could be made. Risk analysis, which is a subset of safety analysis, requires consideration of the probability of an accident occurring and its subsequent consequences.

Risk involves the occurrence or potential occurrence of some accident consisting of an event or sequence of events. The event is defined by its complete description, such as a motor vehicle accident occurring under given circumstances. For each event that occurs, there are several types of ensuing consequences. For example, in a motor vehicle accident, the consequences may include death, a variety of injuries, property damage, and other possible effects. We frequently measure risk, as we shall see later, by the frequency of an accident and the damage measured in the number of lives or dollars lost.

In this book, we shall be interested in what might be termed "societal" or "collective" risk analysis, as opposed to "individual" risk analysis. That is, we wish to explore methods used, and some of the results already obtained, in the analysis of potentially large-scale accidents arising from the operation of nuclear power systems (which includes the fuel supply and waste disposal, as well as reactor operation). Thus we are concerned with the public health risk from people's exposure to ionizing radiation that results from acute releases of radioactive substances arising from a failure or malfunction in one step of the nuclear fuel cycle.

In answer to the question "Why perform risk analysis?" at least two reasons easily come to mind. First, as engineers, we have a moral obligation to design the safest possible system (with the lowest probability of system failure and minimal consequences if the system does fail) within the given set of engineering constraints (such as overall cost, available materials, available personnel and equipment, etc.). By performing a risk analysis, we may obtain sufficient information about the system to redesign it and lower the probability of the occurrence of an accident or mitigate the ensuing consequences. Alternatively, it may be possible to show that the probability of occurrence is negligibly small.

Second, we perform a risk analysis of an engineering system to obtain information useful in a comparative risk assessment. One example is the risks of the competing technologies of coal and nuclear power to produce the same end product—electricity; another example is a nuclear reactor and an oil refinery, which produce energy that may have complementary end uses. Risk analyses performed on competing technologies may help influence a choice of one energy source over another. Risk analyses on complementary technologies may help in the design of a balanced set of safety regulations from one industry to another and result in decreased risk to the public from all energy sources. (Of course, one must be wary of overregulation against risks, which could lead to "risk feedback" caused by the resulting attitude of engineers to do as little as necessary to mitigate risks.)

Anyone familiar with the nuclear power industry recognizes a compelling reason for performing risk analyses: there are certain governmental regulations mandating such evaluations. Thus there is an incentive for some students to study risk analysis methods and example analyses so that they can better educate themselves for future employment.

1-2 AN OVERVIEW OF RISK ANALYSIS

Three general aspects of risk analysis are discussed in this book: the methods used to analyze accident initiation and propagation, risk analyses for nuclear power applications, and other risk assessments.

1-2-1 METHODS

Accident initiation is concerned with the probability of an event occurring. If the initiation is nondeliberate, such as a failure arising because of some natural catastrophe or from normally occurring phenomena such as wear-out due to abrasion, then the problem is one of safety engineering, and principles of reliability engineering are required to study the initiation phase. Reliability engineering encompasses (1) the causes of a system component failure, and the probability of such a failure over a period of time and (2) the way in which the probability of failure of a component of a device influences the probability of failure of the device itself. In its broadest interpretation, reliability engineering encompasses the analysis of reliability, availability, and maintainability (sometimes abbreviated as RAM). Because the study of risk analysis is so integrally tied to that of reliability engineering, and because of the importance of reliability engineering concepts to the operation and maintenance of nuclear facilities, these two topics have been combined in this book.

If the initiation is deliberate (or "man-made"), the problem is one of safeguards engineering; then threat assessments, involving social, economic, and political considerations, are required to study the initiation phase. Safeguards evaluations include both "overt" attacks on a component or system, and "covert" attacks, the distinction being whether the initiating attack group is part of the system. Safeguards will not be considered directly in this book, but many of the principles of risk analysis have applications to safeguards as well.

Chapters 2 through 7 summarize some of the more important concepts from the first few weeks of a course in reliability engineering as typically taught on most university campuses in mechanical or industrial engineering departments. Basic manipulations with probability functions are briefly covered in Chapter 2. Surprisingly, many engineers seem to never have received such an introduction; others may find it to be merely a review. The basic definitions of reliability engineering to be used in the successive chapters are introduced at the end of this chapter.

Chapters 3 through 5 are devoted to the reliability of a single device or component. Chapter 3 contains some probability distributions encountered most frequently in reliability and risk studies. Also included is a brief introduction to extreme-value distributions; if nothing else is accomplished by this latter material, it may alert the reader to seek out appropriate mathematics references to learn more about the probability distribution that (in principle) well describes the occurrence of earthquakes of a magnitude less than some maximum (i.e., extreme) value on the Richter scale.

An introductory overview of data manipulation concepts is presented in Chapter 4. Its purpose is to give the reader a beginning cookbook of recipes necessary to collect and assemble data into a useful format for reliability engineering calculations. Whenever possible, however, it is desirable to use updated versions of data provided by some of the data sources listed in Chapter 5. An appendix contains a brief listing of demand failure probabilities for discrete-time operations and hazard (or instantaneous failure) rates for continuously operating devices.

Chapters 6 through 8 treat the reliability of linked components comprising a system. The laws of probability are used in Chapter 6 to combine units in series and parallel to obtain the reliability of simple systems. Generalizations of these relations to cover other pedagogical examples from reliability engineering are also contained in Chapter 6, and the reader more interested in "applications" than "methods" may wish to skip over the rest of this chapter.

In Chapter 7 we examine the simplest approach to the analysis of the availability of a system when one or more components can be repaired. This introduction to methods developed for incorporating the effects of maintenance of systems permits a comparison of the reliability of a system without repair to its availability, i.e., reliability with repair.

Fault tree analysis is the most widely used technique for analyzing the failure probability of complex systems. The basic concepts of fault tree analysis are treated in Chapter 8, along with simple examples.

Chapter 9 covers the basic concepts of event trees, which are a means of ordering the various stages in the propagation of an accident. Some examples from the nuclear power industry are included in this chapter as well. Chapter 10 completes the material on the methods used in reliability and risk analyses and covers computer programs used in fault tree analysis.

Detailed analysis of the different stages of a system when a major accident unfolds typically involves standard engineering principles such as heat, mass, and momentum transport, strength of materials, and electronics and electrical circuit theory. This subject matter is what the average student has already spent years learning, so these topics will be omitted as a part of the methods of risk analysis. Likewise, the fundamentals of propagation of radionuclides in air, water, and the ground are omitted since most of this material is normally treated in other basic nuclear engineering courses.

1-2-2 Nuclear Power Risks

This section begins with an introduction in Chapter 11 to the definitions and procedures used to conduct risk analyses. As might be expected from the preceding discussion, both event tree and fault tree analyses typically play an important role.

Chapter 12 provides a summary of the Reactor Safety Study, also called the Rasmussen Report or WASH-1400. This study is a generic (and not site-specific) investigation of risks from operation of a boiling water reactor or a pressurized water reactor. Although the study is now somewhat dated (it was published in 1975 and is based on nuclear plants that were operating in 1972), nonetheless, it is still of major importance to engineers interested in risks from nuclear power since the study established the methods used in later investigations, and it is still the most comprehensive one to date. Furthermore, the results of WASH-1400 serve as a "benchmark" against which to compare the results of other risk analyses. Chapter 13 is devoted to a summary of risks from two other types of reactors: the liquid metal fast breeder and the high-temperature gas reactors.

Besides the nuclear reactor, there is a "front-end" and a "back-end" of the nuclear fuel cycle. All parts of the cycle are linked by a transportation system, and the risks from material transport are considered in Chapter 14. The risks from the mining of uranium dioxide are really not "societal," but rather individual risks; therefore they are not covered in a separate chapter.

In Chapter 15, the risks from nuclear waste storage systems are briefly considered. The analysis of such risks is complicated by the fact that very long-range projections of system performance are required because of the long half-lives of a few of the fission products; thus there are larger uncertainties associated with the predictions of risks. Furthermore, each study must be done for a specific site because of the importance of geological considerations.

1-2-3 Other Risk Assessments

Certainly nuclear power applications are not the only ones that create risk to the general public. Chapter 16 covers some risk comparisons of nuclear systems with competing and complementary nonnuclear energy systems. Chapter 17 contains an introduction to risk-benefit analyses and gives an example of the possible use of such studies for deciding how to best allocate a finite amount of money to improve safety systems that will reduce risk; a few nonnuclear examples also are provided.

Finally, Chapter 18 provides an introduction to the important, and not yet well developed, area of risk acceptance analysis. Certainly any engineer or scientist who has tried to argue, with a certain segment of the general public, the relative safety of one engineering system over another knows that certain people are not convinced by facts and statistics. Many risk analysts now feel there is a pressing need to obtain a better understanding of the perception of risks and their acceptance; we hope such knowledge can be used to convince the public that financial allocations to reduce risks should be made proportionally over all industries. A few closing comments about this subject are included in Chapter 19.

GENERAL REFERENCES

References specific to each chapter are provided at the end of each. In this section are summarized the more comprehensive references that the reader may wish to consult. An asterisk is used to point out those references of special interest to students.

Reliability Engineering

B. L. Amstadter, "Reliability Mathematics." McGraw-Hill, New York, 1971.

G. E. Apostolakis, Mathematical Methods of Probabilistic Safety Analysis. Univ. of California at Los Angeles, Rep. UCLA-ENG-7464 (1974).

R. E. Barlow and F. Proschan, "Mathematical Theory of Reliability." Wiley, New York, 1965.

I. Bazovsky, "Reliability Theory and Practice." Prentice-Hall, Englewood Cliffs, New Jersey, 1961.

R. Billinton, R. J. Ringlee, and A. J. Wood, "Power-System Reliability Calculations." MIT Press, Cambridge, Massachusetts, 1973.

* A. J. Bourne and A. E. Green, "Reliability Technology." Wiley (Interscience), New York, 1972.

Conf. Reliability and Fault Tree Anal. Univ. of California at Berkeley, 1974.

* J. B. Fussell and J. S. Arendt, System reliability engineering methodology: A discussion of the state of the art, *Nucl. Safety* **20,** 541 (1979).

* G. J. Hahn and S. S. Shapiro, "Statistical Models in Engineering." Wiley, New York, 1967.

W. Hammer, "Handbook of System and Product Safety." Prentice-Hall, Englewood Cliffs, New Jersey, 1972.

IEEE Std 352, General Principles of Reliability Analysis of Nuclear Power Generating Station Protection Systems (1975).

IEEE Trans. Reliability, recent issues of the monthly journal.

A. K. S. Jardine, "Maintenance, Replacement, and Reliability." Halsted, Wiley, New York, 1973.

A. Kaufmann, D. Grouchko, and R. Cruon, "Mathematical Models for the Study of the Reliability of Systems." Academic Press, New York, 1977.

N. R. Mann, R. E. Schafer, and N. D. Singpurwalla, "Methods for Statistical Analysis of Reliability and Life Data." Wiley, New York, 1974.

Microelec. and Reliability, recent issues of the bi-montly journal.

* *Reliability and Maintainability Symp.,* Annual Proceedings.

N. H. Roberts, "Mathematical Methods in Reliability Engineering." McGraw-Hill, New York, 1964 (out of print).

* W. E. Vesely, F. F. Goldberg, N. H. Roberts, and D. F. Haasl, Fault Tree Handbook U.S. Nuclear Regulatory Commission Rep. NUREG-0492 (1981).

* M. L. Shooman, "Probabilistic Reliability: An Engineering Approach." McGraw-Hill, New York, 1968.

C. O. Smith, "Introduction to Reliability in Design." McGraw-Hill, New York, 1976.

Risk Analysis for Nuclear Power Systems

A. W. Bartsell, V. Joksimovic, and F. A. Silady, An assessment of HTGR accident consequences, *Nucl. Safety* **18,** 761 (1977); see also HTGR Accident Initiation and Progression

Analysis Status Rep., Energy Research and Development Administration Rep. GA-A-13617 (volumes issued from 1975 to 1977).

* J. B. Fussell, and G. R. Burdick (eds.), "Nuclear Systems Reliability Engineering and Risk Assessment." Society for Industrial and Applied Mathematics, Philadelphia, Pennsylvania, 1977.

Nucl. Safety, recent issues of the bi-monthly journal.

H. B. Piper, L. L. Conradi, A. R. Buhl, P. J. Wood, and D. E. W. Leaver, Clinch River breeder reactor plant safety study, *Nucl. Safety* **19**, 316 (1978); see also Clinch River Breeder Reactor Plant (CRBRP) Safety Study—Assessment of Accident Risks in CRBR, Rep. CRBRP-1, Vols. 1 and 2, 1977.

Probabilistic analysis of nuclear reactor safety, *Proc. ANS Topical Meeting, May 8–10*. American Nuclear Society, LaGrange Park, Illinois, 1978.

* Reactor Safety Study—An Assessment of Accident Risks in U.S. Commercial Nuclear Power Plants. Nuclear Regulatory Commission Rep. WASH-1400 (NUREG 75/014) (1975).

Reliability of Nuclear Power Plants, *IAEA Symp. Innsbruck, April 14–18*. International Atomic Energy Agency (1975).

Risks Associated With Nuclear Power: A Critical Review of the Literature, Summary and Synthesis Chapter. National Academy of Sciences (1979).

Other Risk Assessments and Concepts

H. Ashley, R. L. Rudman, and C. G. Whipple (eds.), "Energy and the Environment—A Risk Benefit Approach." Pergamon, Oxford, 1976.

* "CANVEY, An Investigation of Potential Hazards from Operations in the Canvey Island/Thurrock Area." Her Majesty's Stationery Office, London, 1978.

Comparative Risk-Cost-Benefit Study of Alternative Sources of Electrical Energy. Atomic Energy Commission Rep. WASH-1224 (1974).

* H. Inhaber, "Risk of Energy Production," Atomic Energy Control Board Rep. AECB-1119/Rev 3, Ottawa, Ontario, 1980.

E. W. Lawless, "Technology and Social Shock." Rutgers Univ. Press, New Jersey, 1977.

W. W. Lowrance, "Of Acceptable Risk." William Kaufman, Los Altos, California, 1976.

* E. P. O'Donnell and J. J. Mauro, A cost-benefit comparison of nuclear and nonnuclear health and safety protective measures and regulations, *Nucl. Safety* **20**, 525 (1979).

* D. Okrent, Risk-benefit evaluation for large technological systems, *Nucl. Safety* **20**, 148 (1979); see also, A General Evaluation Approach to Risk-Benefit for Large Technological Systems and Its Application to Nuclear Power. Univ. of California at Los Angeles Rep. UCLA-ENG-7777 (1978).

D. Okrent (ed.), Risk-Benefit Methodology and Application: Some Papers presented at the *Eng. Foundation Workshop, September 22–25, Asilomar, California*. Univ. of California at Los Angeles, UCLA-ENG-7598 (December 1975).

W. D. Rowe, "An Anatomy of Risk." Wiley, New York, 1977.

* C. Starr, R. Rudman, and C. Whipple, Philosophical basis for risk analysis, "Annual Review of Energy," No. 1, pp. 629–662, Annual Reviews, Palo Alto, California, 1976.

M. M. R. Williams (ed.), "Nuclear Safety." Pergamon, Oxford, 1979.

Breipohl, Arthur - Probabilistic Reliability McGraw-Hill

NUREG 0492 - Fault Tree Analysis - NRC Gov't Printing Ofc.

PART I

Methods

CHAPTER 2

Probability Concepts

2-1 INTERPRETATIONS AND LAWS OF PROBABILITY

2-1-1 INTERPRETATIONS

The two most common interpretations of probability are the relative frequency approach and the axiomatic or subjective approach. The relative frequency interpretation requires that a sample space S be defined, with event A being a member of S. If event A occurs X number of times out of a number n of repeated experiments whose outcomes are described by S, then the probability $P(A)$ of the outcome of event A is defined as

$$P(A) = \lim_{n \to \infty} (X/n). \tag{2-1}$$

For fixed n, the quantity X/n is the relative frequency of occurrence of A.

Since it is impossible to actually conduct an infinite number of trials so that $n \to \infty$, usually $P(A)$ is just approximated by (X/n). The law of large numbers and the central limit theorem [1–4] provide a justification that improved estimates of $P(A)$ will be obtained by increasing n.

The interpretation of Eq. (2-1) is clear enough for experiments that can be repeated. There are many occasions in which the knowledge available is less precise, especially when the engineer deals with rarely occurring events that form the basis of many risk evaluations. Then it is necessary to resort to the axiomatic or subjective approach to the concept of probability, which we shall use from now on.

The axiomatic interpretation begins with the broad view that probability is nothing more than a measure of uncertainty about the likelihood of an event. Stated more precisely, "a probability assignment is a numerical encoding of a state of knowledge" [5, 6]. Beginning with such a broad definition means that we shall have to impose some constraints before obtaining something that can be used in quantitative analyses. Examples of several kinds of knowledge are [5]:

(a) *Symmetry* Sometimes it is known that a system is symmetrical, as in the case of honest dice or coins. For example, even if an experiment consisting of 100 flips of a coin gave 56 heads and 44 tails, the probability of the event that heads will appear would be assigned a probability of 0.5 because it would be believed that an insufficient number of flips had been performed to give the outcome 0.5.

(b) *Frequencies* Sometimes historical data concerning a system are known, e.g., how many years the annual rainfall in a particular area exceeds a given amount. Even if the rainfall in a given year exceeds the expected amount, the frequency of occurrence in past years would be used in making predictions about rainfall in future years—unless there were reason to believe that a meteorological change had occurred.

(c) *Averages* Sometimes the average result of what has occurred in the past is known. For example, the average annual rainfall at a particular location may be known, so this would be used as an estimate of the expected amount of rainfall in the next year, again, provided no meteorological changes occur.

With the axiomatic approach, the analyst must at least assign probabilities in a "coherent" manner, which requires that such probabilities obey the axioms and laws of probability [7]. The axiomatic approach is formally developed in a deductive way by the use of three axioms [5]. The first axiom is that the probability for the outcome of event A is a number between zero and unity,

$$0 \leq P(A) \leq 1. \tag{2-2}$$

The probability of a certain event is assigned the value of unity, and the probability of an impossible event is zero.

The second axiom deals with two mutually exclusive events, A and the complement of A, "NOT A"; the latter event can be denoted as event $\bar{A}$. The *addition rule for probabilities* then states that

$$P(A) + P(\bar{A}) = 1, \tag{2-3}$$

a result that follows since either A or $\bar{A}$ is certain to occur.

2-1-2 Intersection of Events

It is natural to talk about the intersection of two events, A_1 and A_2, which is sometimes denoted as $A_1 \cap A_2$ or more simply A_1A_2. (Note that this latter symbol is still "A_1 AND A_2," *not* "A_1 times A_2.") The *product rule for probabilities* is the third and last axiom and may be stated as

$$P(A_1A_2) = P(A_1 | A_2)P(A_2) = P(A_2 | A_1)P(A_1). \tag{2-4}$$

In Eq. (2-4), the concept of a "conditional probability" $P(A_1 \mid A_2)$ has been introduced and is defined as the probability of event A_1 given that event A_2 has occurred. In the special case that events A_1 and A_2 are independent, so that the probability that event A_1 occurs is independent of the occurrence of event A_2, then $P(A_1 \mid A_2) = P(A_1)$ and $P(A_1A_2) = P(A_1)P(A_2)$. A second special case occurs if events A_1 and A_2 are mutually exclusive (i.e., "disjoint"), in which case $P(A_1|A_2) = 0$ and $P(A_1A_2) = 0$.

The results for the product rule may be easily extended to cover more than two events. If there are N events $A_1, A_2, \cdots, A_N$, then in general

$$P(A_1A_2 \cdots A_N) = P(A_1)P(A_2|A_1) \cdots P(A_N|A_1A_2 \cdots A_{N-1}). \quad (2\text{-}5)$$

If the events are independent, then

$$P(A_1A_2 \cdots A_N) = P(A_1)P(A_2) \cdots P(A_N), \quad (2\text{-}6)$$

whereas if the events are mutually exclusive

$$P(A_1A_2 \cdots A_N) = 0. \quad (2\text{-}7)$$

For the probabilities in risk analysis, often only an upper bound of $P(A_1A_2 \cdots A_N)$ is needed. For example, in Eq. (2-4), since all probabilities and conditional probabilities must satisfy Eq. (2-2), it follows that $P(A_1A_2)$ has an upper bound given by the minimum of $P(A_1)$ or $P(A_2)$,

$$P(A_1A_2) \leq \min[P(A_1), P(A_2)]. \quad (2\text{-}8)$$

Equation (2-8) can be extended to treat any number of events, so that

$$P(A_1A_2 \cdots A_N) \leq \min[P(A_1), P(A_2), \cdots, P(A_N)]. \quad (2\text{-}9)$$

It also is possible to obtain an estimate for the probability of three events in terms of the probabilities of pairs of events, as given by

$$P(A_1A_2A_3) \leq \min[P(A_1A_2), P(A_1A_3), P(A_2A_3)]. \quad (2\text{-}10)$$

This equation follows by a procedure similar to that used to obtain Eq. (2-8) and may be termed a *double-event bound*. Thus it follows, for example, that a *triple-event bound* would be

$$P(A_1A_2A_3 \cdots A_N) \leq \min[\text{probabilities of all triple combinations}], \quad N \geq 4. \quad (2\text{-}11)$$

2-1-3 Union of Events

Another type of event is the union of two events A_1 and A_2. This is sometimes denoted as $A_1 \cup A_2$, or it may be denoted as $A_1 + A_2$; both symbols mean "or" so that $A_1 + A_2$ means "A_1 OR A_2." It should be remembered that two events cannot be added together; only probabilities or other numbers can be added.

The general equation expressing the probability of the union of two events, $A_1 + A_2$, is

$$P(A_1 + A_2) = P(A_1) + P(A_2) - P(A_1A_2). \tag{2-12}$$

The right-hand side may be interpreted as the sum of the probabilities of the two events considered independently, with the third term to eliminate the possible double counting arising from the "overlap" caused by the intersection of the two events. Of course, if the two events are independent, then from Eqs. (2-6) and (2-12) it follows that

$$P(A_1 + A_2) = P(A_1) + P(A_2) - P(A_1)P(A_2). \tag{2-13}$$

On the other hand, if the two events are mutually exclusive, then

$$P(A_1 + A_2) = P(A_1) + P(A_2). \tag{2-14}$$

The preceding equations can be generalized to the case of more than two events, where in general

$$P(A_1 + A_2 + \cdots + A_N) = \sum_{n=1}^{N} P(A_n) - \sum_{n=1}^{N-1} \sum_{m=n+1}^{N} P(A_nA_m) + - \cdots + (-1)^{N-1}P(A_1A_2 \cdots A_N). \tag{2-15}$$

The rth term on the right-hand side of Eq. (2-15) contains

$$\binom{N}{r} = \frac{N!}{r!(N-r)!}$$

probabilities for all possible combinations of the N events A_n considered r at a time. From Eq. (2-7) it follows that if the events are all mutually exclusive, then only the first term on the right-hand side of Eq. (2-15) is nonvanishing. If all the events are independent, then the form of the equation can be improved by collecting terms and rearranging to obtain a product of factors on the right-hand side,

$$1 - P(A_1 + A_2 + \cdots + A_N) = \prod_{n=1}^{N} [1 - P(A_n)]. \tag{2-16}$$

The alternating signs in the series in Eq. (2-15) immediately suggest the bounds for $P(A_1 + A_2 + \cdots + A_N)$. Indeed, it may be shown that

$$P(A_1 + A_2 + \cdots + A_N) \leq \sum_{n=1}^{N} P(A_n), \tag{2-17}$$

$$P(A_1 + A_2 + \cdots + A_N) \geq \sum_{n=1}^{N} P(A_n) - \sum_{n=1}^{N-1} \sum_{m=n+1}^{N} P(A_nA_m). \tag{2-18}$$

2-1-4 Decomposition of an Event

Frequently it is useful to analyze the probability of occurrence of an event A_1 by using the conditional probabilities $P(A_1|A_2)$ and $P(A_1|\bar{A}_2)$. The decomposition rule

$$P(A_1) = P(A_1|A_2)\,P(A_2) + P(A_1|\bar{A}_2)\,P(\bar{A}_2) \tag{2-19}$$

follows from Eqs. (2-4) and (2-14); the first term on the right-hand side is $P(A_1)$ if A_2 occurs, while the second is $P(A_1)$ if $\bar{A}_2$ occurs. Equation (2-19) can be generalized to provide a more detailed decomposition, if necessary; e.g.,

$$\begin{aligned} P(A_1) = {} & P(A_1|A_2A_3)\,P(A_2A_3) + P(A_1|A_2\bar{A}_3)\,P(A_2\bar{A}_3) \\ & + P(A_1|\bar{A}_2A_3)\,P(\bar{A}_2A_3) + P(A_1|\bar{A}_2\bar{A}_3)\,P(\bar{A}_2\bar{A}_3). \end{aligned} \tag{2-20}$$

The number of terms when $P(A_1)$ is expressed in terms of conditional probabilities $P(A_1|A_2A_3 \cdots A_N)$, etc., is 2^{N-1}.

If the events $A_2A_3 \cdots A_N$ are all independent, then Eq. (2-7) may be used. For example, Eq. (2-20) then becomes

$$\begin{aligned} P(A_1) = {} & P(A_1|A_2A_3)\,P(A_2)P(A_3) + P(A_1|A_2\bar{A}_3)\,P(A_2)P(\bar{A}_3) \\ & + P(A_1|\bar{A}_2A_3)\,P(\bar{A}_2)P(A_3) + P(A_1|\bar{A}_2\bar{A}_3)\,P(\bar{A}_2)P(\bar{A}_3). \end{aligned} \tag{2-21}$$

2-1-5 Use of the Equations

Equations (2-4) through (2-21) are a terse listing of the calculus of probabilities needed in reliability analyses of systems with more than one component [8, 9]. They are presented here for reference purposes for the student who has not already encountered them. We shall defer their application until needed in the study of reliability engineering.

It is important to note that all the theorems of probabilities also hold for conditional probabilities; e.g.,

$$P(A_1 + A_2|B) = P(A_1|B) + P(A_2|B) - P(A_1A_2|B) \tag{2-22}$$

is the form analogous to Eq. (2-12). Indeed, some authors [5, 6] insist that all probabilities are conditional in the sense that they are based upon certain hypotheses or assumptions H about a system. In such cases, for example, Eqs. (2-12) and (2-22) would appear as

$$P(A_1 + A_2|H) = P(A_1|H) + P(A_2|H) - P(A_1A_2|H) \tag{2-23}$$

$$P(A_1 + A_2|BH) = P(A_1|BH) + P(A_2|BH) - P(A_1A_2|BH). \tag{2-24}$$

The use of such a convention does serve to remind the risk analyst to check that the operating environment for the device is the same as that for which the failure probability data have been generated.

Before we leave this section, it is appropriate to discuss a special case that arises very often in risk analyses: the evaluation of the occurrence of independent and highly infrequent (or rare) events. In such cases, the probability of two or more events is small compared to the probability for any single event. Then from Eq. (2-15),

$$P(A_1 + A_2 + \cdots + A_N) \approx \sum_{n=1}^{N} P(A_n). \tag{2-25}$$

Also Eq. (2-6) remains applicable, so

$$P(A_1A_2 \cdots A_N) = P(A_1)P(A_2) \cdots P(A_N). \tag{2-26}$$

These two equations comprise the *rare-events approximation*.

To summarize:

1. Probability is a numerical encoding of a state of knowledge.
2. There are three fundamental axioms of probability: the range of values is to be in the interval [0, 1], the addition rule, and the product rule.
3. A probability is either conditional or unconditional.
4. The general probability expressions are simplified if the events are independent or mutually exclusive.
5. Equations for bounding a probability are available.

2-2 THE BAYES EQUATION

A basic result for conditional probabilities follows by rewriting Eq. (2-4), the product axiom for probabilities:

$$\begin{aligned} P(A_nB) &= P(A_n)\, P(B|A_n) \\ &= P(B)\, P(A_n|B). \end{aligned} \tag{2-27}$$

Here A_n denotes the nth of N mutually exclusive hypotheses or events, $n = 1, 2, \cdots, N$, while B is some other hypothesis or event. To calculate the probability for event A_n that incorporates the additional evidence from the occurrence of B, the two equations (2-27) are solved to give

$$P(A_n|B) = P(A_n)\left[\frac{P(B|A_n)}{P(B)}\right]. \tag{2-28}$$

This equation is one form of the Bayes equation: the left-hand side of the equation represents the posterior probability of A_n when B is known, while the first factor on the right-hand side is the prior probability of A_n and the second factor represents the relative change in the probability of A_n when B becomes known.

From the addition axiom (2-3), it follows that for mutually exclusive events

$$\sum_{n=1}^{N} P(A_n|B) = 1. \tag{2-29}$$

Therefore if this equation is multiplied by $P(B)$, then

$$P(B) = \sum_{n=1}^{N} P(B)P(A_n|B)$$

$$= \sum_{n=1}^{N} P(A_nB), \tag{2-30}$$

where the product axiom has been used in deriving the second form of the equation. Applying the product rule again results in still another way of expressing $P(B)$ as

$$P(B) = \sum_{n=1}^{N} P(B|A_n)\, P(A_n). \tag{2-31}$$

This equation is the *extension rule* of probabilities and allows $P(B)$ to be expressed in terms of the previously known probabilities $P(A_n)$ and all the conditional probabilities $P(B|A_n)$. Finally, substitution of Eq. (2-31) in Eq. (2-28) gives the final form for the Bayes equation:

$$P(A_n|B) = [P(A_n)P(B|A_n)] \Big/ \left[\sum_{m=1}^{N} P(A_m)P(B|A_m)\right], \qquad n = 1 \text{ to } N. \tag{2-32}$$

The high degree of symmetry in the last equation shows that once the entire set of probabilities $P(B|A_n)$ becomes known, then the calculation of the posterior $P(A_n|B)$ becomes straightforward.

The Bayes equation is an important tool, when one specifies by the subjective approach the possibility of rarely occurring events, because it enables one to "reverse" the order of information gathering about a failure process. That is, in some instances it may be more convenient to do a sampling of event B, given each of the events A_n, rather than the other way around; Eq. (2-32) then shows how the desired probabilities $P(A_n|B)$ are obtained.

Example 2-1 We consider an elementary nuclear reactor core monitoring system (CMS) that consists of an uncompensated ionization chamber (IC), a temperature sensor (TS), and a pressure sensor (PS). From the manufacturers' data, these subsystems are known to have a probability of failure over the time period of interest and at the reactor operating conditions of 0.02, 0.04, and 0.01, respectively. All three systems must work for the core monitoring system to be considered functional. After installation of the units and reactor operation, however, suppose one observes that the monitoring system is not good and presumes that this is because one of the three subsystems has failed.

To determine the probability that it was the TS that failed during reactor operation, we initiate a testing program. Suppose we learn that when IC fails, the CMS fails with probability 0.10; when TS fails, the CMS fails with probability 0.15; and when PS fails, the CMS fails with probability 0.10.

The solution is straightforward since it is known from the data that $P(\text{IC}) = 0.02, P(\text{TS}) = 0.04$, and $P(\text{PS}) = 0.01$; from the testing program, we recognize that $P(\text{CMS}|\text{IC}) = 0.10$, $P(\text{CMS}|\text{TS}) = 0.15$, and $P(\text{CMS}|\text{PS}) = 0.10$. Use of Eq. (2-32) now gives

$$P(\text{TS})|\text{CMS})$$

$$= [0.04(0.15)]/[(0.02)(0.10) + 0.04(0.15) + 0.01(0.10)] = 0.667. \quad \diamond$$

The Bayes equation (2-32) is often used as a means of revising failure data. When applied in this way, it serves as an important link between axiomatic probability and relative frequency probability: the data tend to modify axiomatic probabilities $P(A_n)$ to yield posterior probabilities $P(A_n|B)$ closer to the relative frequency.

If *nothing* is known about the $P(A_n)$, $n = 1$ to N, in Eq. (2-32) then the *principle of insufficient reason* requires that we pick equal probabilities for each event A_n, i.e., that we select the uniform prior distribution $P(A_n) = 1/N$. In such a case, the prior probability distribution is diffuse and the posterior probabilities $P(A_n|B)$ tend to be more influenced by data in the form of $P(B|A_n)$. On the other hand, when the prior probabilities $P(A_n)$ are dramatically different, the resulting $P(A_n|B)$ tend to differ less from the $P(A_n)$.

A recent use of the Bayes equation, in an application of some concern to the nuclear power industry, was to the rail transport of spent nuclear fuel [10].

Example 2-2 It is desirable to use the record of past spent fuel shipments (as of 1979, about 4000 without a single release of radioactivity) to assess the probability of release per shipment. Let B stand for the statement "we have 4000 shipments with no releases," and let A_1 represent the hypothesis that the probability of release per shipment is 10^{-3}. Other hypotheses A_n $(n = 2$ to $6)$ are identified in Table 2-1.

Table 2-1 *Bayesian Calculations for Example 2-2*

	n					
	1	2	3	4	5	6
A_n	10^{-3}	10^{-4}	10^{-5}	10^{-6}	10^{-7}	10^{-8}
$P(B\|A_n)$	0.0183	0.6703	0.9608	0.9960	0.9996	0.99996
Uniform prior						
$P(A_n)$	0.1667	0.1667	0.1667	0.1667	0.1667	0.1667
$P(A_n\|B)$	0.004	0.1443	0.2068	0.2144	0.2152	0.2153
Nonuniform prior[a]						
$P(A_n)$	0.01	0.2	0.4	0.3	0.08	0.01
$P(A_n\|B)$	0.0002	0.1475	0.4228	0.3287	0.0880	0.0110

[a] From S. Kaplan and B. J. Garrick, On the use of a Bayesian reasoning in safety and reliability decisions—three examples, *Nucl. Technol.* **44**, 231 (1979).

If the probability of release per shipment were 10^{-3}, the probability of B, i.e., of 4000 release-free trips, would be

$$P(B|A_1) = (1 - 10^{-3})^{4000} = 0.0183.$$

This result is derived by using Eq. (2-3) to calculate the probability of a single safe shipment as $1 - 10^{-3}$, and then assuming that all shipments are independent so that Eq. (2-6) may be used. Similar calculations give the other $P(B|A_n)$ values in Table 2-1; these numbers contain the new data to be combined with the $P(A_n)$ to give the revised probabilities $P(A_n|B)$.

If a uniform prior distribution is assumed for the $P(A_n)$, so that each probability equals $\frac{1}{6}$, the results of using Eq. (2-32) to obtain the $P(A_n|B)$ in Table 2-1 show that the probability is almost certainly not 10^{-3}, corresponding to hypothesis A_1, and is more likely to be 10^{-5} or smaller.

Of course the $P(A_n|B)$ results depend upon the prior distribution selected. Thus if we had been more realistic in our first estimate of the probabilities $P(A_n)$, as with the nonuniform prior values shown in Table 2-1, then the introduction of information B would have had only a small effect on our degree of confidence in the propositions A_3, A_4, A_5, and A_6, increasing these probabilities by approximately 10%. On the other hand, information B reduces $P(A_2)$ by about 26% and virtually nullifies $P(A_1)$.

We conclude that the experience of 4000 release-free shipments is not sufficient to distinguish between probabilities of 10^{-5} and less. However, it is sufficient to substantially reduce our belief that the probability is on the order of 10^{-4} and nearly rule out any belief that the probability could be 10^{-3} or greater. ◇

2-3 PROBABILITY DISTRIBUTION FUNCTIONS

For many engineering problems, and particularly in reliability analyses where failure phenomena are studied as a function of time, the outcome of an experiment may be one of an infinite number of mutually exclusive events. Hence instead of considering discrete events A_1, A_2, etc., we really want to use a continuous random variable, say x, to consider the events. Because the probability for an event to occur between x and $x + \Delta x$ depends upon the size of Δx, we define $p(x)\,dx$ as the probability that outcome x for an experiment occurs within dx about x. This means that $p(x)$ is actually a probability *density*, or probability *per unit dx* for the outcome x.

The analog to the summation of probabilities $\sum_{n=1}^{N} P(A_n)$ for mutually exclusive events A_n becomes the cumulative probability, written as $P(X)$ or $P(x \leq X)$, which is defined as

$$P(X) = \int_{x_{\min}}^{X} p(x)\,dx. \tag{2-33}$$

If $X = x_{\max}$, then

$$P(x_{\max}) = 1, \tag{2-34}$$

since x is certain to occur within the range of $x_{\min}$ to $x_{\max}$. From the last two equations it is evident that

$$0 \leq p(x)\,dx \leq 1. \tag{2-35}$$

Care must be exercised when changing the variables of a probability density function $p(x)$ to $p(y)$ because it is necessary to include the Jacobian of the transformation, so that

$$p(y) = \left|\frac{dx}{dy}\right| p(x)\Big|_{x=y}. \tag{2-36}$$

Example 2-3 If a probability density function $f_n(\tau)$ is normally expressed in the τ-variable, and we desire to express it in terms of t, where $\tau = \ln t$, then the new probability density function $f_{\ln}(t)$ is

$$f_{\ln}(t) = t^{-1} f_n(\tau)\big|_{\tau=\ln t}. \quad \diamond$$

Generalizations of probability distributions to more than a single continuous random variable can be made, but such extensions usually are more involved than necessary to introduce the mathematical theory of reliability, where the continuous variable of interest is time t.

To summarize:

1. Probabilities can be used to treat discrete events, say A_1, A_2, etc., or a continuous random variable such as time.
2. Since a probability P is dimensionless, a probability density $p(x)$ has units of the inverse of x.

2-4 PROBABILITY CONCEPTS FOR FAILURE ANALYSES

Two adjectives describing failures are *instantaneous* and *degradation.* Simple examples of an instantaneous failure are the fracture of the filament of an incandescent light bulb and the rupture of the casing for a set of bearings. Examples of degradation failures are the gradual wearing out of bushings, used in lieu of bearings, and the rusting away of iron surfaces.

One of the problems associated with the study of degradation failures is in defining when the "failure" actually occurs; many times a component will be marginally serviceable and hence replacement or repair will not be required. In such cases, it is convenient to define the failure of an item undergoing degradation as occurring when unscheduled maintenance or repair actions are initiated; in other situations, failure is defined as occurring when performance parameters lie outside the specified limits of tolerance.

Chapter 5 contains information about classifying failures by mode of failure. For now we shall be content to presume that any failure of interest occurs instantly, and focus our attention on the time of failure.

For failure analyses as a function of time, discrete probabilities and probability distribution functions take on specific meanings, and so a special vocabulary is used. For this reason, it is very important to understand the concepts in the remainder of this chapter before continuing the study of the next several chapters.

There are two types of systems to be considered: those that operate on demand, and those that operate continuously. Demand failures occur in a system during its intermittant, possibly repetitive operation: either the system operates at the nth demand, event D_n, or does not operate, event $\bar{D}_n$. The probability $P(W_{n-1})$ that the system works for each of $(n - 1)$ operations is

$$P(W_{n-1}) = P(D_1 D_2 D_3 \cdots D_{n-1}). \tag{2-37}$$

Just because the system works for $(n - 1)$ operations, this does not mean that the system will operate at the nth demand. That is, $P(D_n|W_{n-1})$ is the conditional probability that the system will operate at the nth demand, given that the system works for $(n - 1)$ demands, while $P(\bar{D}_n|W_{n-1})$ is the corresponding conditional probability of failure. By Eq. (2-4), the probability that a system will not operate on the nth demand when it worked for all previous demands is

$$P(\bar{D}_n W_{n-1}) = P(\bar{D}_n|W_{n-1})P(W_{n-1}). \tag{2-38}$$

Using Eq. (2-5), the last equation also may be written as

$$\begin{aligned} &P(D_1D_2 \cdots D_{n-1}\bar{D}_n) \\ &\quad = P(\bar{D}_n|D_1D_2 \cdots D_{n-1})P(D_{n-1}|D_1D_1 \cdots D_{n-2}) \\ &\qquad \cdots P(D_2|D_1)P(D_1). \end{aligned} \tag{2-39}$$

For demand-type failures, one ideally would like to have a complete tabulation of all the probabilities in Eq. (2-39) for every intermittantly operating component in a system. Usually it is necessary, because of limitations in the experimental data available, to assume that the demand events are identical and independent; then any failure is assumed to be *random* so that $P(\bar{D}_n|W_{n-1}) = P(\bar{D})$ and $P(D_n|W_{n-1}) = P(D)$. In such a case, Eq. (2-39) reduces to

$$P(D_1D_2 \cdots D_{n-1}\bar{D}_n) = P(\bar{D})[P(D)]^{n-1} = P(\bar{D})[1 - P(\bar{D})]^{n-1} \tag{2-40}$$

so only the demand failure probability $P(\bar{D})$ need be tabulated. The demand failure probability $P(\bar{D})$ is frequently denoted by the symbol Q_d.

Note that Eqs. (2-39) and (2-40) describe the probability of failure on the nth demand, which differs from the probability that a repairable system will undergo a failure sometime during n demands. For example, for random failures, the latter probability would be n times the former since the failure could occur on any one of the n demands.

Example 2-4 A light switch fails randomly with a demand failure probability of 10^{-4}. On the average, the switch is used 20 times per week. What is the probability that the switch will fail at the end of a 3-year period? What is the probability it could fail exactly once during the 3 years if it were immediately repaired after failure?

Over a 3-year period, the switch could be used $20 \times 52 \times 3 = 3120$ times, so the probability of failure on the 3120th demand is

$$10^{-4}(1 - 10^{-4})^{3119} = 0.732 \times 10^{-4}.$$

The probability it could fail once anytime during the 3 years is

$$3120(0.732 \times 10^{-4}) = 0.228. \ \diamond$$

For systems in continuous operation, which do not undergo repair, the analog to Eq. (2-38) is given as

$$f(t)\,dt = \lambda(t)\,dt[1 - F(t)]. \tag{2-41}$$

Here

$f(t)\,dt$ = the probability for failure in dt about t,
$\lambda(t)\,dt$ = the probability for failure in dt about t, given that it survived to time t,
$1 - F(t)$ = the probability that the device did not fail prior to time t.

Another way of saying the same thing is

$$f(t) = \lambda(t)\,[1 - F(t)], \tag{2-42}$$

where $f(t)$ is the failure probability density, i.e., the probability of failure in dt about t per unit time. The term $\lambda(t)$ is the conditional failure rate and is often called the hazard rate; the units of $\lambda(t)$ are inverse time. It should be pointed out the $\lambda(t)$ does not obey an equation like Eq. (2-36).

For a system that is known to be functioning at time $t = 0$, the cumulative probability for failure between 0 and t, $F(t)$, is related to the probability density for failure by the equation

$$F(t) = \int_0^t f(t')\,dt'. \tag{2-43}$$

This result is consistent with Eq. (2-2) since every system is initially operable, so

$$F(0) = 0. \tag{2-44}$$

(If $f(t) = 0$ for $0 \le t \le T$, the system continues to remain "as good as new" so $F(T) = 0$). For every device that must eventually fail,

$$F(\infty) = 1, \tag{2-45}$$

and this equation serves as a normalization condition for the failure probability density $f(t)$. There are devices, however, that need not eventually fail even though they do not undergo repairs. This could occur, for example, if there were a threshold loading required for failure; then care must be exercised in evaluating $F(t)$ as $t \to \infty$. Such complications are beyond the scope of interest here.

Equation (2-43) may be differentiated to give

$$f(t) = dF(t)/dt, \tag{2-46}$$

and this result can be used in Eq. (2-42) to show that

$$\lambda(t) = \frac{dF(t)/dt}{1 - F(t)} = -d\ln[1 - F(t)]/dt. \tag{2-47}$$

Reliability $R(t)$ is defined as the probability that a specified fault event has not occurred in a system for a given period of time t and under specified operating conditions. Stated another way, reliability is the probability a system performs a specified function or mission under given conditions for a prescribed time. Reliability is just the complementary probability to $F(t)$, i.e.,

$$R(t) = 1 - F(t). \tag{2-48}$$

In other words, $F(t)$ is the unreliability, the probability that the device or system will fail at some time between 0 and t, and $R(t)$ is the probability that it will not fail during that time period.

Using the last equation, the preceding five equations may be written as

$$R(t) = \int_t^\infty f(t')\,dt', \tag{2-49}$$

$$R(0) = 1, \tag{2-50}$$

$$R(\infty) = 0, \tag{2-51}$$

$$f(t) = -\frac{dR(t)}{dt}, \tag{2-52}$$

$$\lambda(t) = -\frac{d \ln R(t)}{dt} = -\frac{1}{R(t)}\frac{dR(t)}{dt} = \frac{f(t)}{R(t)}. \tag{2-53}$$

Equation (2-53) may be integrated, and then Eq. (2-50) used, to obtain the very useful equation

$$R(t) = \exp\left[-\int_0^t \lambda(t')\,dt'\right]. \tag{2-54}^*$$

A combination of Eqs. (2-53) and (2-54) then leads to the important result

$$f(t) = \lambda(t) \exp\left[-\int_0^t \lambda(t')\,dt'\right]. \tag{2-55}$$

Other equations relating $f(t)$, $F(t)$, $R(t)$, and $\lambda(t)$ are given in Table 2-2. Although there are four parameters, only one is independent. Generally $\lambda(t)$ is the parameter tabulated because it is the one measured experimentally and because it tends to vary less rapidly with time than the other parameters.

Once the conditional failure rate $\lambda(t)$ is completely specified, Eq. (2-54) gives the probability that a device or system will survive to a time of interest. The converse calculation also can be done to obtain the minimum

* The symbol exp is used interchangeably with e to denote an exponential.

Table 2-2 *A Summary of Equations Relating* $\lambda(t)$, $R(t)$, $F(t)$, *and* $f(t)$

Word description	Symbol =	First relationship =	Second relationship =	Third relationship
Hazard rate	$\lambda(t)$	$-(1/R)\, dR/dt$	$f(t)/(1 - F(t))$	$f(t)/R(t)$
Reliability	$R(t)$	$\int_t^\infty f(\tau)\, d\tau$	$1 - F(t)$	$\exp\left[-\int_0^t \lambda(\tau)\, d\tau\right]$
Cumulative failure probability	$F(t)$	$\int_0^t f(\tau)\, d\tau$	$1 - R(t)$	$1 - \exp\left[-\int_0^t \lambda(\tau)\, d\tau\right]$
Failure probability density	$f(t)$	$dF(t)/dt$	$-dR(t)/dt$	$\lambda(t)R(t)$

time T the system will operate before its cumulative probability of failure exceeds a specified value; then T must be determined from

$$\int_0^T \lambda(t)\, dt = -\ln R(T) = -\ln[1 - F(T)]. \tag{2-56}$$

Example 2-5 Calculate the reliability of a device for which the conditional failure rate is defined by

$$\begin{aligned} \lambda(t) &= 0, & 0 \le t \le a, \\ &= \lambda, & a \le t \le b, \\ &= \lambda e^{(t-b)/c}, & t \ge b, \quad c > 0. \end{aligned}$$

From Eq. (2-54), it is found that $R(t) = 1$, $0 \le t \le a$, and

$$\begin{aligned} R(t) &= R(a) \exp\left[-\int_a^t \lambda(t')\, dt'\right] \\ &= \exp[-\lambda(t - a)], \qquad a \le t \le b. \end{aligned}$$

Likewise it follows that

$$\begin{aligned} R(t) &= R(b) \exp\left[-\int_b^t \lambda(t')\, dt'\right] \\ &= \exp\{-\lambda[b - a - c + c \exp(t - b)/c]\}, \qquad t \ge b. \diamond \end{aligned}$$

Example 2-6 For Example 2-5, in which $a = 10$ hr, $b = 1000$ hr, $c = 500$ hr, and $\lambda = 10^{-4}$ per hr, calculate the minimum time T in hr before the cumulative failure probability equals 0.2.

For $F(T) = 0.2$, it is obvious that $T > 10$ hr, so we first assume $10 < T < 1000$ hr and use Eq. (2-56) for this problem to find

$$\lambda(T - a) = -\ln[1 - F(T)].$$

After substituting the numerical values, we obtain $T = 2241$ hr. Since $T > 1000$ hr, the assumption was incorrect, so Eq. (2-56) is again used to find

$$\lambda[(b - a - c) + c \exp(T - b)/c] = -\ln[1 - F(T)].$$

From this equation, we conclude that $T =$ ~~1851~~ 1624 hr. ◇

For many devices, the behavior of $\lambda(t)$ follows the classic "bathtub curve" of Fig. 2-1. Early in life, $\lambda(t)$ for such a device is high because of "wear-in failures" or failures arising because of poor quality control practices. During the middle portion of useful life, failures occur at a rather uniform rate corresponding to random failures. Finally, late in life, $\lambda(t)$ begins to increase because of "wear-out failures." As one might expect, there is a difference in the shape of the curves for different devices.

For a typical major safety system in a complicated device such as a nuclear reactor, early life failures tend to be random because of high quality control requirements, while maintenance and replacement of components help mitigate against wear-out failures. Besides, when devices fail infrequently and are sufficiently complex and costly so that many tests cannot be performed to characterize patterns of failure, only an estimate of $\lambda(t)$ is available. Thus the usual procedure in many reliability and risk analyses is to assume that failures are random so that

$$\lambda(t) = \lambda. \tag{2-57}$$

The assumption of random failures for all times means that the reliability

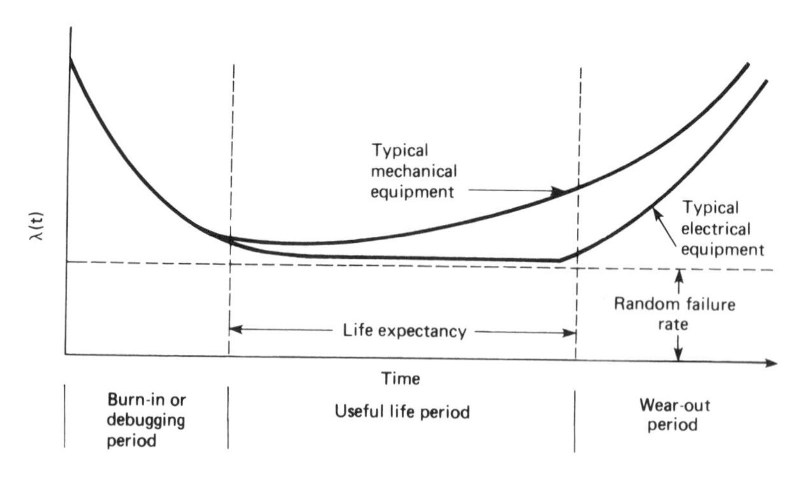

Fig. 2-1 *Time dependence of conditional failure (hazard) rate. (From H. E. Lambert, Lawrence Livermore Laboratory Report UCRL-51829, 1975.)*

of the device may be written simply as

$$R(t) = e^{-\lambda t}. \tag{2-58}$$

One additional definition is helpful when discussing quantities associated with reliability. The first moment of the failure probability density $f(t)$ for a system is defined as the mean time to failure (MTTF). Thus

$$\text{MTTF} = \int_0^\infty tf(t)\,dt \Big/ \int_0^\infty f(t)\,dt = \int_0^\infty tf(t)\,dt. \tag{2-59}$$

Here the equation has been simplified by using Eq. (2-45). The MTTF also can be expressed in terms of $R(t)$ if Eq. (2-52) is used in Eq. (2-59) and the result is integrated once by parts. If the integral in Eq. (2-59) is defined, which always happens if $tR(t) \to 0$ as $t \to \infty$, then

$$\text{MTTF} = \int_0^\infty R(t)\,dt. \tag{2-60}$$

The MTTF is especially simple in the case of random failures since

$$\text{MTTF} = 1/\lambda. \tag{2-61}$$

It should be emphasized that Eqs. (2-41) through (2-61) are for analyzing the probability of a device that cannot be repaired. If repairs are possible, then the system may remain operable for $t \to \infty$, for at least part of the time. In such cases, it is more useful to consider the system *instantaneous availability* $A(t)$, which is defined as the probability a system performs a specified function or mission under given conditions *at* a prescribed time. The probability $1 - A(t)$ is termed the instantaneous unavailability. The instantaneous availability is bounded such that

$$R(t) \leq A(t) \leq 1 \tag{2-62}$$

since $A(t) = R(t)$ for a device that does not undergo repair.

An important difference between $A(t)$ and $R(t)$ is their behavior for large times. As t becomes large, $R(t)$ approaches zero, whereas availability functions reach some steady-state value. Further discussion of the differences between $R(t)$ and $A(t)$ is given in Chapter 7.

To summarize:

1. There are two conditional probabilities of failures of interest, failure probabilities per demand for intermittently operated systems and hazard rates for systems undergoing continuous operation.
2. The conditional failure rate $\lambda(t)$ contains all the information needed to describe failures of a system; when $\lambda(t)$ is not known precisely, it is frequently approximated by a constant λ.

3. The reliability $R(t)$ can be used to describe the time-dependent availability of a system that can eventually fail, while the instantaneous availability $A(t)$ is used for systems that undergo repair.

EXERCISES

2-1 Write out $P(A_1 + A_2 + A_3 + A_4)$ for the case in which the events A_n are not mutually exclusive.

2-2 Repeat the calculations of Example 2-2 when A_n ($n = 1$ to 8) represents the hypothesis that the probability of release is $10^{-(n+1)}$ for (a) $P(A_1) = P(A_7) = 0.1$, $P(A_8) = 0.05$, with the remaining five probabilities 0.15, and (b) $P(A_n) = 0.125$.

2-3 Repeat the calculations of Example 2-2 when A_n ($n = 1$ to 6) represents the hypothesis that the probability of release is 10^{-n} for (a) the uniform $P(A_n)$ in Table 2-1, and (b) the nonuniform $P(A_n)$ in Table 2-1.

2-4 The horn on an automobile operates on demand 99.96% of the time. Each event is independent of all the others. How many times would you expect to be able to honk the horn with a 50% probability of not having a single failure?

2-5 It is assumed that the hazard rate for a pressure valve is given by $\lambda(t) = 1/(t + 2)$. (a) What is the cumulative probability of failure $F(t)$? (b) What is the probability density $f(t)$ for failure at time t?

2-6 A hazard rate $\lambda(t)$ is piecewise continuous according to the relation

$$\lambda(t) = \lambda, \qquad 0 \leq t < t_1$$
$$\lambda(t) = \lambda + k(t - t_1), \qquad t \geq t_1.$$

(a) What is the probability that the system has not failed during time t? (b) What is the failure probability density?

2-7 A light bulb operates continuously to illuminate a display panel. The conditional failure rate of the bulb $\lambda(t) = 5 \times 10^{-5}t$, with t in days. (a) What is the mean time to failure (MTTF) for the bulb? (b) What is the MTTF if $\lambda(t) = 5 \times 10^{-7}t$?

2-8 The failure probability density for a device is given by $f(t) = kE_2(at)$, where $E_n(x)$ is the exponential integral function having properties defined in Appendix A, k is a normalization constant, and a is another constant. (a) Calculate k by using Eq. (2-45). (b) Obtain the cumulative probability of failure $F(t)$. (c) Obtain the reliability $R(t)$. (d) Obtain the hazard rate $\lambda(t)$. (e) Obtain the mean time to failure MTTF.

REFERENCES

1. A. Papoulis, "Probability, Random Variables, and Stochastic Processes." McGraw-Hill, New York, 1965.
2. W. Feller, "An Introduction to Probability Theory and Its Applications," Vol. 1, 3rd ed. Wiley, New York, 1968.
3. N. R. Mann, R. E. Schafer, and N. D. Singpurwalla, "Methods for Statistical Analysis of Reliability and Life Data." Wiley, New York, 1974.
4. R. Von Mises, "Probability, Statistics and Truth," 2nd rev. English ed. Macmillan, New York, 1957.
5. M. Tribus, "Rational Descriptions, Decisions and Designs." Pergamon, Oxford, 1969.
6. R. T. Cox, Probability, frequency and reasonable expectation, *Am. J. Phys.* **14,** 1–13 (1946).
7. G. Apostolakis, Probability and risk assessment: The subjectivistic viewpoint and some suggestions, *Nucl. Safety* **19,** 305 (1978).
8. M. L. Shooman, "Probabilistic Reliability: An Engineering Approach." McGraw-Hill, New York, 1968.
9. A. J. Bourne and A. E. Green, "Reliability Technology." Wiley (Interscience), New York, 1972.
10. S. Kaplan and B. J. Garrick, On the use of a Bayesian reasoning in safety and reliability decisions—three examples, *Nucl. Technol.* **44,** 231 (1979).

Probability Distributions for Describing Failures

Chapters 3 through 5 treat the failures of systems comprised of single components. Chapter 3 contains some of the probability distributions used to describe such failures, and Chapter 4 describes the manipulation of data for component failures to obtain the parameters needed in the probability distributions. Chapter 5 provides information on sources of data and some representative values.

Before considering the single-component systems, however, we shall examine systems whose failures can be described by discrete distributions.

3-1 DISCRETE DISTRIBUTIONS

In Section 2-4 we saw that system operation may be either intermittent or continuous. This section contains a discussion of two of the most useful discrete probability distributions used for failure analyses, the binomial and Poisson distributions [1–5].

Two parameters of interest for any discrete probability distribution $P(r)$ of random variable r are the mean m and the variance σ^2. For n possible outcomes, the mean is defined as

$$m = \sum_{r=0}^{n} rP(r), \tag{3-1}$$

while the variance, which measures the deviation of values about the mean, is

$$\sigma^2 = \sum_{r=0}^{n} (r - m)^2 P(r). \tag{3-2}$$

These two parameters provide measures of the differences between various discrete distributions.

3-1-1 Binomial Distribution

In the simplest of systems there are only two outcomes of events, either the system functions on demand, which can be denoted by D, or it fails ($\bar{D}$). It follows from Eq. (2-3) that the two probabilities are related by

$$P(D) = 1 - P(\bar{D}). \quad (3\text{-}3)$$

Suppose that the performance of the system is not known, so that an experiment consisting of n demands or trials is to be performed, where n is fixed. The demands are specified to be independent (or Bernoulli trials) such that $P(\bar{D})$ is constant for each trial. In order to describe the experiment with the binomial distribution, it is necessary that the ordering of the events not affect the result of the experiment. The possible outcomes of the experiment correspond to the different terms in the binomial expansion of the equation

$$[P(D) + P(\bar{D})]^n = 1. \quad (3\text{-}4)$$

Now introduce the discrete random variable r, defined to be the number of demands for which the system fails. The random variable r obeys the binomial distribution, with parameter $P(\bar{D})$ and index n. The probability of r is obtained by selecting the proper term from the binomial expansion of Eq. (3-4) and it has the form

$$P(r) = \binom{n}{r} [P(\bar{D})]^r[P(D)]^{n-r} = \frac{n!}{r!(n-r)!} [P(\bar{D})]^r[P(D)]^{n-r}. \quad (3\text{-}5)$$

It may be shown for the binomial distribution that

$$m = n\, P(\bar{D}) \quad (3\text{-}6)$$

$$\sigma^2 = n\, P(\bar{D})\, P(D). \quad (3\text{-}7)$$

Another probability distribution obtainable from Eq. (3-5) is the cumulative probability $P(\leq x)$ that the system fails for x *or fewer* demands. It follows by addition of the appropriate terms in the expansion of Eq. (3-4) that

$$P(\leq x) = \sum_{r=0}^{x} P(r). \quad (3\text{-}8)$$

Thus the probability that the system fails for $(x + 1)$ or more demands would be the complement of $P(\leq x)$,

$$P(>x) = 1 - \sum_{r=0}^{x} P(r). \quad (3\text{-}9)$$

For a large enough n, the calculation of the binomial distribution can be approximated by a normal distribution with the same mean m and variance σ^2. This approximation gives good results if $nP(\bar{D})$ and $nP(D)$ are both at least 5.

The binomial distribution is used in reliability engineering calculations for a single component that operates on demand and can be repaired to an "as good as new" state immediately after it fails. Then r is the number of failures in n demands and $P(r)$ is the probability that the component will fail on r demands. Thus Eq. (3-5) is a generalization of Eq. (2-40) in the case in which the unit operated on demand can undergo repair.

The binomial distribution treats the case in which there are only two possible outcomes from any demand, $\bar{D}$ and D. If there are more than two possibilities, then the *multinomial* distribution must be used [1–3].

Example 3-1 A switch that can be repaired has a failure rate of 10^{-4}/demand. Calculate the probability that in 1000 operations the switch will fail exactly two times and the probability that it will fail two or more times.

Since $P(\bar{D}) = 10^{-4}$ and $n = 1000$, Eq. (3-5) gives

$$P(2) = \frac{1000!}{2!(1000 - 2)!}(10^{-4})^2(1 - 10^{-4})^{1000-2} = 0.0045.$$

To evaluate the probability the switch will fail two or more times, from Eq. (3-5) we calculate $P(1) = 0.0905$ and $P(0) = 0.9048$ and use Eq. (3-9) with $x = 1$ to obtain

$$P(>1) = 1 - P(0) - P(1) = 0.0047. \ \diamond$$

A second interpretation of the binomial distribution for failure analyses involves the case of n identical units that initially function, in which case $P(\bar{D})$ becomes the probability that a single unit fails. Then $P(r)$ describes the probability that r of the n units in the system will fail.

Example 3-2 A system has four identical components that operate simultaneously and independently; two must remain operating or the system will fail. If the failure of each component is 0.02 over the design life of the system, determine the probability of system failure.

Since $P(\bar{D}) = 0.02$ and $n = 4$, from Eq. (3-5) we find $P(0) = 0.9223$ and $P(1) = 0.0753$, so from Eq. (3-9)

$$P(>1) = 1 - P(0) - P(1) = 0.0024. \ \diamond$$

3-1-2 Poisson Distribution

The Poisson distribution is like the binomial distribution in that it describes phenomena for which the *average* probability of an event is constant, independent of the number of previous events. In this case, however, the system undergoes transitions *randomly* from one state with n occurrences of an event to another with $(n + 1)$ occurrences, in a process that is *irreversible*. That is, the ordering of the events cannot be interchanged. Another distinction between the binomial and Poisson distributions is that for the Poisson process the number of possible events should be large.

The Poisson distribution may be inferred from the identity

$$e^{-\mu}e^{\mu} = 1, \tag{3-10}$$

where the most probable number of occurrences of the event is μ. If the factor e^{μ} is expanded in a power series expansion, the probability $P(r)$ that exactly r random occurrences will take place can be inferred as the rth term in the series, i.e.,

$$P(r) = \frac{e^{-\mu}\mu^{r}}{r!}. \tag{3-11}$$

This probability distribution leads directly to the interpretation that:

$$\begin{aligned} e^{-\mu} &= \text{the probability that an event will not occur,} \\ \mu e^{-\mu} &= \text{the probability that an event will occur exactly once,} \\ (\mu^2/2!)e^{-\mu} &= \text{the probability that an event will occur exactly twice, etc.} \end{aligned}$$

The mean and the variance of the Poisson distribution follow from Eq. (3-11) and from Eqs. (3-1) and (3-2) as $n \rightarrow \infty$. From the definition of μ,

$$m = \mu, \tag{3-12}$$

although it is less obvious that

$$\sigma^2 = \mu. \tag{3-13}$$

The cumulative Poisson probability that an event will occur x times or less is

$$P(\leq x) = \sum_{r=0}^{x} P(r). \tag{3-14}$$

Of course the probability that the event will occur $(x + 1)$ or more times would be the complement of $P(x)$.

The Poisson distribution is useful for analyzing the failure of a system that consists of a large number of identical components that, upon failure, cause irreversible transitions in the system. Each component is assumed to fail independently and randomly. Then μ is the most probable number of system failures over the lifetime. The distribution is the limiting form of a binomial distribution for a system of n identical components for which the probability $P(\bar{D})$ of a component failure is small and the number of components n is large; in this case, μ becomes $nP(\bar{D})$.

Example 3-3 A given nuclear reactor is fueled with 200 assemblies, each of which can fail if the cladding on a fuel rod fails. If each assembly fails in an independent and random manner over the exposure time, calculate the probability of 3 assemblies failing if, on the average, 1% of the fuel assemblies are known to fail.

The mean number of assembly failures is $\mu = 2$, so using Eq. (3-11) for $r = 3$ gives

$$P(3) = (2^3/3!)e^{-2} = 0.1804.$$

As a check, we can use the probability of a single assembly failing, $P(\bar{D}) = 0.01$, and the binomial distribution of Eq. (3-5) with $n = 200$ to obtain

$$P(3) = \frac{200!}{3!(200 - 3)!}(0.01)^3(0.99)^{200-3} = 0.1814. \quad \diamond$$

To summarize:

1. The binomial distribution is useful for systems with two possible outcomes of events (failure–no failure) in cases where there is a known, finite number of (Bernoulli) trials and the ordering of the trials does not affect the outcome.
2. The Poisson distribution treats systems in which randomly occurring phenomena cause irreversible transitions from one state to another.

3-2 CONTINUOUS DISTRIBUTIONS

This section contains a discussion of five of the most useful probability distributions for describing failures as a function of time: the Erlangian, exponential, gamma, lognormal, and Weibull distributions [1–5]. For failure analyses, only times in the range $0 \leq t \leq \infty$ are of interest, a constraint that is satisfied by all the distributions. In this case, the mean value of a distribution is given by

$$m = \int_0^\infty tf(t)\,dt, \tag{3-15}$$

and the variance is

$$\sigma^2 = \int_0^\infty (t - m)^2 f(t)\, dt. \tag{3-16}$$

It should be observed that m is the mean time to failure (MTTF) as defined in Eq. (2-59). The mean and variance are the two lowest moments of a set that are sometimes useful when characterizing the differences among different failure probability distributions.

3-2-1 Erlangian and Exponential Distributions

The Erlangian distribution is the time-dependent form of the Poisson discrete distribution. The Erlangian distribution arises frequently in reliability engineering calculations involving random failures, i.e., those failures for which the hazard rate $\lambda(t)$ is a constant λ. To derive the distribution from Eq. (3-11), we recognize that the mean number of failures μ is the product of λ and time t. The probability of exactly r failures occurring in time t is then given by

$$P(r, t) = e^{-\lambda t}(\lambda t)^r / r!, \tag{3-17}$$

and the cumulative probability of x or fewer failures is

$$P(\leq x, t) = \sum_{r=0}^{x} \frac{e^{-\lambda t}(\lambda t)^r}{r!}. \tag{3-18}$$

Equation (3-17) is useful since it permits calculation of the failure probability density $f(t)$ for the rth failure in dt about t. What is required, of course, is for the system to have undergone $(r - 1)$ prior failures so that it is ready to fail for the rth time with a conditional probability λ. Thus the Erlangian distribution follows from Eq. (3-17) as

$$f(t) = \lambda P(r - 1, t) = \frac{\lambda(\lambda t)^{r-1} e^{-\lambda t}}{(r - 1)!}, \quad \lambda > 0, \quad r \geq 1. \tag{3-19}$$

The Erlangian distribution is valid for an integer number of failures r. The most important special case is for $r = 1$, in which case the exponential distribution is obtained,

$$f(t) = \lambda e^{-\lambda t}. \tag{3-20}$$

The cumulative failure probability for the exponential distribution is

$$F(t) = 1 - e^{-\lambda t}, \tag{3-21}$$

and the two moments are

$$m = 1/\lambda \tag{3-22}$$

$$\sigma^2 = 1/\lambda^2. \tag{3-23}$$

The use of this distribution for analyzing the (first) purely random failure of a device characterized by a constant hazard rate was discussed in Chapter 2.

Both the Erlangian and exponential distributions are special cases of the gamma distribution, so we need not treat their applications separately in reliability engineering calculations.

3-2-2 Gamma Distribution

The gamma distribution failure probability density obeys the equation

$$f(t) = \frac{\lambda(\lambda t)^{r-1}e^{-\lambda t}}{\Gamma(r)}, \qquad \lambda > 0, \qquad r > 0, \tag{3-24}$$

where parameter r need not be an integer. The gamma function $\Gamma(r)$ is defined in Appendix A and is tabulated in standard references to mathematical functions [6]. In the special case that r is an integer, the Erlangian distribution is recovered; in the special case that $\lambda = 0.5$ and $r = 0.5\eta$, where η is the number of degrees of freedom, the gamma distribution becomes the *chi-square* distribution.

The cumulative failure probability $F(t)$ corresponding to Eq. (3-24) is

$$F(t) = \frac{1}{\Gamma(r)}\int_0^{\lambda t} y^{r-1}e^{-y}\,dy$$

$$= \frac{1}{\Gamma(r)}\gamma(r, \lambda t). \tag{3-25}$$

The incomplete gamma function $\gamma(r, \lambda t)$, discussed in Appendix A, tends to the (complete) gamma function $\Gamma(r)$ as $t \to \infty$, a result that is consistent with $F(\infty) = 1$.

The mean and variance of the gamma distribution follow from Eqs. (3-15), (3-16), and (3-24) as

$$m = r/\lambda \tag{3-26}$$

$$\sigma^2 = r/\lambda^2. \tag{3-27}$$

The gamma distribution is especially appropriate for systems subjected to an environment of repetitive, random shocks generated according to the Poisson distribution; thus the failure probability depends upon how many shocks the device has suffered, i.e., its age. As another application, if the mean rate of wear of a device is a constant, but the rate of wear is subject to random variations, then the gamma function should be used.

The two parameters are the shape parameter r and the scale parameter λ. As we can see from Fig. 3-1, the shape of $f(t)$ depends significantly upon the value of r. Figure 3-2 illustrates the impact on the hazard rate $\lambda(t)$.

For some devices, such as those for which corrosion of metals is important, it may be appropriate to modify the two-parameter gamma distribution by introducing a time delay τ before the onset of failures begins. Then Eq. (3-24) is modified to read as

$$f(t) = \frac{\lambda^r (t-\tau)^{r-1} e^{-\lambda(t-\tau)}}{\Gamma(r)}, \qquad t \geq \tau \tag{3-28}$$

$$= 0, \qquad t < \tau.$$

In such a case, the mean value of Eq. (3-26) becomes

$$m = \tau + r/\lambda. \tag{3-29}$$

Example 3-4 Data manipulation (of the type to be discussed in Chapter 4) has shown that a device subjected to repetitive random shocks satisfies a gamma distribution with parameters $r = 3$ and $\lambda = 10^{-3}$/hr, and that no

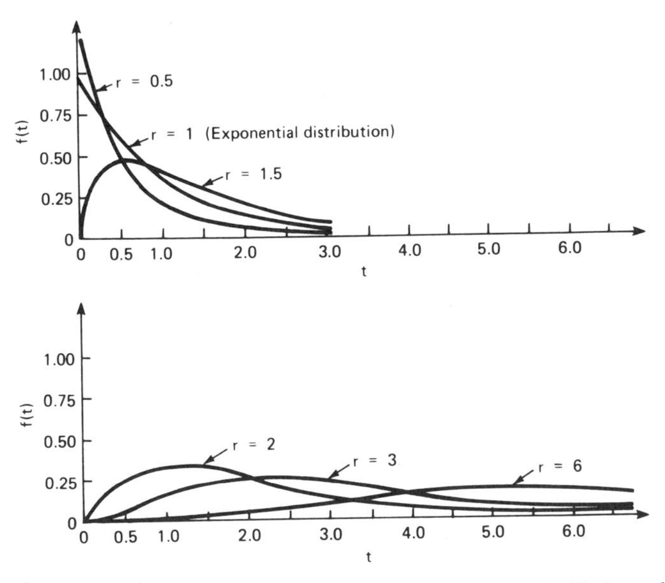

Fig. 3-1 *Gamma distribution failure probability density. (From G. A. Wadsworth and J. G. Bryan, "Introduction to Probability and Random Variables." McGraw-Hill, New York, 1960.)*

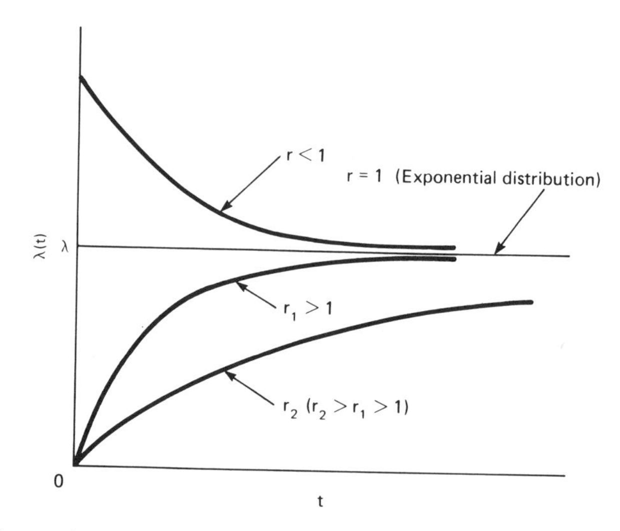

Fig. 3-2 *Gamma distribution hazard rate.* [*From G. E. Apostolakis, Mathematical methods of probabilistic safety analysis, UCLA-ENG-7464 (September 1974).*]

failures can occur until 200 hr have passed. Estimate (a) the probability of failure after the device has operated for $t = 4500$ hr and (b) its mean time to failure.

In this problem, the time displacement is $\tau = 200$ hr. Integrating Eq. (3-28) from 0 to t and using Eq. (3-25) gives the cumulative failure probability

$$F(4500) = \frac{1}{\Gamma(3)} \gamma[3,\ 10^{-3}(4500 - 200)] = \frac{1}{\Gamma(3)} \gamma(3,\ 4.3),$$

and from Appendix A it can be estimated that $F(4500 \text{ hr}) \approx 0.8$. Using Eq. (3-29) gives

$$\text{MTTF} = 200 + 3/10^{-3} = 3200 \text{ hr.} \ \diamond$$

3-2-3 Lognormal Distribution

The lognormal distribution (sometimes spelled out as the logarithmic-normal distribution) of a random variable t is one for which the logarithm of t follows a normal or Gaussian distribution. The equation describing the failure probability distribution can be written as

$$f(t) = \frac{1}{\sqrt{2\pi}\,\alpha t} \exp\left[-\frac{[\ln(t/\beta)]^2}{2\alpha^2}\right], \qquad \alpha,\ \beta > 0. \qquad (3\text{-}30)$$

The shape parameter α (which is dimensionless) and the scale parameter or "characteristic life" β (in units of time) are sufficient to specify the shape of $f(t)$; often the symbol $\beta' = \ln \beta$ (in units of ln-time) is used instead of β, so one should always check the units of the scale parameter.

The failure probability density is depicted in Fig. 3-3 where we can see that the distrubition is skewed to the right as compared to the Gaussian distribution, which is symmetric about its mean value; the skewness increases with increasing values of α. The lognormal distribution is related to the normal distribution in the manner explained in Example 2-3.

The cumulative failure probability follows by integrating Eq. (3-30) to obtain

$$\begin{aligned} F(t) &= \pi^{-1/2} \int_{-\infty}^{z} e^{-u^2}\, du \\ &= \tfrac{1}{2}[1 - \operatorname{erf}|z|], \qquad t < \beta \\ &= \tfrac{1}{2}[1 + \operatorname{erf} z], \qquad t > \beta, \end{aligned} \tag{3-31}$$

where z is defined as

$$z = \frac{\ln(t/\beta)}{\sqrt{2}\alpha} \tag{3-32}$$

and where the error function erf z defined in Appendix A is tabulated [6].

The mean and variance of the lognormal distribution, as defined by Eqs. (3-15) and (3-16), are

$$m = \beta \exp(\alpha^2/2) \tag{3-33}$$

$$\sigma^2 = \beta^2 \exp \alpha^2(\exp \alpha^2 - 1). \tag{3-34}$$

From the last equation, it is obvious that β^2 for the lognormal distribution cannot be interpreted as the variance σ^2, in contrast to the case for the

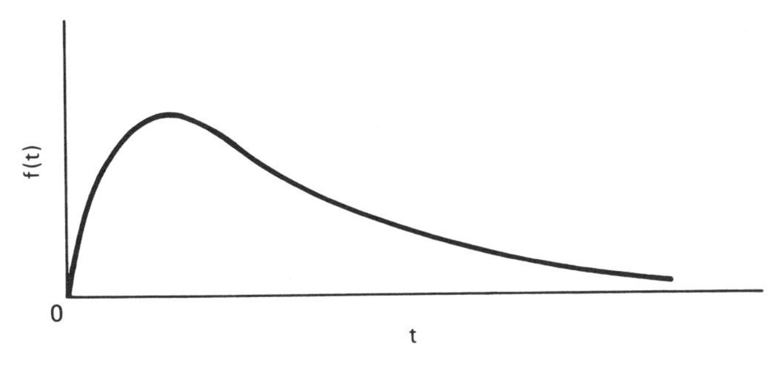

Fig. 3-3 *Lognormal distribution failure probability density.* [*From G. E. Apostolakis, Mathematical methods of probabilistic safety analysis, UCLA-ENG-7464 (September 1974).*]

normal distribution. Procedures to estimate the parameters α and β from data will be developed in Chapter 4.

The lognormal distribution arises in processes in which the change in a random variable at the nth step is a random proportion of the variable at the $(n - 1)$st step. Another way of saying the same thing is that the lognormal distribution is needed when factors or percentages characterize the variation. Thus if X represents a quantity that can vary by factors in its error, having a possible range between X_0/f and $X_0 f$, where X_0 is some midpoint reference value and f an error factor, then a lognormal is the natural distribution for describing the phenomenon.

One of the reasons that the lognormal distribution is frequently suitable for describing failures in reliability and risk analysis is that data for rarely occurring events may not be extensive, so that component failure rates may vary by factors. For example, a failure rate estimated at 10^{-6}/hr can vary from 10^{-5}/hr to 10^{-7}/hr if the error factor is 10. When the failure rate is expressed as 10^{-X}, where X is some exponent, use of the lognormal distribution implies that the exponent satisfies a normal distribution. Thus we can view the lognormal distribution as one for situations in which there is considerable uncertainty in the failure parameters.

Another feature of the lognormal distribution is that the skewness to higher times incorporates the general behavior of the data for unlikely phenomena since the skewness accounts for the occurrence of infrequent but large deviate values, such as abnormally high failure rates due to batch defects, environmental degradation, and other causes.

To complete the discussion of the lognormal distribution, it should be pointed out that the two-parameter form can be generalized to a three-parameter form which incorporates a time-delay parameter τ if t in Eq. (3-32) is replaced by $t - \tau$. Then

$$m = \tau + \beta \exp(\alpha^2/2) \tag{3-35}$$

and σ^2 is given in Eq. (3-34). The three-parameter lognormal distribution replaces the two-parameter version whenever there is no possibility of failure for $0 \leq t \leq \tau$.

Example 3-5 Data manipulation (of the type to be discussed in Chapter 4) has shown that the limited failure information about a particular pump manufactured by XYZ Company fits a lognormal distribution with a mean time to failure of 2.5×10^4 hr. It is also known that 40% have failed by $T = 10^4$ hr. Calculate the probability of failure for a pump intended for $T_* = 5 \times 10^4$ hr of service.

From Eq. (3-33),

$$\ln \text{MTTF} = \ln \beta + \alpha^2/2.$$

Since $F(T) < 0.5$, Eq. (3-31) gives

$$2F(T) = 1 - \text{erf}\left|\frac{\ln T - \ln \beta}{\sqrt{2}\,\alpha}\right|$$

$$= 1 - \text{erf}\left|\frac{\ln T - \ln \text{MTTF} + \alpha^2/2}{\sqrt{2}\,\alpha}\right|, \qquad T < \beta.$$

Substitution of $F(T) = 0.4$ and the values for T and MTTF gives

$$\text{erf}\left|\frac{-0.9163 + \alpha^2/2}{\sqrt{2}\,\alpha}\right| = 0.2,$$

and a table of error functions yields

$$\left|\frac{-0.9163 + \alpha^2/2}{\sqrt{2}\,\alpha}\right| = 0.179.$$

If we remove the absolute value sign from the left-hand side of the equation by assuming the argument is positive, we obtain the root $\alpha = 1.632$ and a negative root that does not satisfy the constraint $\alpha \geq 0$. The values of α and MTTF then lead to $\beta = 6622$ hr, an answer that must be incorrect since $\beta < T$. Therefore we remove the absolute value sign by taking the negative argument and obtain $\alpha = 1.125$, which leads to $\beta = 1.328 \times 10^4$ hr. Finally, Eq. (3-31) is used with $T_* > \beta$ to give

$$F(T_*) = \frac{1}{2}\left[1 + \text{erf}\left(\frac{\ln(T_*/\beta)}{\sqrt{2}\,\alpha}\right)\right]$$

$$= \tfrac{1}{2}[1 + \text{erf}(0.833)] = 0.78. \ \diamond$$

3-2-4 WEIBULL DISTRIBUTION

The Weibull is a very general and popular failure distribution that has been shown to apply to a large number of diverse situations. The three-parameter form of the distribution is

$$f(t) = \frac{\alpha}{\beta}\left(\frac{t-\tau}{\beta}\right)^{\alpha-1} \exp\left[-\left(\frac{t-\tau}{\beta}\right)^{\alpha}\right],$$

$$\alpha > 0, \qquad \beta > 0, \qquad 0 \leq \tau \leq t \leq \infty, \tag{3-36}$$

and

$$F(t) = 1 - \exp\{-[(t-\tau)/\beta]^{\alpha}\} \tag{3-37}$$

$$m = \tau + \beta\Gamma(1 + \alpha^{-1}), \tag{3-38}$$

$$\sigma^2 = \beta^2\{\Gamma(1 + 2\alpha^{-1}) - [\Gamma(1 + \alpha^{-1})]^2\}. \tag{3-39}$$

The shape of the distribution depends primarily on the shape parameter α, as we see from Fig. 3-4. The scale parameter or "characteristic life" is β, and the time-delay parameter is τ. For $\alpha = 1$, the exponential distribution is obtained, with hazard rate $\lambda = \beta^{-1}$. Furthermore, as α increases, the Weibull distribution tends to the normal distribution; indeed, for $\alpha \geq 4$, Eq. (3-36) and the normal distribution are almost indistinguishable! Another special case of the Weibull model is the *Rayleigh* distribution, for which $\alpha = 2$.

The applications of the Weibull distribution perhaps are most easily understood by observing that the hazard rate for the Weibull model is

$$\lambda(t) = \frac{\alpha}{\beta}\left(\frac{t - \tau}{\beta}\right)^{\alpha-1}. \tag{3-40}$$

That is, the Weibull model is the appropriate one for fitting any data for which the conditional probability of failure $\lambda(t)$ satisfies a power law as a function of time. Hazard rates of this type are depicted in Fig. 3-5.

Another application of the Weibull distribution is that it is the distribution for the failure of a device that consists of a large number of identical components, each of which can fail independently according to the gamma distribution and all of which must function for the device to not fail. An example of such a device is a fuel assembly of a nuclear reactor, which consists of a large number of (presumably identical) fuel rods; if the cladding fails for any rod, then the assembly can be deemed to have "failed."

Example 3-6 An analysis of 39 Mark-IA driver fuel assemblies for the EBR-II reactor has shown [see N. J. Olson *et al.*, *Nucl. Technol.* **28**, 134

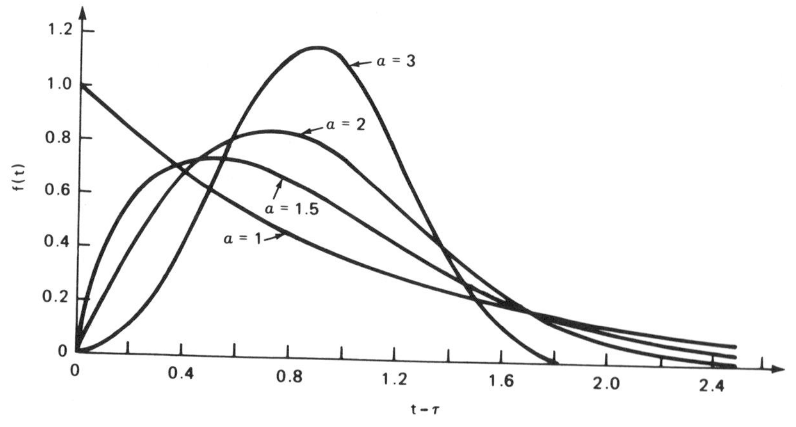

Fig. 3-4 *Weibull distribution failure probability density. (From N. H. Roberts, "Mathematical Methods in Reliability Engineering." McGraw-Hill, New York, 1964.)*

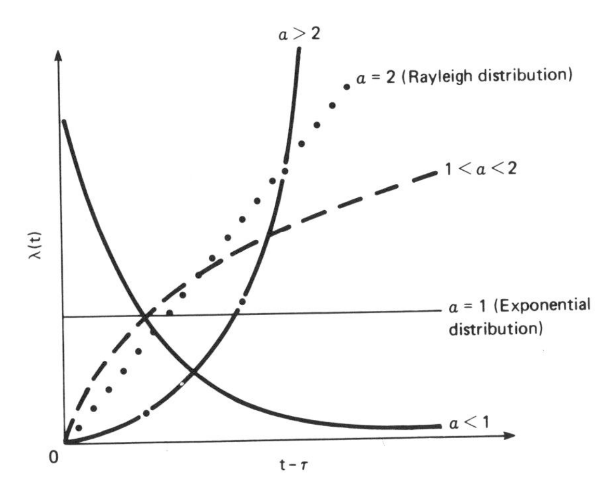

Fig. 3-5 *Weibull distribution hazard rate. [From G. E. Apostolakis, Mathematical methods of probabilistic safety analysis, UCLA-ENG-7464 (September 1974).]*

(1976)] that the cumulative failure probability $F(BU_{max})$ as a function of maximum percent burnup BU_{max} at normal EBR-II operating conditions is

$$F(BU_{max}) = 1 - \exp\left[-\left(\frac{BU_{max} - 3.0}{0.674}\right)^{5.91}\right], \qquad BU_{max} \geq 3.0.$$

Calculate the mean percent burnup m at the time of failure.

Comparison of $F(BU_{max})$ with Eq. (3-37) shows that $\alpha = 5.91$, $\beta = 0.674$, and $\tau = 3.0\%$. From Eq. (3-38) it follows that

$$m = 3.0 + 0.674\Gamma(1 + 1/5.91),$$

and use of a table of gamma functions gives $m = 3.62\%$. ◇

3-2-5 Which Distribution Should Be Used?

To summarize the preceding material in this section:

1. The gamma distribution is a natural extension of the Poisson discrete distribution and encompasses the Erlangian and exponential distributions as special cases; it is frequently useful for characterizing fatigue failures arising from repetitive shocks.

2. The lognormal distribution is like the normal distribution except that the independent variable is ln t rather than t; it is frequently used to describe very infrequent failures for which the uncertainties in the failure rates vary according to (possibly noninteger) powers of ten.

3. The Weibull distribution includes both the exponential and normal distributions as special cases; it is widely used because it encompasses all cases in which the hazard rate varies according to a power of t.
4. The parameters for these distributions are determined from the first and second moments of $f(t)$, i.e., the mean and variance, and from the time of the beginning of possible failures.

In general, a model for a continuous distribution function is chosen on the basis that the physical nature of the problem fits most or all of the underlying assumptions associated with a particular distribution. Frequently, however, insufficient theoretical grounds are available for selecting a distribution, so it may be necessary to infer the distribution by trial-and-error curve fitting of data for $F(t)$ versus t, as described in Chapter 4.

Another procedure for determining which distribution is appropriate uses estimates of the ratios of higher moments of $f(t)$ [5]. If we denote the moments about the mean as

$$\mu_n = \int_0^\infty (t - m)^n f(t)\, dt, \qquad n = 2, 3, \ldots, \tag{3-41}$$

so that μ_2 is the variance, then the third moment is *skewness,* which is a measure of the asymmetry of the distribution, and $n = 4$ gives the *kurtosis,* which is related to the peakedness of $f(t)$. The *coefficient of skewness* $\beta_1^{1/2}$ is defined as

$$\beta_1^{1/2} = \mu_3/(\mu_2)^{3/2} \tag{3-42}$$

and measures the skewness of the distribution relative to its degree of spread. The *coefficient of kurtosis* β_2 is

$$\beta_2 = \mu_4/\mu_2^2. \tag{3-43}$$

The use of dimensionless coefficients β_1 and β_2 to distinguish between different distributions is illustrated by the plot of β_2 vs β_1 in Fig. 3-6. For example, we see that the exponential distribution is the point for $\beta_1 = 4$ and $\beta_2 = 9$. All gamma distributions have parameters such that $\beta_2 - 1.5\beta_1 = 3$, which is the straight line connecting the point for the exponential distribution [for r of Eq. (3-24) equal to unity] with that point at $\beta_1 = 0$ and $\beta_2 = 3$ for the normal distribution (for $r \to \infty$).

The lognormal distribution in Fig. 3-6 is nearly a straight line given approximately by $\beta_2 - [16(1 + \epsilon)/9]\,\beta_1 = 3$, where $\epsilon \le 0.1$. The Weibull distribution lies along two distinct curves depending upon whether the shape parameter α exceeds 3.6, for which $\beta_1 = 0$. For $1 < \alpha \le 11.5$, the curve always lies somewhat below that for the gamma distribution, while

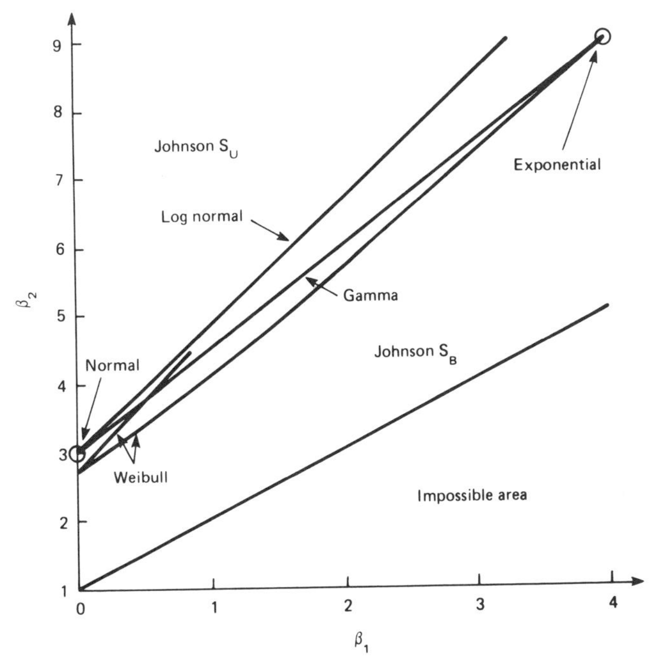

Fig. 3-6 *Regions in* (β_1, β_2) *plane for various distributions.*

for $\alpha > 11.5$, the curve lies above that for the gamma distribution and rapidly approaches that for the lognormal distribution [7].

An estimate of the parameters β_1 and β_2 can be used to plot a point on Fig. 3-6. If the plotted point is reasonably close to a point or curve corresponding to one of the standard distributions, then that distribution can be selected to represent the data. Only rarely will the plotted point agree precisely with a standard distribution, however, because estimates of β_1 and β_2 are subject to fluctuations in sampling and the moments μ_3 and μ_4 are especially sensitive to a few extreme observations.

If the plotted point from the estimated parameters β_1 and β_2 lies far from a point or curve corresponding to one of the standard distributions, this suggests that a more general distribution may be required. The *Johnson* distribution is a three-parameter distribution (actually a four-parameter distribution if there is a time-delay τ) that spans the entire possible area in Fig. 3-6. There are three alternate forms of the Johnson distribution: the Johnson S_L form is just the lognormal distribution, the Johnson S_U distribution spans the area above the lognormal curve in the figure, and the Johnson S_B distribution spans the possible area below the lognormal curve. This entire area also may be fitted to one of the forms of the *Pearson* distribution. The interested reader is urged to consult Hahn and Shapiro [5] for details about these more general distributions.

3-3 SYNTHESIZED DISTRIBUTIONS

Three categories of synthesized distributions for a device are considered. A mixed distribution failure model is a linear combination of two or more probability densities for all times, and a composite distribution failure model is a piecewise-continuous-with-time failure probability density. A convoluted distribution arises for a device that has replacement units in standby ready for sequential use as each unit fails.

3-3-1 Mixed Distribution Models

If $f_i(t)$ is a failure probability density with hazard rate $\lambda_i(t)$, $i = 1$ to I, then the corresponding density $f(t)$ for the mixed distribution of a single-component device can be written as

$$f(t) = \sum_{i=1}^{I} k_i f_i(t). \tag{3-44}$$

The mixing parameters k_i, $i = 1$ to I, must be such that

$$0 \leq k_i \leq 1, \tag{3-45}$$

$$\sum_{i=1}^{I} k_i = 1. \tag{3-46}$$

Example 3-7 Construct a mixed distribution failure probability density for a combination of an exponential distribution and a gamma distribution.

From Eqs. (3-20), (3-24), and (3-44) through (3-46),

$$f(t) = k\,\lambda_1 \exp(-\lambda_1 t) + (1 - k)\frac{\lambda_2^r\, t^{r-1} \exp(-\lambda_2 t)}{\Gamma(r)}.$$

Figure 3-7a shows the exponential distribution (curve 1), the gamma distribution (curve 2), and the sum (curve 3) for $k = 0.1$. Figure 3-7b illustrates the instantaneous failure rate $\lambda(t)$ for the mixed model. For the case shown, there initially is a period of diminishing hazard rate ($0 \leq t \leq t_H$) during which "weak" items in a large population would be expected to fail; at later times, phenomena such as wear cause the rate to increase. ◇

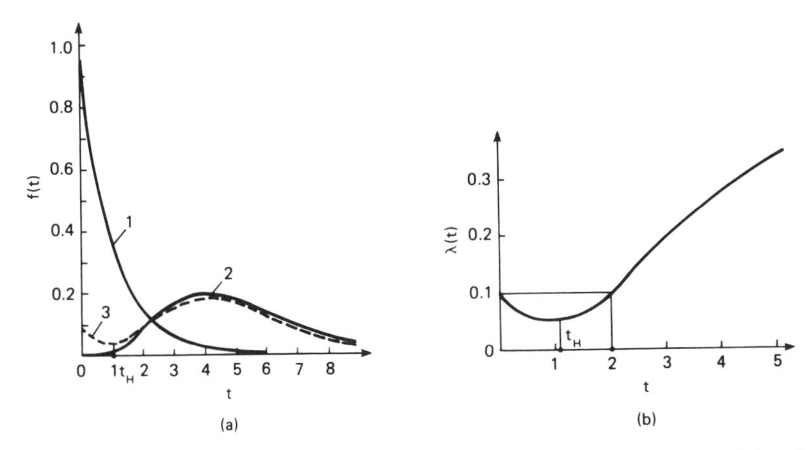

Fig. 3-7 *Mixed exponential and gamma distribution with* $\lambda_1 = \lambda_2 = 1$, $r = 5$, and $k = 0.1$. *(From I. B. Gertsbakh and K. B. Kordonskiy, "Models of Failure." Springer-Verlag, New York, 1969.)*

Example 3-8 Failures of a given device can be classified as either sudden (catastrophic) or delayed (wear-out). Develop a mixed distribution model for the cumulative failure probability of the device.

Catastrophic failures may occur as soon as the device is exposed to an operating environment outside the maximum tolerances for operation; then the Weibull distribution with a location parameter $\tau_1 = 0$ and a shape parameter $\alpha_1 < 1$ is an appropriate model. Wear-out failures are due to aging of the device; a Weibull model with a location parameter $\tau_2 > 0$ and a shape parameter $\alpha_2 > 1$ is an appropriate failure model. From Eq. (3-37)

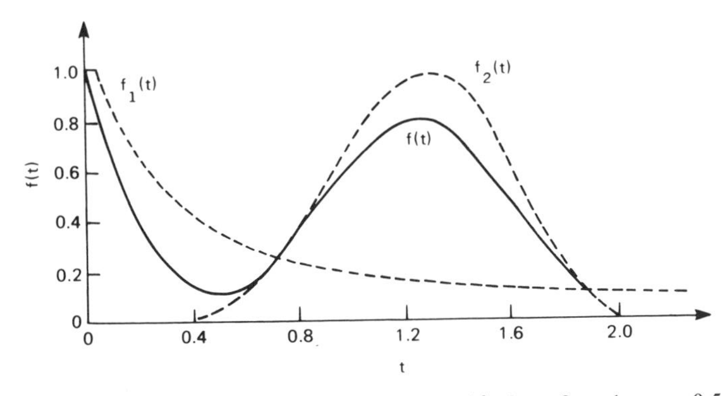

Fig. 3-8 *Mixed Weibull failure probability density with* $\beta_1 = \beta_2 = 1$, $\alpha_1 = 0.5$, $\alpha_2 = 3$, $\tau_1 = 0$, $\tau_2 = 0.4$, *and* $k = 0.2$. *(From N. R. Mann,* et al. *"Methods for Statistical Analysis of Reliability and Life Data." Wiley, New York, 1974.)*

and Eqs. (3-44) through (3-46), the cumulative failure probability is

$$F(t) = k\left\{1 - \exp\left[-(t/\beta_1)^{\alpha_1}\right]\right\} + (1 - k)\left\{1 - \exp\left[-\left(\frac{t - \tau_2}{\beta_2}\right)^{\alpha_2}\right]\right\}$$

for β_1 and $\beta_2 > 0$ and $0 \le k \le 1$. A special case of the corresponding $f(t)$ is shown in Fig. 3-8. ◇

3-3-2 Composite Distribution Models

A composite failure model for a one-component system can be constructed by linking together different failure probability densities for different time intervals. Then $f_j(t)$ denotes the composite probability density function for time interval $T_{j-1} \le t \le T_j$, where the times T_{j-1} and T_j are the partition parameters for the jth interval.

A special case of a composite distribution exists for any device that cannot fail for a finite period of time τ. In such situations, the device is not sensitive to any load to which it is subjected, so $f_1(t) = 0$ for $0 \le t \le \tau$. A composite distribution has already been illustrated in Example 2-5.

Example 3-9 A device is known to always fail in a random fashion while in a *phased mission* mode of operation consisting of three stages: $0 \le t < T_1$, $T_1 \le t < T_2$, and $t \ge T_2$. Obtain the cumulative failure probability for the device.

Three nonsynchronous hazard rates, denoted as λ_j, $j = 1$ to 3, characterize the hazard rate, which can be written as

$$\lambda(t) = \lambda_1 + (\lambda_2 - \lambda_1)H(t - T_1) + (\lambda_3 - \lambda_2)H(t - T_2),$$

where $H(x) = 1, x \ge 0$, and $H(x) = 0, x < 0$. Using Eqs. (2-48) and (2-54), the cumulative failure probability is

$$\begin{aligned} F(t) &= 1 - \exp(-\lambda_1 t), \qquad 0 \le t < T_1, \\ &= 1 - \exp[-\lambda_1 t - (\lambda_2 - \lambda_1)(t - T_1)], \qquad T_1 \le t \le T_2, \\ &= 1 - \exp[-\lambda_1 t - (\lambda_2 - \lambda_1)(t - T_1) \\ &\quad - (\lambda_3 - \lambda_2)(t - T_2)], \qquad t \ge T_2. \end{aligned}$$ ◇

A composite model has the advantage that it can sometimes provide flexibility in fitting and explaining failure data. It is really nothing more than the well-known method of approximating a function by dividing it into a number of regions. Intuitively, the greater the number of segments taken, the more accurate the approximation becomes, but engineering judgment must be exercised to balance goodness of fit and computational complexity.

3-3-3 Convoluted Distribution Models

A device that has replacement units in standby can continue to operate provided at least one of its units has not failed. The first unit operates until failure at $t = t_1$; the jth unit fails at $t = t_j$. In this section, it is assumed that there is perfect switching between units; cases in which this assumption is not made are discussed in Chapter 6.

The failure probability density for the ith *and all prior* units, $f_{12\cdots i}(t)$, may be expressed in terms of that for the $(i - 1)$th unit and all prior units, $f_{12\cdots(i-1)}(t)$, as the convolution of two failure probability densities:

$$f_{12\cdots i}(t) = \int_0^t f_i(t - t')\, f_{12\cdots(i-1)}(t')dt'. \tag{3-47}$$

In this equation, the failure probability density for the ith unit, $f_i(t - t')$, accounts for the system failure probability density for the time $(t - t')$ during which the ith unit is in operation, while the $f_{12\cdots(i-1)}(t')dt'$ accounts for the failure probability of the $(i - 1)$th unit in dt' about time t' after all other units j, $j < (i - 1)$, have failed. The integration over the time of failure t' of the $(i - 1)$th unit ranges from 0 to t because the actual time of the ith failure is not known.

Equation (3-47) can be rewritten in the form of nested integrals by recursively applying the equation. The result is

$$f_{12\cdots i}(t) = \int_0^t dt_{i-1} f_i(t - t_{i-1}) \int_0^{t_{i-1}} dt_{i-2} f_{i-1}(t_{i-1} - t_{i-2}) \cdots \times \int_0^{t_2} dt_1 f_2(t_2 - t_1) f_1(t_1). \tag{3-48}$$

Thus, for example, for a three-unit system, with initially two replacement units in standby, ready for use, the system probability density for failure is given by

$$f_{123}(t) = \int_0^t dt_2 f_3(t - t_2) \int_0^{t_2} dt_1 f_2(t_2 - t_1) f_1(t_1). \tag{3-49}$$

Any distribution discussed in Section 3-2 may be substituted for $f_i(t)$ in Eqs. (3-47) through (3-49). The $f_i(t)$ need not be identical for all i, but the system must be capable of functioning with any unit in operation.

Example 3-10 Calculate the failure probability density for a system consisting of i identical units, all having identical constant hazard rates λ. The units are used successively.

Equation (3-20) for the exponential failure model and Eq. (3-47) can be combined to give

$$f_{12\cdots i}(t) = \lambda \frac{(\lambda t)^{i-1}}{(i-1)!} e^{-\lambda t}.$$

This result should not be surprising since it is precisely the same as for the ith occurrence in the Poisson process, as seen from Eq. (3-19). ◇

To summarize the results of this section:

1. A mixed distribution consists of the addition of failure distributions with variable mixing coefficients for the different constituent probability densities.
2. A composite distribution is obtained from a set of piecewise-continuous hazard rates, each valid over a finite interval of time; if the composite hazard rate is continuous, so also is the composite failure probability density.
3. A convoluted distribution arises for a multi-unit system with one or more replacement units in standby that can be switched into service instantaneously with no switch failures.

3-4 EXTREME-VALUE DISTRIBUTIONS

Extreme-value distributions, or weakest-link distributions [8], deal with the probability of occurrence $F^*(x_*)$ of either the maximum or minimum x_* when a large number of independent events are sampled from an initial distribution $F(x)$ [3, 5, 8, 9]. [Here the asterisk is used to emphasize that the distribution $F^*(x_*)$ differs from the parent distribution $F(x)$ and x_* differs from x.] A classification system has been devised according to the behavior of $F(\infty)$ and to whether the minimum or maximum values are selected. Results of the classification scheme are shown in Table 3-1, along with a few natural phenomena for which the extreme value distributions apply. By restricting our attention to cases in which the number of events in the sample is large, we need consider only the *asymptotic* distributions for maximum or minimum values.

As an example, consider the daily discharge x from a river, which varies according to $F(x)$, over a year's time. If the maximum discharge x_* in a year is termed a flood, the distribution of sizes of floods $F^*(x_*)$ over a large number of years is a Type I asymptotic distribution of maximum values. For such a case, the cumulative probability is

$$F^*(x_*) = \exp[-e^{-\alpha(x_*-\beta)}],$$

$$-\infty < x_* < \infty, \qquad \alpha < 0, \qquad -\infty < \beta < \infty. \tag{3-50}$$

$\alpha > 0$

Table 3-1 *Classification Scheme and Applications for Extreme-value Distributions*[a]

Values sampled	Initial distribution sampled from, $F(x)$	Extreme-value distribution type, $F^*(x_*)$	Some applications
Maximum	Gamma	Type I (maximum values)	Sea wave height River level
	Lognormal	Type I (maximum values)	Flood damage magnitude Earthquake magnitude and frequency
Minimum	Normal (Gaussian)	Type I (minimum values)	Material fracture strength
Minimum	Gamma	Type III (identical to the Weibull)	Drought occurrence Wind speed minimum

[a] Type II distributions are not of interest for reliability and risk analyses.

Differentiation of this equation with respect to x_* reveals that the probability density function for the Type I asymptotic distribution of maximum values is

$$f^*(x_*) = \alpha \exp[-\alpha(x_* - \beta) - e^{-\alpha(x_* - \beta)}]. \tag{3-51}$$

The mean value of x_* and the variance are

$$m = \beta + 0.577\alpha^{-1} \tag{3-52}$$

$$\sigma^2 = 1.645\alpha^{-2}, \tag{3-53}$$

while the hazard function in units of x_*^{-1} is

$$\lambda^*(x_*) = \frac{\alpha e^{-\alpha(x_* - \beta)}}{\exp[e^{-\alpha(x_* - \beta)}] - 1}. \tag{3-54}$$

Another quantity of interest is the return period $T(x_*)$ of the extremes of magnitude at least x_*; this return period is given as

$$T(x_*) = [1 - F^*(x_*)]^{-1}, \tag{3-55}$$

where $F^*(x_*)$ is given in Eq. (3-50). For example, if x is the rainfall in a year, it takes an average of $T(x_*)$ years for an annual maximum rainfall of at least x_* to occur once.

Floods, tornadoes, hurricanes, and earthquakes are examples of natural disasters that can be fitted with a Type I asymptotic distribution of maximum values. The data for the probability of occurrence and the severity of each of these phenomena are generally quite "site-specific" because they depend upon the locality under consideration. Therefore detailed meteorological studies are required to use a Type I distribution for any risk analysis.

Example 3-11 An example of flooding data for a specific geographic location is shown in Fig. 3-9. The probability per year that a flood daily discharge rate will not exceed a specified rate is plotted versus the magnitude of the flood. The return period is also given as an axis. The upper and lower lines about the straight line define a control band that contains all of the observations.

This application of a Type I distribution assumes that the values of the daily discharges follow an exponential distribution, that 365 days in a year is sufficiently large for the asymptotic theory to be applicable, and that the daily discharges are independent [9]. The last assumption clearly does not hold, but probability plots of actual data have suggested that the Type I distribution provides a reasonable representation. ◇

The Type I asymptotic distribution of minimum values is a distribution giving the probability $F^*(x_*)$ of the minimum magnitude x_*. Such a distribution is obtained, for example, when one studies the fracture strength of a material that has a Gaussian distribution of crack sizes. Then the probability $F^*(x_*)$ that the minimum material strength is x_* is given by

$$F^*(x_*) = 1 - \exp\{-\exp[\alpha(x_* - \beta)]\}, \qquad -\infty < x_* < \infty$$

$\alpha > 0$ ~~$\alpha < 0,$~~ $\quad -\infty < \beta < \infty. \qquad$ (3-56)

The other functions of interest for the Type I asymptotic distribution of minimum values are

$$f^*(x_*) = \alpha \exp\{\alpha(x_* - \beta) - \exp[\alpha(x_* - b)]\}, \tag{3-57}$$

$$m = \beta - 0.577\alpha^{-1}, \tag{3-58}$$

$$\sigma^2 = 1.645\alpha^{-2}, \tag{3-59}$$

$$\lambda^*(x_*) = \alpha \exp[\alpha(x_* - \beta)]. \tag{3-60}$$

A comparison of Eqs. (3-50) through (3-60) shows that the two Type I extreme-value distributions are closely related, although the values of $\lambda^*(x_*)$ shown in Fig. 3-10 do display a dramatic difference.

The Type III extreme-value distribution of minimum values is the Weibull distribution. As mentioned before, the Weibull distribution is suitable for describing the probability of failure of a nuclear reactor fuel assembly made up of identical rods provided the failure of an assembly occurs whenever one rod fails and the lifetime of each rod is given by the gamma distribution as the parent distribution.

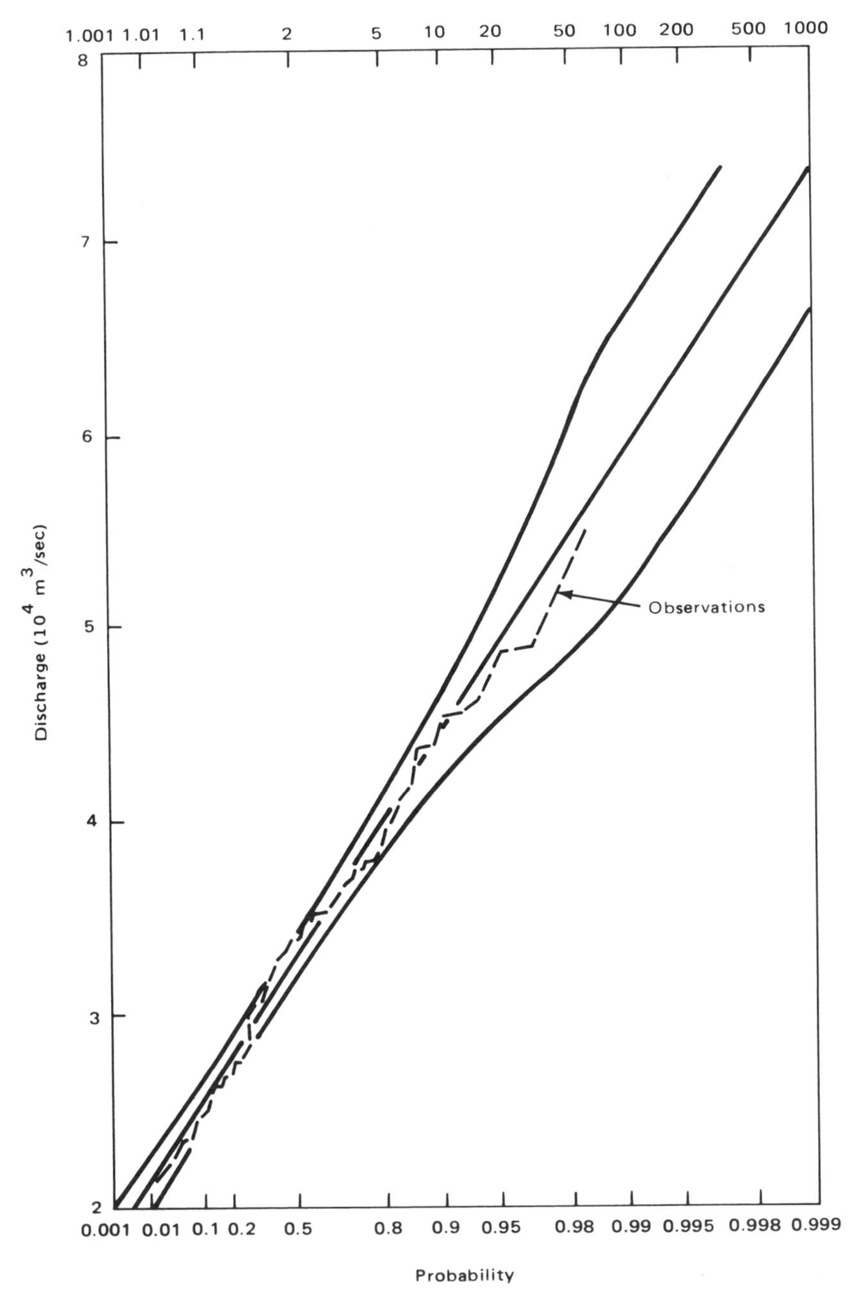

Fig. 3-9 *Mississippi River floods, Vicksburg, Mississippi, 1898–1949. (From E. J. Gumbel, "Statistics of Extremes." Columbia Univ. Press, New York, 1958.)*

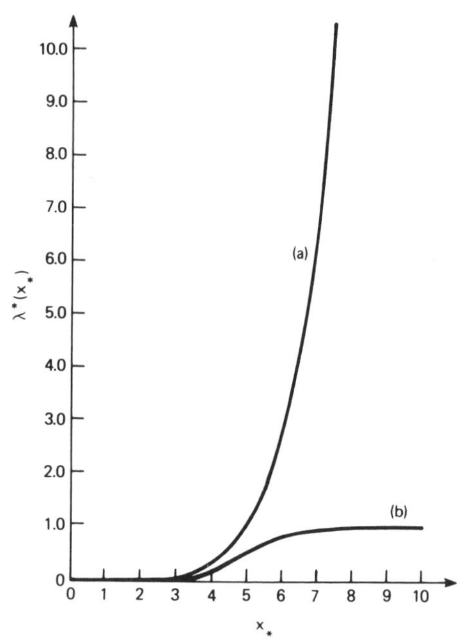

Fig. 3-10 *Type I extreme-value distribution hazard functions for* $\beta = 5$ *and* $\alpha = 1$ *(a) of minimum values and (b) of maximum values. (From G. J. Hahn and S. S. Shapiro, "Statistical Models in Engineering." Wiley, New York, 1967.)*

In summary, the distributions of maximum or minimum values in large samples drawn from a gamma distribution, and the maximum value in large samples drawn from the lognormal distribution, are extreme-value distributions.

EXERCISES

3-1 If a reactor has 3 coolant loops, each of which has a failure probability of 0.05 over the lifetime of the plant, what is the probability that at least one loop will fail?

3-2 A slot machine has 3 reels, each with 10 different symbols. What is the probability of obtaining 2 lemons (a) exactly twice in 20 trials? (b) more than twice in 20 trials?

3-3 Generalize the binomial distribution of Eq. (3-5) by replacing $P(\bar{D})$ by Θ, r by a (noninteger) variable $(\alpha - 1)$, and $n - r$ by a variable $(\beta - 1)$

to obtain

$$f(\Theta) = \frac{\Gamma(\alpha+\beta)}{\Gamma(\alpha)\Gamma(\beta)}\Theta^{\alpha-1}(1-\Theta)^{\beta-1}, \qquad \alpha > 0, \qquad \beta > 0, \qquad 0 \le \Theta \le 1$$

$$= 0 \text{ otherwise.}$$

This is the *beta* probability distribution [3] that is used when a random variable Θ takes values in the interval [0, 1]. For the variable Θ' defined for the interval $[\Theta_1, \Theta_2]$, (a) develop the linear transformation to map Θ' to Θ, and (b) obtain an equation for $f(\Theta')$.

3-4 The probability of an event occurring per trial is 0.1. (a) What is the probability of exactly 12 events occurring out of 100 trials? (b) What is this probability if it is calculated by using the Poisson approximation to the binomial distribution?

3-5 The electric motors in a widget factory suffer a collective failure rate of 1.37 failures/yr, based upon the records from the last 20 years. What is the probability that 3 or more motors fail in a single year?

3-6 It is found that the number of system breakdowns occurring with a constant rate in a given length of time has a mean value of 2 breakdowns. What are the probabilities, in the same length of time, of (a) no breakdowns, (b) 1 breakdown, (c) 2 breakdowns, (d) 10 breakdowns?

3-7 Some control units have times to their first failures that are exponentially distributed. The mean of this distribution is 5000 hr. What is the probability that a unit that has survived (a) 1000 hr, (b) 5000 hr, and (c) 10,000 hr will survive an additional 1000 hr?

3-8 The probability that a system with constant hazard rate λ will fail to survive for a mission of 100 λ is 0.5. What are the probabilities that (a) it survives for 500 λ, (b) it fails within 1000 λ?

3-9 A device fails with a hazard rate given by

$$\lambda(t) = at, \qquad t \le T,$$
$$= aT^2/t, \qquad t > T,$$

where a is a constant and $aT^2 > 1$. If $a = 10^{-2}$ hr^{-2} and $T = 15$ hr, (a) calculate the time τ after the device is placed in service before the probability of its failure is 0.95; (b) derive the equation for MTTF and accurately determine its numerical value using published tables.

3-10 Given the probability density function $f(t) = Ct^a e^{-t}$, where $a \ge 0$, calculate (a) the normalization constant C that makes $f(t)$ satisfy the constraint $F(\infty) = 1$, (b) and the mean and variance of $f(t)$.

3-11 A large number of valve mechanisms are found to have times to failure that follow a Weibull distribution. The parameters of this distribution are $\beta = 10$ years and $\alpha = 0.5$. What is the probability that such a mechanism will survive (a) 1 year, (b) 5 years, and (c) 10 years without failure, and (d) what is the mean time to failure?

3-12 An analysis reveals that a particular batch of components fails in such a way that the distribution of failures may be described by a two-parameter Weibull function with $\beta = 10^5$ hr and $\alpha = 2.3$. Find (a) the probability that one of these components will not fail in 10,000 hr, (b) the mean value for the distribution.

3-13 The manager of a trucking firm has determined that the trucks in the fleet require major overhauls according to a Weibull distribution, with $\beta = 250{,}000$ km and $\alpha = 1.7$. What are the probabilities that a truck will not break down in (a) 100,000 km, (b) 500,000 km?

3-14 For a system with $\lambda(t) = 0.007\, t^{-0.3}$/yr, find (a) the probability the system will fail in 5 years, (b) the mean time to failure.

3-15 Calculate the cumulative failure probability for the composite distribution model of Fig. E3-15 if $a_2 = a_3$.

3-16 A device has a continuous hazard rate specified by the equations

$$\begin{aligned} \lambda(t) &= a\, t^{-0.5}, \qquad & 0 \le t \le t_1 \\ \lambda(t) &= \lambda, & t_1 \le t \le t_2 \\ \lambda(t) &= b(t - t_1)^2, & t \ge t_2 \end{aligned}$$

(a) Evaluate the constants a and b in terms of λ. (b) Give the equations for the reliability of the device as a function of time.

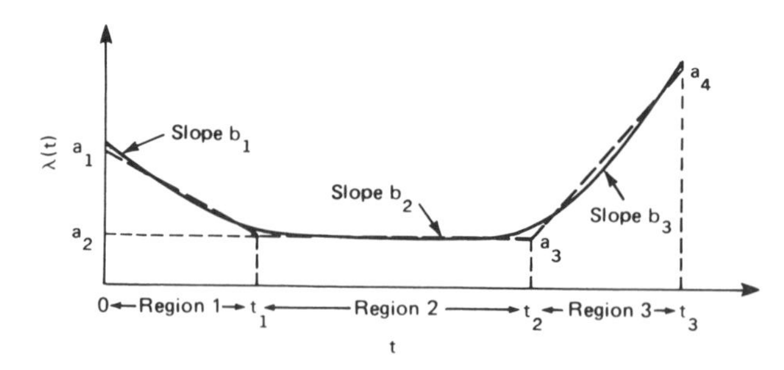

Fig. E3-15 *Piecewise-linear model hazard rate. (From M. L. Shooman, "Probabilistic Reliability: An Engineering Approach." McGraw-Hill, New York, 1968.)*

3-17 Another model, different from those developed in Section 3-3, can be developed by approximating the hazard rate for a device by the power series expension

$$\lambda(t) = \sum_{i=1}^{I} K_i t^i$$

for constant K_i values. (a) Calculate the cumulative failure probability $F(t)$. (b) For $I = 4$, fit the model to the four points a_i, $i = 1$ to 4, in Fig. E3-15 and solve the four resulting equations for the constants K_i in terms of the a_i, $i = 1$ to 4 and t_1, t_2, and t_3.

REFERENCES

1. A. J. Bourne and A. E. Green, "Reliability Technology." Wiley, New York, 1972.
2. M. L. Shooman, "Probabilistic Reliability: An Engineering Approach." McGraw-Hill, New York, 1968.
3. N. R. Mann, R. E. Schafer, and N. D. Singpurwalla, "Methods for Statistical Analysis of Reliability and Life Data." Wiley, New York, 1974.
4. C. Lipson and N. J. Sheth, "Statistical Design and Analysis of Engineering Experiments." McGraw-Hill, New York, 1973.
5. G. J. Hahn and S. S. Shapiro, "Statistical Models in Engineering." Wiley, New York, 1967.
6. M. Abramowitz and I. A. Stegun (eds.), "Handbook of Mathematical Functions." National Bureau of Standards, Washington, D.C., 1964; reprinted by Dover, New York.
7. D. N. Rousu, Weibull skewness and kurtosis as a function of shape parameters, *Technometrics* **15,** 927 (1973).
8. W. G. Ireson, "Reliability Handbook." McGraw-Hill, New York, 1966.
9. E. J. Gumbel, "Statistics of Extremes." Columbia Univ. Press, New York, 1958.

CHAPTER 4

Data Manipulation Concepts

4-1 CURVE FITTING OF DATA

When plotting cumulative failure probabilities to fit data to the different probability distributions of Chapter 3, it is necessary to decide what probability value to assign to the time of each failure. If we knew the exact percent of the population failing before each failure, that percent would be the true rank of each failure. Since we do not know the true rank when testing a small number of samples, however, we must estimate it.

We use an estimate such that, in the long run, the positive and negative errors of the estimate cancel each other. In other words, half the time the rank will be too high and the other half the rank will be too low. A rank with this property is called a *median* or 50% rank. Table 4-1 is a table of such median ranks for different sample sizes, ordered as a function of the failure order number; more extensive tables of median ranks also are available [1]. Whenever the tables are not available, or when the sample size n exceeds that in Table 4-1, the median ranks are given approximately by $(j - 0.3)/(n + 0.4)$, where j is the failure order number.

Example 4-1 Ten identical devices are tested until failure at times t_i, $i = 1$ to 10, with $t_{i+1} \geq t_i$. What are the predicted cumulative probabilities of failure $F(t_i)$ for the 10 data points?

Because the sample size $n = 10$ is so small, we should not set $F(t_1) = 0.10$, $F(t_2) = 0.20$, etc. From Table 4-1, the cumulative failure probabilities are assigned values of $F(t_1) = 0.0670$, $F(t_2) = 0.1632$, etc., up to $F(t_{10}) = 0.9330$. ◇

When fitting data, it is advantageous to use the proper probability paper since the data can then be fitted to a straight line [2–6]. For example, the paper of Fig. 4-1 should be used when fitting data to the exponential distribution,

$$F(t) = 1 - e^{-\lambda(t-\tau)}, \qquad t \geq \tau \tag{4-1}$$
$$= 0, \qquad t < \tau.$$

Table 4-1 *Table of Median Ranks r for Different Sample Sizes n*

r	Sample size n: 1	2	3	4	5	6	7	8	9	10
1	0.5000	0.2929	0.2063	0.1591	0.1294	0.1091	0.0943	0.0830	0.0741	0.0670
2		0.7071	0.5000	0.3864	0.3147	0.2655	0.2295	0.2021	0.1806	0.1632
3			0.7937	0.6136	0.5000	0.4218	0.3648	0.3213	0.2871	0.2594
4				0.8409	0.6853	0.5782	0.5000	0.4404	0.3935	0.3557
5					0.8706	0.7345	0.6352	0.5596	0.5000	0.4519
6						0.8909	0.7705	0.6787	0.6065	0.5481
7							0.9057	0.7979	0.7129	0.6443
8								0.9170	0.8194	0.7406
9									0.9259	0.8368
10										0.9330

r	Sample size n: 11	12	13	14	15	16	17	18	19	20
1	0.0611	0.0561	0.0519	0.0483	0.0452	0.0424	0.0400	0.0378	0.0358	0.0341
2	0.1489	0.1368	0.1266	0.1178	0.1101	0.1034	0.0975	0.0922	0.0874	0.0831
3	0.2366	0.2175	0.2013	0.1873	0.1751	0.1644	0.1550	0.1465	0.1390	0.1322
4	0.3244	0.2982	0.2760	0.2568	0.2401	0.2254	0.2125	0.2009	0.1905	0.1812
5	0.4122	0.3789	0.3506	0.3263	0.3051	0.2865	0.2700	0.2553	0.2421	0.2302
6	0.5000	0.4596	0.4253	0.3958	0.3700	0.3475	0.3275	0.3097	0.2937	0.2793
7	0.5878	0.5404	0.5000	0.4653	0.4350	0.4085	0.3850	0.3641	0.3453	0.3283
8	0.6756	0.6211	0.5747	0.5347	0.5000	0.4695	0.4425	0.4184	0.3968	0.3774
9	0.7634	0.7018	0.6494	0.6042	0.5650	0.5305	0.5000	0.4728	0.4484	0.4264
10	0.8511	0.7825	0.7240	0.6737	0.6300	0.5915	0.5575	0.5272	0.5000	0.4755
11	0.9389	0.8632	0.7987	0.7432	0.6949	0.6525	0.6150	0.5816	0.5516	0.5245
12		0.9439	0.8734	0.8127	0.7599	0.7135	0.6725	0.6359	0.6032	0.5736
13			0.9481	0.8822	0.8249	0.7746	0.7300	0.6903	0.6547	0.6226
14				0.9517	0.8899	0.8356	0.7875	0.7447	0.7063	0.6717
15					0.9548	0.8966	0.8450	0.7991	0.7579	0.7207
16						0.9576	0.9025	0.8535	0.8095	0.7698
17							0.9600	0.9078	0.8610	0.8188
18								0.9622	0.9126	0.8678
19									0.9642	0.9169
20										0.9659

Then the time delay τ is the value at which the straight line intersects the abscissa, and the estimate of the hazard rate, $\hat{\lambda}$, is obtained from

$$\hat{\lambda} = (t^* - \tau)^{-1}, \tag{4-2}$$

where t^* is the time for which the cumulative percentage of failures $F(t^*) = 0.632$, corresponding to $\ln[1 - F(t^*)] = -1$.

Table 4-1 (continued)[a]

	Sample size n									
r	21	22	23	24	25	26	27	28	29	30
1	0.0330	0.0315	0.0301	0.0288	0.0277	0.0266	0.0256	0.0247	0.0239	0.0231
2	0.0797	0.0761	0.0728	0.0698	0.0670	0.0645	0.0621	0.0599	0.0579	0.0559
3	0.1264	0.1207	0.1155	0.1108	0.1064	0.1023	0.0986	0.0951	0.0919	0.0888
4	0.1731	0.1653	0.1582	0.1517	0.1457	0.1402	0.1351	0.1303	0.1259	0.1217
5	0.2198	0.2099	0.2009	0.1927	0.1851	0.1781	0.1716	0.1655	0.1599	0.1546
6	0.2665	0.2545	0.2437	0.2337	0.2245	0.2159	0.2081	0.2007	0.1939	0.1875
7	0.3132	0.2992	0.2864	0.2746	0.2638	0.2538	0.2445	0.2359	0.2279	0.2204
8	0.3599	0.3438	0.3291	0.3156	0.3032	0.2917	0.2810	0.2711	0.2619	0.2533
9	0.4066	0.3884	0.3718	0.3566	0.3425	0.3295	0.3175	0.3063	0.2959	0.2862
10	0.4533	0.4330	0.4145	0.3975	0.3819	0.3674	0.3540	0.3415	0.3299	0.3191
11	0.5000	0.4776	0.4572	0.4385	0.4212	0.4053	0.3905	0.3767	0.3639	0.3519
12	0.5466	0.5223	0.5000	0.4795	0.4606	0.4431	0.4270	0.4119	0.3979	0.3848
13	0.5933	0.5669	0.5427	0.5204	0.5000	0.4810	0.4635	0.4471	0.4319	0.4177
14	0.6400	0.6115	0.5854	0.5614	0.5393	0.5189	0.5000	0.4823	0.4659	0.4506
15	0.6867	0.6561	0.6281	0.6024	0.5787	0.5568	0.5364	0.5176	0.5000	0.4835
16	0.7334	0.7007	0.6708	0.6433	0.6180	0.5946	0.5729	0.5528	0.5340	0.5164
17	0.7801	0.7454	0.7135	0.6843	0.6574	0.6325	0.6094	0.5880	0.5680	0.5493
18	0.8268	0.7900	0.7562	0.7253	0.6967	0.6704	0.6459	0.6232	0.6020	0.5822
19	0.8735	0.8346	0.7990	0.7662	0.7361	0.7082	0.6824	0.6584	0.6360	0.6151
20	0.9202	0.8792	0.8417	0.8072	0.7754	0.7461	0.7189	0.6936	0.6700	0.6480
21	0.9669	0.9238	0.8844	0.8482	0.8148	0.7840	0.7554	0.7288	0.7040	0.6808
22		0.9684	0.9271	0.8891	0.8542	0.8218	0.7918	0.7640	0.7380	0.7137
23			0.9698	0.9301	0.8935	0.8597	0.8283	0.7992	0.7720	0.7466
24				0.9711	0.9329	0.8976	0.8648	0.8344	0.8060	0.7795
25					0.9722	0.9354	0.9013	0.8696	0.8400	0.8124
26						0.9733	0.9378	0.9048	0.8740	0.8453
27							0.9743	0.9400	0.9080	0.8782
28								0.9752	0.9420	0.9111
29									0.9760	0.9440
30										0.9768

[a] From L. G. Johnson, "The Statistical Treatment of Fatigue Experiments." Elsevier, Amsterdam, 1964.

There is no general probability paper for fitting the gamma distribution to all values of the shape parameter r, but paper is available for the cases in which $r = 0.5, 1, 1.5, 2, 3, 4$, or 5 and is known as chi-square probability paper for $2r$ degrees of freedom (see Appendix A). The paper with $r = 1$ is the probability paper of Fig. 4-1.

For the lognormal distribution, the probability paper of Fig. 4-2 should be used. This probability paper is the same as that for the normal distribution, except that the logarithm of the time-value is used along the abscissa.

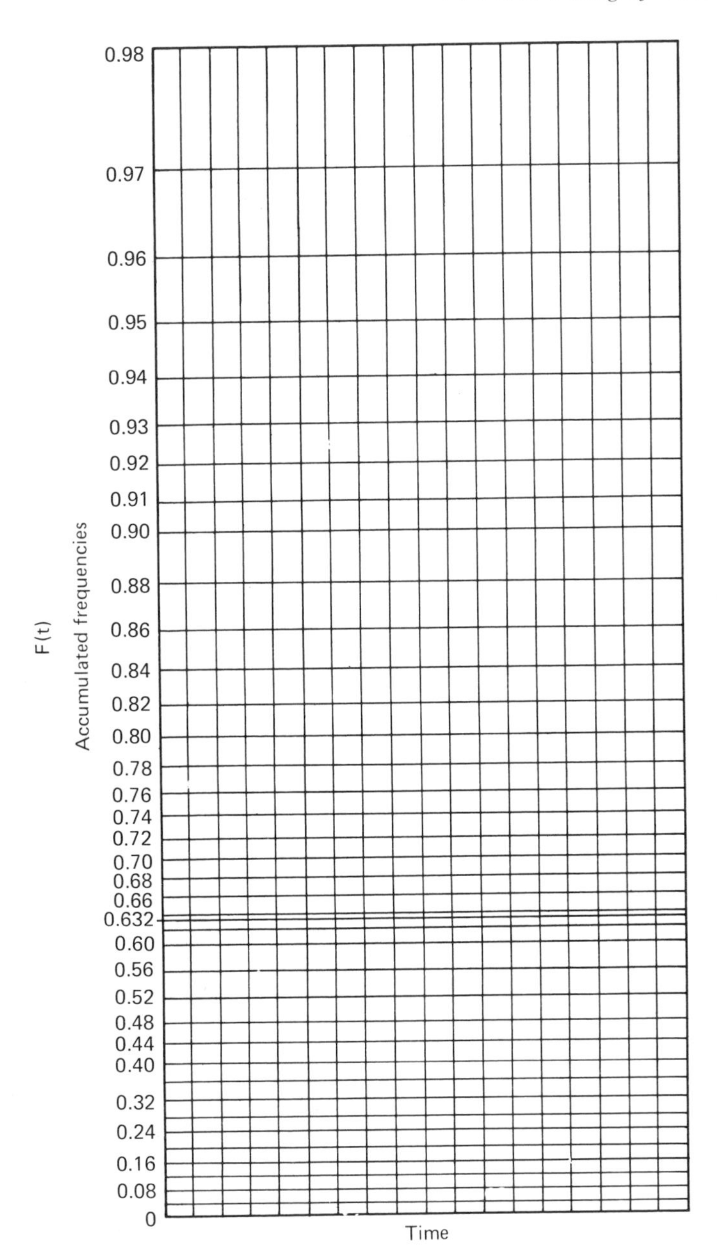

Fig. 4-1 *Exponential distribution probability paper. (From I. B. Gertsbakh and K. B. Kordonskiy, "Models of Failure." Springer-Verlag, New York, 1969.)*

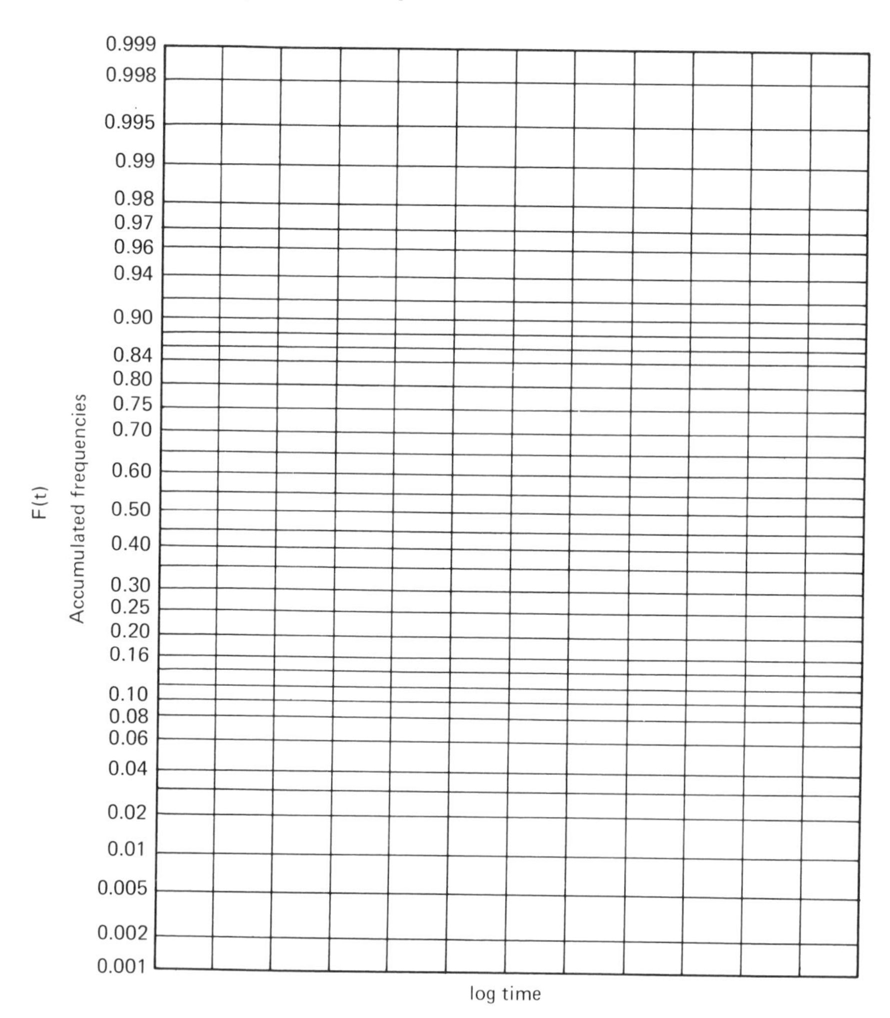

Fig. 4-2 *Lognormal distribution probabilty paper. (From I. B. Gertsbakh and K. B. Kordonskiy, "Models of Failure." Springer-Verlag, New York, 1969.)*

The parameter β in Eq. (3-30) is estimated to be the time at the abscissa corresponding to a cumulative probability of 0.5, and the parameter α is estimated as the difference between the ln-time abscissa values for which $F(t) = 0.5$ and $F(t) = 0.159$. Use of lognormal paper is described in more detail elsewhere [2, 5].

Because of the widespread use of the Weibull distribution for describing failures of devices, it is worthwhile to describe in more detail how the parameters α and β are obtained by fitting failure data. This is done by taking logarithms of Eq. (3-37) twice to obtain

$$\ln \ln[1 - F(t)]^{-1} = -\alpha \ln \beta + \alpha \ln(t - \tau), \tag{4-3}$$

which is the equation for a stright line on paper for which the ordinate is a loglog scale and the abscissa is a log scale. An example of such paper is shown in Fig. 4-3.

The paper is used by plotting the cumulative percent failures $F(t)$ versus

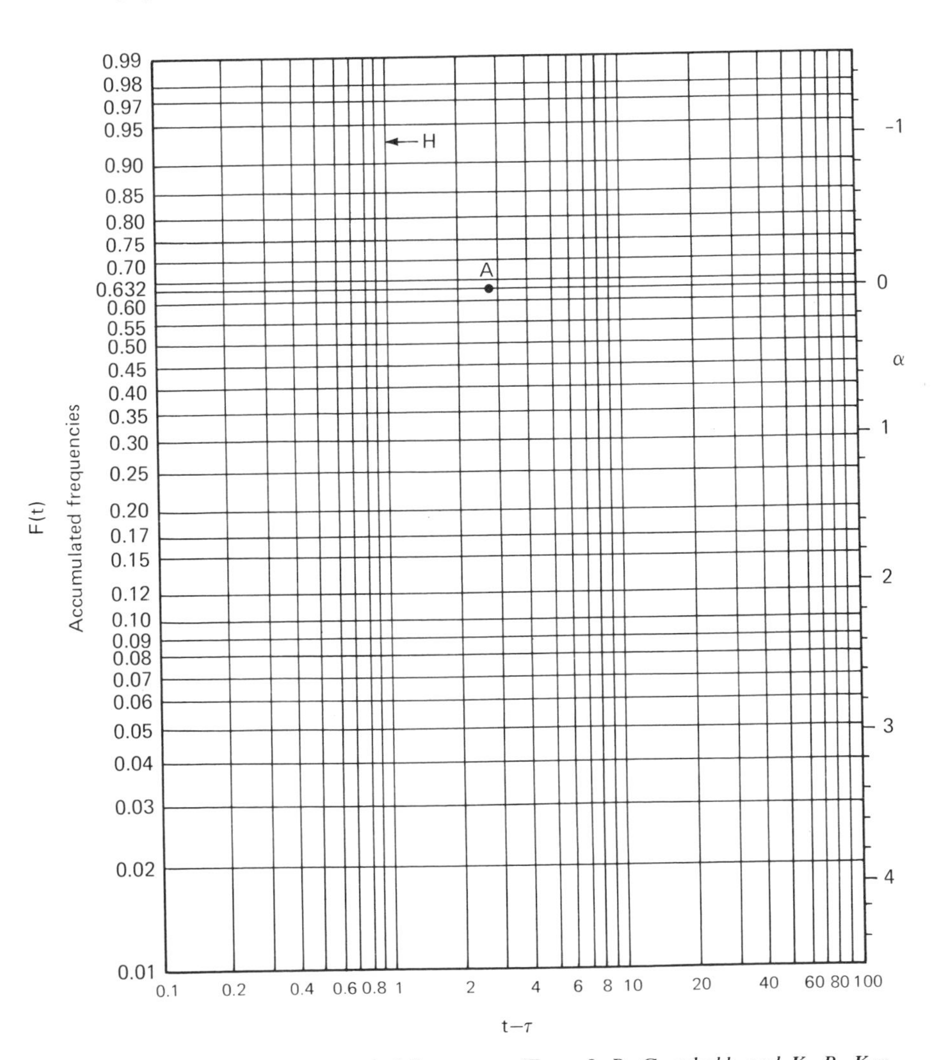

Fig. 4-3 *Weibull distribution probability paper. (From I. B. Gertsbakh and K. B. Kordonskiy, "Models of Failure." Springer-Verlag, New York, 1969.)*

the time of each failure on the abscissa $(t - \tau)$ and then drawing a best straight line through the data: the estimate of the shape factor β, denoted as $\hat{\beta}$, is obtained from the value of $t - \tau$ corresponding to

$$F(t) = 1 - e^{-1} = 0.632, \tag{4-4}$$

since, for this value of $F(t)$, the left-hand side of Eq. (4-3) vanishes. The value of τ, unfortunately, must be obtained by trial-and-error; a good starting value, of course, is $\tau = 0$.

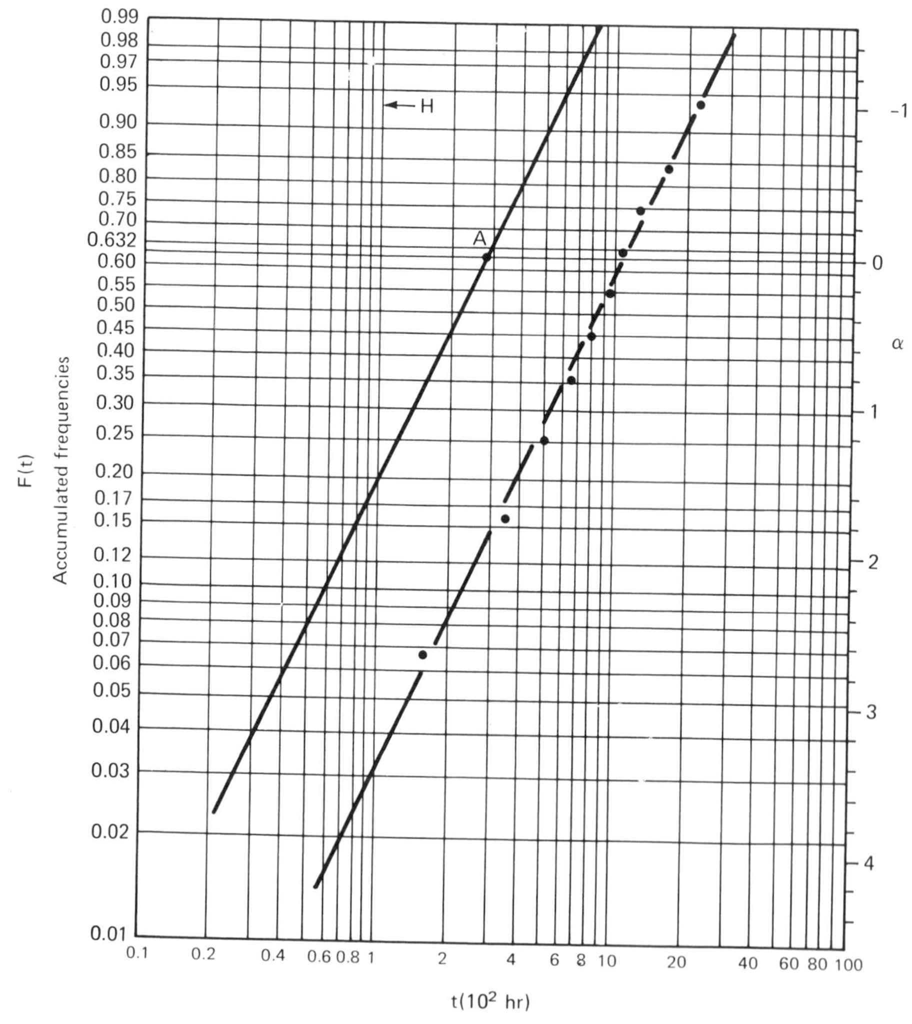

Fig. 4-4 *Cumulative failure probability versus time for Example 4-2.*

To obtain $\hat{\alpha}$, the estimate of α, a line parallel to that through the data is constructed through the special point A in Fig. 4-3, which has coordinates (2.718, 0.632). The intercept of the line through A with the vertical line H gives the value $\hat{\alpha}$, which may be read from the right-hand scale of α-values in Fig. 4-3.

Example 4-2 [3] Ten identical devices are tested with failures occurring at 1.7, 3.5, 5.0, 6.5, 8.0, 9.6, 11, 13, 18, and 22 ($\times 10^2$) hr. It is believed that the data may be fitted with a Weibull distribution.

Use of the median ranks from Exercise 4-1 to plot these values versus the times of failure results in Fig. 4-4. Following the procedure just described, the values of $\hat{\alpha} = 1.5$ and $\hat{\beta} = 1100$ hr are obtained for $\tau = 0$.◇

Once the parameters have been estimated by the plotting method, one may wish to answer the question, "How well does the distribution, using the parameters just determined, fit the data?" A "goodness-of-fit" test, such as the classical *Kolmogorov-Smirnov* or *Cramér-von Mises* test, may be used; other specially developed tests for particular distributions also are available [5, 7]. Such statistical tests are beyond the scope of an introduction to reliability and risk analysis, however.

In the remainder of this chapter, we shall survey data manipulation techniques that do not involve curve fitting.

4-2 INTRODUCTION TO ESTIMATION THEORY

Sampling theory is a study of relationships that exist between a population and samples drawn from that population. Sampling in which each member of the population may be chosen more than once is called *sampling with replacement;* sampling in which each member may not be chosen more than once is termed *sampling without replacement.* A finite population in which sampling is with replacement theoretically can be considered to be infinite in size, since any number of samples can be drawn.

For each sample of size N that can be drawn randomly (either with or without replacement) from a given population, we can compute a statistic such as the mean, standard deviation, or higher-order moments. For example, the arithmetic mean $\bar{X}$ of a set of N numbers $X_1, X_2, \ldots, X_N$ is

$$\bar{X} = N^{-1} \sum_{j=1}^{N} X_j, \tag{4-5}$$

while the variance s^2 (i.e., the square of the standard deviation) is defined as

$$s^2 = N^{-1} \sum_{j=1}^{N} (X_j - \bar{X})^2. \tag{4-6}$$

[The higher moments about the mean, such as the skewness and kurtosis, have the same form as Eq. (4-6) except the power two on the right-hand side is replaced by three and four, respectively.] The statistics in Eqs. (4-5) and (4-6) will vary from sample to sample, giving rise to a sampling distribution of the mean and the variance, respectively.

The estimation of the population mean m and variance σ^2 from the corresponding sample statistics (i.e., sample mean and variance) is a part of statistical inference. If the mean of the sampling distribution of a statistic equals the corresponding population parameter, then the statistic is said to be an "unbiased estimator" of the parameter, and the value of the statistic is an "unbiased estimate"; otherwise the statistic is a "biased estimator" whose value is a "biased estimate." For example, the sample mean $\bar{X}$ given in Eq. (4-5) is an unbiased estimate $\hat{m}$ of the population mean m, so

$$\hat{m} = \bar{X}, \tag{4-7}$$

but to obtain an unbiased estimate of the population variance σ^2, denoted by $\hat{\sigma}^2$, it is necessary to use the equation [4–7]

$$\hat{\sigma}^2 = \frac{N}{N - 1} s^2, \tag{4-8}$$

where s^2 is defined in Eq. (4-6).

Now suppose all possible samples of size N are drawn without replacement from a finite population of size $N_P > N$. If the mean and standard deviation of the sampling distribution of the mean are denoted by $m_{\bar{X}}$ and $\sigma_{\bar{X}}$, and the population mean and standard deviation by m and σ, respectively, then

$$m_{\bar{X}} = m, \tag{4-9}$$

$$\sigma_{\bar{X}} = \frac{\sigma}{\sqrt{N}} \sqrt{\frac{N_p - N}{N_p - 1}}. \tag{4-10}$$

If the population is infinite, or if the sampling is done with replacement, then the last equation simplifies to

$$\sigma_{\bar{X}} = \sigma/\sqrt{N}. \tag{4-11}$$

With large enough N, say $N \geq 30$, the sampling distribution of means is approximately a normal distribution with mean $m_{\bar{X}}$ and standard deviation $\sigma_{\bar{X}}$, irrespective of the size of N_p (provided $N_p \geq 2n$). That is, the sampling distribution is said to be *asymptotically normal.*

Example 4-3 Five hundred ball bearings have a mean mass of 5.02 g and a standard deviation of 0.30 g. If a sample of 100 ball bearings is chosen from this group, $m_{\bar{X}} = 5.02$ g and

$$\sigma_{\bar{X}} = \frac{0.30\ \text{g}}{\sqrt{100}}\sqrt{\frac{500 - 100}{500 - 1}} = 0.027\ \text{g}.$$

If the sample of 100 ball bearings is chosen with replacement, $m_{\bar{X}} = 5.02$ g and

$$\sigma_{\bar{X}} = 0.30\ \text{g}/\sqrt{100} = 0.03\ \text{g}. \quad \diamond$$

Equations (4-7) and (4-8) give single numbers for the unknown parameters m and σ^2; these estimates are referred to as *point estimates. Interval estimates* are a second kind of estimate for which ranges of the unknown parameters are calculated. Such estimates are used to indicate the precision or accuracy of a point estimate and, for this reason, are sometimes referred to as confidence estimates. For example, for the sample mean $\bar{X}$, defined in Eq. (4-5), the 95% confidence coefficient for estimation of the population mean m for a sample of large size obeying a normal distribution is given by $\bar{X} \pm 1.96\ \sigma_{\bar{X}}$. Another way of writing the 95% confidence limit is

$$P(\bar{X} - 1.96\ \sigma_{\bar{X}} < m < \bar{X} + 1.96\ \sigma_{\bar{X}}) = 0.95. \tag{4-12}$$

More generally, the confidence limits are given by $\bar{X} \pm t\sigma_{\bar{X}}$, where t must be obtained from a table of t-values from *Student's* t*-distribution* for two-sided confidence interval estimation. This table is given in Table 4-2 for the case of a large sample size N; more complete tables are available for cases when the sample size is small [1, 8].

Example 4-4 For the sample of 100 ball bearings drawn without replacement in Example 4-3,

$$\begin{aligned} &P(\bar{X} - t\ \sigma_{\bar{X}} < m < \bar{X} + t\ \sigma_{\bar{X}}) \\ &\quad = P(5.02 - 0.027t < 5.02 < 5.02 + 0.027t). \end{aligned}$$

Thus, for example, for $P(4.98 < m < 5.06)$, we calculate $t = 1.48$ so that from Table 4-2 we find

$$P(4.98 < m < 5.06) \simeq 0.85.$$

Conversely, with 95% confidence the population mean m is within $4.967 < m < 5.073$ since

$$\begin{aligned} &P(\bar{X} - 1.96\ \sigma_{\bar{X}} < m < \bar{X} + 1.96\ \sigma_{\bar{X}}) \\ &= P(4.967 < m < 5.073) = 0.95. \quad \diamond \end{aligned}$$

Table 4-2 *Confidence Levels for the Mean of a Normal Distribution (for a Large Sample Size)*

Two-sided confidence level (%)	One-sided confidence level (%)	t
99.73	99.86	3.00
99	99.5	2.58
98	99	2.33
96	98	2.05
95.45	97.72	2.00
95	97.5	1.96
90	95	1.645
80	90	1.28
68.27	84.14	1.00
50	75	0.6745

A one-sided interval estimate also may be calculated. Because the normal distribution is symmetric, the one-sided estimates corresponding to Eq. (4-12) are given by

$$P(m < \bar{X} + 1.96\,\sigma_{\bar{X}}) = P(m > \bar{X} - 1.96\,\sigma_{\bar{X}}) = 0.975. \qquad (4\text{-}13)$$

The one-sided confidence levels from Student's t-distribution are also listed in Table 4-2 for a sample of large size.

Interval estimates for reliability and risk analyses using the lognormal distribution follow from the approach just used for the normal distribution except that a ln-time variable is used instead of time. The determination of interval estimates for the gamma and Weibull distributions is considerably more complicated [7, 9] and is beyond the scope of coverage here. For our purposes, it will suffice that one is aware of the differences between point and interval estimates and knows that the curve fitting in Section 4-1 and the analytical methods in Section 4-3 are used to obtain point estimates.

4-3 POINT ESTIMATES FOR CONTINUOUS DISTRIBUTIONS

The equations used to obtain point estimates depend upon the type of environment the samples are subjected to during a *life test*. For example, if the life test is terminated at time t_s before all N items have failed, then Type I censoring of the life test has been done; on the other hand, Type II censoring occurs when the life test is terminated at the time of a particular failure K, where $K \leq N$. In this section, we shall focus mostly on the simplest case of sampling where all available units are tested to failure so that there is no censoring.

Only the moment estimator, maximum likelihood, and maximum entropy methods will be considered here for obtaining point estimates of the parameters in the failure probability distributions of Section 3-2; least-squares estimates are known more commonly by the name *method of least squares* and are discussed elsewhere [1, 10].

4-3-1 Moment Estimators

The moment method of point estimation is a straightforward application of the concepts in the previous section. If the numbers $t_1, t_2, \ldots, t_N$ represent a set of data for the actual failure times of a set of N identical units, then an unbiased estimator for the mean time to failure, defined by

$$m = \int_0^\infty t f(t)\, dt, \tag{4-14}$$

is

$$\hat{t} = \frac{1}{N} \sum_{n=1}^{N} t_n. \tag{4-15}$$

For the variance

$$\sigma^2 = \int_0^\infty (t - m)^2 f(t)\, dt, \tag{4-16}$$

the estimator is

$$\hat{\sigma}^2 = \frac{1}{N-1} \sum_{n=1}^{N} (t_n - \hat{t}\,)^2. \tag{4-17}$$

Here the caret continues to denote the estimator value. [Note that the factor of $(N - 1)^{-1}$ appears on the right-hand side of Eq. (4-17) instead of N^{-1}, in order to keep the estimator unbiased.]

Equations (4-14) through (4-17) may be combined to give two equations for calculating two unknown parameters for the failure probability distributions. The results summarized in Table 4-3 are collected from the appropriate equations in Section 3-2.

Example 4-5 For the ten devices described in Example 4-2, estimate the parameters α and β for the Weibull distribution using moment estimators.

The first step is to calculate $\hat{t}$ and $\hat{\sigma}^2$ using Eqs. (4-15) and (4-17), respectively, using the ten failure times. It is found from the data that $\hat{t} = 983$ hr and $\hat{\sigma}^2 = 4.114 \times 10^5$ hr^2. From Table 4-3 it is necessary to

Table 4-3 *Moment Estimators for Failure Probability Distributions*[a]

Distribution name	Distribution $f(t)$	$\hat{t} = \frac{1}{N} \Sigma_{n=1}^{N} t_n$	$\hat{\sigma}^2 = \frac{1}{N-1} \Sigma_{n=1}^{N} (t_n - \hat{t})^2$
exponential	$\lambda e^{-\lambda t}$	$1/\hat{\lambda}$	not needed
gamma	$\frac{\lambda(\lambda t)^{r-1} e^{-\lambda t}}{\Gamma(r)}$	$\hat{r}/\hat{\lambda}$	$\hat{r}/\hat{\lambda}^2$
lognormal	$\frac{1}{\sqrt{2\pi}\,\alpha t} \exp\left[-\frac{[\ln(t/\beta)]^2}{2\alpha^2}\right]$	$\hat{\beta}\exp(\hat{\alpha}^2/2)$	$\hat{\beta}^2 \exp \hat{\alpha}^2 (\exp \hat{\alpha}^2 - 1)$
2-parameter Weibull	$\frac{\alpha}{\beta}\left(\frac{t}{\beta}\right)^{\alpha-1} \exp[-(t/\beta)^\alpha]$	$\hat{\beta}\Gamma(1 + 1/\hat{\alpha})$	$\hat{\beta}^2\{\Gamma(1 + 2/\hat{\alpha}) - [\Gamma(1 + 1/\hat{\alpha})]^2\}$

[a] All N units tested to failure.

solve for $\hat{\alpha}$ and $\hat{\beta}$ from

$$\hat{\beta}\Gamma(1 + 1/\hat{\alpha}) = \hat{t}.$$

$$\hat{\beta}^2\{\Gamma(1 + 2/\hat{\alpha}) - [\Gamma(1 + 1/\hat{\alpha})]^2\} = \hat{\sigma}^2.$$

These two equations must be solved simultaneously by an iterative method in order to obtain the $\hat{\alpha}$ and $\hat{\beta}$. Since $\hat{t}$ and $\hat{\alpha}^2$ are merely numbers, one simple iterative procedure is to rewrite the two equations as

$$\hat{\beta}_1 = \hat{t}/\Gamma(1 + 1/\hat{\alpha})$$

$$\hat{\beta}_2 = \{[\hat{\sigma}^2 + \hat{t}^2]/\Gamma(1 + 2/\hat{\alpha})\}^{1/2}$$

and graphically search for the $\hat{\alpha}$ for which $\hat{\beta}_1 = \hat{\beta}_2 = \hat{\beta}$. The result is that $\hat{\alpha} = 1.58$ and $\hat{\beta} = 1095$ hr, in close agreement with the results of Example 4-2. ◇

It should be remembered that the form of estimator (4-15) assumes that all N units available are tested to failure. In the event that R units fail out of a total of N and the testing is stopped at $t = t_s$ (Type I censoring), then Eq. (4-15) should be replaced by

$$\hat{t} = \frac{1}{R}\sum_{n=1}^{R} t_n + \left(\frac{N - R}{R}\right) t_s. \tag{4-18}$$

In testing schemes such as *accelerated life testing* or other sophisticated schemes, equations other than Eqs. (4-15) and (4-18) should be used [7].

4-3-2 Maximum Likelihood Estimators

The maximum likelihood method requires the use of the "likelihood function" L, which is defined for N data points as [7, 10]

$$L(t_1, t_2, \cdots, t_N; \theta_1, \theta_2, \cdots, \theta_M) = \prod_{n=1}^{N} f(t_n; \theta_1, \theta_2, \cdots, \theta_M). \tag{4-19}$$

Here the function $f(t_n; \theta_1, \theta_2, \cdots, \theta_M)$ is the failure probability density for the selected distribution at the time of each failure t_n, $n = 1$ to N. The values of θ_1, θ_2, etc., are the M parameters to be estimated. For example, for the gamma distribution in Table 4-3, the two θ-parameters are r and λ. Likewise, for the lognormal and two-parameter Weibull distributions, the parameters are α and β.

In the event that the tests of N units are terminated at t_s before the last failure, so that only R units fail at $t_1, t_2, \cdots, t_R$, then the terminated-testing form of L must be used. This is

$$L(t_1, t_2, \cdots, t_R; \theta_1, \theta_2, \cdots, \theta_M) = \frac{N!}{(N-R)!} \prod_{n=1}^{R} f(t_n; \theta_1, \theta_2, \cdots, \theta_M)[1 - F(t_s)]^{N-R}, \tag{4-20}$$

where $F(t_s)$ is the cumulative distribution function of the assumed probability density $f(t_s; \theta_1, \theta_2, \cdots, \theta_M)$. For the remainder of the discussion, we shall assume that all units are tested until failure.

The estimate of each parameter θ_m is obtained from Eq. (4-19) by maximizing $\ln L$ with respect to all the parameters. Since

$$\ln L = \sum_{n=1}^{N} \ln f(t_n), \tag{4-21}$$

the conditions for obtaining the $\hat{\theta}_m$ follow from the simultaneous solution of the set of equations [7, 10]

$$\frac{\partial \ln L}{\partial \theta_m} = 0, \qquad m = 1 \text{ to } M. \tag{4-22}$$

Each estimate $\hat{\theta}_m$ is a random variable, so different samples will, in general, yield different values. The variance of $\hat{\theta}_m$ is calculated from

$$\operatorname{var} \hat{\theta}_m = -\frac{1}{\dfrac{\partial^2 \ln L}{\partial \theta_m^2}} \tag{4-23}$$

if N is large [7, 10].

Example 4-6 Derive the two equations for a maximum likelihood fit of data to a gamma distribution.

From the form of $f(t)$ in Table 4-3 and Eq. (4-21), it follows that

$$\begin{aligned}\ln L &= \sum_{n=1}^{N} \ln\left[\frac{\lambda^r t_n^{r-1} e^{-\lambda t_n}}{\Gamma(r)}\right] \\ &= \sum_{n=1}^{N} [r \ln \lambda - \ln \Gamma(r) + (r-1) \ln t_n - \lambda t_n] \\ &= N[r \ln \lambda - \ln \Gamma(r)] + (r-1) \sum_{n=1}^{N} \ln t_n - \lambda \sum_{n=1}^{N} t_n.\end{aligned}$$

We first calculate

$$\frac{\partial \ln L}{\partial \lambda} = N\, r/\lambda - \sum_{n=1}^{N} t_n,$$

and thus from Eq. (4-22) we find that the estimate of λ is obtained implicitly in terms of the estimate of r as

$$\hat{r}/\hat{\lambda} = N^{-1} \sum_{n=1}^{N} t_n.$$

This result agrees with one equation obtained by the moment method.

To obtain a second equation for estimating r and λ of the gamma distribution, we next calculate

$$\frac{\partial \ln L}{\partial r} = N[\ln \lambda - \psi(r)] + \sum_{n=1}^{N} \ln t_n,$$

where $\psi(r)$ is the digamma function defined in Appendix A. Using Eq. (4-22), we obtain the result

$$\psi(\hat{r}) - \ln \lambda = N^{-1} \sum_{n=1}^{N} \ln t_n.$$

This second equation for the gamma distribution differs from that for the moment method and is sufficiently complicated that the two equations must be solved iteratively to obtain estimates for r and λ. ◇

The equations from the maximum likelihood method for estimating the desired parameters of the exponential, gamma, lognormal, and two-parameter Weibull distributions are given in Table 4-4. Also shown in the table are equations obtained from the maximum entropy method, which is to be discussed after the next example.

Example 4-7 For the devices described in Example 4-2, estimate the parameters α and β for the Weibull distribution using maximum likelihood estimators.

Table 4-4 *Maximum Likelihood Estimators (MLE) and Maximum Entropy Estimators (MEE) for Failure Probability Distributions Defined in Table 4-3*

Distribution name	Data collected[a]	=	MLE[b]	MEE[b]
exponential	$N^{-1}\sum_{n=1}^{N} t_n$	=	$1/\hat{\lambda}$	same
gamma	$N^{-1}\sum_{n=1}^{N} \ln t_n$	=	$-\ln\hat{\lambda} + \psi(\hat{r})$	same
	$N^{-1}\sum_{n=1}^{N} t_n$	=	$\hat{r}/\hat{\lambda}$	same
lognormal	$N^{-1}\sum_{n=1}^{N} \ln t_n$	=	$\ln\hat{\beta}$	same
	$N^{-1}\sum_{n=1}^{N} (\ln t_n)^2$	=	$\hat{\alpha}^2 + (\ln\hat{\beta})^2$	same
2-parameter Weibull	$N^{-1}\sum_{n=1}^{N} \ln t_n$	=	not used	$\ln\hat{\beta} - \gamma/\hat{\alpha}$
	$N^{-1}\sum_{n=1}^{N} t_n^{\hat{\alpha}}$	=	$\hat{\beta}^{\hat{\alpha}}$	same
	$\hat{\beta}^{-\hat{\alpha}}N^{-1}\sum_{n=1}^{N} t_n^{\hat{\alpha}}\ln t_n - N^{-1}\sum_{n=1}^{N}\ln t_n = \hat{\alpha}^{-1}$			not used

[a] All N units tested to failure.
[b] $\gamma = 0.5772$ and $\psi(\hat{r})$ is the digamma function defined in Appendix A.

The first step is to guess an initial $\hat{\alpha}$ and $\hat{\beta}$. Of course, if a graphical fit of the data has been performed, then these values can be used. The two maximum likelihood estimator MLE equations from Table 4-4 are then evaluated (somewhat laboriously) to see if the initial guess was correct. Iteration eventually gives the estimates $\hat{\alpha} = 1.67$ and $\hat{\beta} = 1102$ hr. ◇

4-3-3 Maximum Entropy Estimators

The maximum entropy estimator method sometimes differs from the maximum likelihood method; it has not been widely studied [11, 12]. The method derives its name from the Shannon entropy function H of information theory; if all N units are tested to failure, H is defined in terms of the selected failure probability density $f(t)$ as

$$H = -\sum_{n=1}^{N} f(t_n) \ln f(t_n). \tag{4-24}$$

[A comparison of Eqs. (4-21) and (4-24) shows that the H function weights the magnitude of each value of $\ln f(t_n)$ by the magnitude $f(t_n)$ rather than by unity.]

The basis for the algorithm for estimating the desired parameters is Jaynes' principle which states that the minimally prejudiced probability distribution is that which maximizes H subject to constraints supplied by moments of the assumed distribution [11]. The algorithm for maximizing H can be conveniently stated by first rewriting the probability density f in the form

$$f(t_n) = \exp[-\theta_0 - \theta_1 T_1(t_n) - \theta_2 T_2(t_n) - \cdots], \tag{4-25}$$

where $\theta_0, \theta_1, \cdots \theta_M$ are constants in terms of the M parameters to be estimated, and the functions $T_1(t), T_2(t), \cdots T_M(t)$ are functions of t. The estimates of the desired parameters follow from the set of equations [11]

$$-\frac{\partial \theta_0}{\partial \theta_m} = N^{-1} \sum_{n=1}^{N} T_m(t_n), \qquad m = 1 \text{ to } M. \tag{4-26}$$

Example 4-8 Derive the two equations for a maximum entropy fit of data to a gamma distribution.

From the form of the gamma distribution in Table 4-3 and Eq. (4-25) it is found that

$$\theta_0 = \ln \Gamma(r) - r \ln \lambda$$

$$\theta_1 = \lambda$$

$$\theta_2 = 1 - r$$

$$T_1(t) = t$$

$$T_2(t) = \ln t.$$

Substituting these parameters into Eq. (4-26) gives

$$-\frac{\partial \theta_0}{\partial \theta_1} = -\frac{\partial}{\partial \lambda}[\ln \Gamma(r) - r \ln \lambda] = \frac{r}{\lambda} = N^{-1} \sum_{n=1}^{N} t_n$$

and

$$-\frac{\partial \theta_0}{\partial \theta_2} = -\frac{\partial \theta_0}{\partial (1 - r)} = \frac{\partial [\ln \Gamma(r) - r \ln \lambda]}{\partial r} = \psi(r) - \ln \lambda = N^{-1} \sum_{n=1}^{N} \ln t_n.$$

Use of the notation $\hat{\lambda}$ and $\hat{r}$ to denote the estimates of λ and r, respectively, leads to the results in Table 4-4 for the maximum entropy method; the equations are precisely the same as those from the maximum likelihood method. ◇

The equations for estimating the parameters of the exponential, gamma, lognormal, and the two-parameter Weibull distributions are given in Table 4-4. We can see that the results for the first three distributions are identical to those from the maximum likelihood method, but differ for the Weibull model.

4-3-4 Comparison of Estimators

The maximum likelihood estimator (MLE) is the normally accepted one because [10]:

1. The MLE is a sufficient estimator if a sufficient estimator exists for the problem.
2. The MLE is efficient (i.e., its variance is small) for N large.
3. The MLE possesses the property of invariance (which means that if w is an estimator for var x, and if w is invariant, then $w^{0.5}$ is an estimator for σ_x).
4. The var $\hat{\theta}_m$ of Eq. (4-23) can be computed and its distribution described in the limit as $N \to \infty$.

Because the MLE is more cumbersome to use than the maximum entropy estimator (MEE) for the Weibull model, a numerical comparison of the results is of interest.

Example 4-9 For the devices described in Example 4-2, estimate the parameters α and β for the Weibull distribution using maximum entropy estimators.

An iteration procedure involving the equations in Table 4-4 eventually leads to the following results; we see that there are some differences in the values of $\hat{\alpha}$ and $\hat{\beta}$. ◇

Method	$\hat{\alpha}$	$\hat{\beta}$, hr
Curve fitting (from Example 4-2)	1.5	1100
Moment (from Example 4-5)	1.58	1095
Maximum likelihood (from Example 4-7)	1.67	1102
Maximum entropy	1.53	1137

EXERCISES

4-1 Twelve identical parts were tested to failure at 1.20, 1.28, 1.39, 1.71, 2.31, 2.50, 3.05, 3.63, 4.02, 5.03, 6.89, and 9.68 ($\times 10^3$) hr. Fit the data to an exponential distribution by determining $\hat{\lambda}$ and $\hat{\tau}$ using the curve-fitting method.

4-2 Four pumps were tested to failure at 8, 14, 21, and 36 ($\times$ 10^3) hr. Fit the data to a lognormal distribution by determining $\hat{\alpha}$ and $\hat{\beta}$ using the curve-fitting method.

4-3 Three air compressors were tested to failure at 400, 475, and 550 hr. Fit the data to a lognormal distribution by determining $\hat{\alpha}$ and $\hat{\beta}$ using the curve-fitting method.

4-4 Sixteen sets of gears were tested until wear-out occurred at the times of 21, 33, 40, 51, 62, 70, 81, 90, 90, 100, 113, 123, 126, 142, 150, and 162 days. Fit the data to a two-parameter Weibull distribution by determining $\hat{\alpha}$ and $\hat{\beta}$ using the curve-fitting method [5].

4-5 Six sets of bearings were tested until wear-out caused failures at 1.3, 2.7, 4.0, 5.2, 6.6, and 9.8 ($\times 10^5$) cycles. Fit the data to a two-parameter Weibull distribution by determining $\hat{\alpha}$ and $\hat{\beta}$ using the curve-fitting method [5].

4-6 Table E4-6 shows the distribution of the diameters of the heads of rivets manufactured by a company. Compute (a) the mean diameter $\bar{X}$, and (b) the variance s^2 [6].

4-7 Measurements of the diameters of a random sample of 200 ball bearings made by a certain machine during one week showed a mean of 0.824 cm and a standard deviation of 0.042 cm. Find (a) 95% and (b) 99% confidence limits for the mean diameter of all the ball bearings [6].

4-8 Find (a) 98%, (b) 90%, and (c) 99.73% confidence limits for the mean diameter of the ball bearings in Exercise 4-7 [6].

4-9 Five gears were tested to failure, and the following failure times were recorded: 0.5, 0.9, 1.7, 2, and 3.2 ($\times 10^5$) sec. For the two-parameter Weibull model, determine $\hat{\alpha}$ and $\hat{\beta}$ using the (a) curve-fitting method, (b) moment method, (c) maximum likelihood method, (d) maximum entropy method.

4-10 Ten shafts were tested to failure, which occurred at the following number of cycles: 3.5, 6.5, 8, 9.2, 13, 14.5, 16.8, 18, 19.5, and 24 ($\times 10^5$).

Table E4-6 *Diameters of Rivet Heads*[a]

Diameter (cm)	Frequency
0.7247–0.7249	2
0.7250–0.7252	6
0.7253–0.7255	8
0.7256–0.7258	15
0.7259–0.7261	42
0.7262–0.7264	68
0.7265–0.7267	49
0.7268–0.7270	25
0.7271–0.7273	18
0.7274–0.7276	12
0.7277–0.7279	4
0.7280–0.7282	1
	Total 250

[a] From M. R. Spiegel, "Statistics." Schaum, New York, 1961.

For the two-parameter Weibull model, determine $\hat{\alpha}$ and $\hat{\beta}$ using the (a) curve-fitting method, (b) maximum likelihood method, (c) maximum entropy method.

4-11 A nonreplacement life test was carried out on a sample of ten pumps, which failed at the following times after commencement of the test: 4.6, 8, 10, 12, 14, 17, 20, 22, 26, and 33 ($\times 10^4$) sec. For the two-parameter Weibull model, determine the $\hat{\alpha}$ and $\hat{\beta}$ using the (a) curve-fitting method, (b) maximum likelihood method, (c) maximum entropy method. Finally, (d) calculate the mean time to failure for each of the three methods.

REFERENCES

1. C. Lipson and N. J. Sheth, "Statistical Design and Analysis of Engineering Experiments." McGraw-Hill, New York, 1973.
2. I. B. Gertsbakh and K. B. Kordonskiy, "Models of Failure." Springer-Verlag, Berlin and New York, 1969.
3. D. J. Smith, "Reliability Engineering." Barnes and Noble (Harper), New York, 1972.
4. C. O. Smith, "Introduction to Reliability in Design." McGraw-Hill, New York, 1976.
5. G. J. Hahn and S. S. Shapiro, "Statistical Models in Engineering." Wiley, New York, 1967.
6. M. R. Spiegel, "Statistics." Schaum (Schaum's Outline Series), New York, 1961.

7. N. R. Mann, R. E. Schafer, and N. D. Singpurwalla, "Methods for Statistical Analysis of Reliability and Life Data." Wiley, New York, 1974.
8. M. Abramowitz and I. A. Stegun, "Handbook of Mathematical Functions." U.S. Government Printing Office, Washington, D.C., 1964, reprinted by Dover, New York.
9. G. H. Lemon and J. B. Wattier, Confidence and 'A' and 'B' allowable factors for the Weibull distribution, *IEEE Trans. Reliability* **R-25,** 16 (1976).
10. M. L. Shooman, "Probabilistic Reliability: An Engineering Approach." McGraw-Hill, New York, 1968.
11. M. Tribus, "Rational Descriptions, Decisions and Designs." Pergamon, Oxford, 1969.
12. N. J. McCormick, Implications of information theory for reliability testing of reactor components, *Trans. Am. Nucl. Soc.* **23,** 226 (1976).

CHAPTER 5

Failure Data

5-1 INTRODUCTION

This chapter provides an introduction to the sources of failure data, some representative numerical values for hardware and human errors, and the problem of common cause failures. Acquisition of good failure data frequently is the crux of a reliability or risk analysis. However, for some purposes of system reliability engineering, a set of precise data are not required. This is fortunate, since present data sources often contain only an *averaged* constant hazard rate rather than detailed time-dependent values.

The state of the art of reliability data can be summarized as follows [1]:

> Reliability data are unlike other data such as nuclear cross sections. The basis of cross-section data is a large finite number of nuclei, all with identical properties. The basis of reliability data is stochastic, involving ever-changing components and environmental conditions. In addition, reliability data have the property of describing a statistical distribution rather than a fixed but unknown value. In other words, even under perfect measurement conditions, reliability data for a single event cannot be determined by a single test but rather must be determined from years of observation of equipment in various environments during which time the equipment designs may change. In addition, it is frequently easy to determine how many components of a certain type failed in a certain way during a known time period. Unfortunately, it is often not known how many identical components did not fail during this time interval; i.e., the population of components in use is not known. Collecting and evaluating reliability data is very frustrating.

The failure of a component is a function not only of its design and quality of construction, but also of the environment in which it is placed. Important considerations are temperature, electrical and mechanical stresses, and cycling-of-operation of the component. One of the reasons that failure data are sometimes not representative of the actual failure probability is that the operating conditions and environment of the device may not always be the same. That is, the device under analysis may not meet the system definition under which its failure data were acquired. It is important to remember that system reliability is defined as the probability of performing a specified function or mission *under given conditions for a prescribed time.*

For quantitative safety analyses, it is not sufficient to know only the number of failures occurring in a given period. Other qualifying information frequently is needed to judge the effect of environmental factors and the quality and frequency of maintenance. Such considerations include [2]:

1. modes of failure experienced,
2. sample size,
3. environmental or special working conditions,
4. the number of successful functions in relation to the failures, particularly in the case of equipment subject to intermittent cycles of operation,
5. true running time during the survey period, particularly for equipment which is run for standby or backup purposes at random times during the survey period,
6. repair time,
7. time intervals between failure,
8. frequency of, or intervals between, periodic inspection or proof tests.

It is especially important to identify the mode of failure of a component. For example, not only can a system fail because of a structural fault, but it may fail to open or close (or to start or stop). Frequently, the probability that a component "fails open," i.e., fails to close, is different from the probability that the component "fails closed," and these probabilities are different from the probability that the component fails under load.

A second major reason that failure data are sometimes in error is because of the limited number of experiments or data points available for some devices. As a rough rule, the lower the cost of the device, the better the failure data information because more tests have been performed.

5-2 SOURCES OF FAILURE DATA

To collect, store, and retrieve failure information effectively for devices of many types is an enormous task and requires an organization that is consistent with defined reliability objectives. Numerous programs have been undertaken to methodically develop parts and equipment reliability data. Up until the past few years, much of the best information from the standpoint of reliability has tended to be of a specialized nature, often confined to the elctronics field and related to military, aeronautical, and space applications.

As a part of the Reactor Safety Study [3], to be discussed in Chapter 12, an effort was made to compile all the hazard rate and demand-failure data that would be useful for reactor components in a unified framework suitable for subsequent updating. The failure data in that study were compiled primarily from the sources in Table 5-1; the Nuclear Regulatory Commission continues to expand the data base developed during the Study.

In the Reactor Safety Study, it was found necessary to forego the reporting of point estimates for failure rates because of the variability of operating conditions and, in some cases, the incompleteness of pertinent data. Therefore, interval estimates were reported. Reporting of failure data over ranges helps to incorporate uncertainties and variations due to different reporting sources, variations in application, fluctuations in the daily operational environmental conditions, and minor variations that can exist from one unit to the next for a given component.

A good data source for the nuclear power industry, listed in Table 5-2, is the IEEE Std 500-1977 [4]. This standard contains reliability data on electrical, electronic, and sensing components found in most nuclear power generating stations and conventionally fueled generators and industrial plants. Similar data compiled for military electronic equipment are in MIL-HDBK-217B [5]. These references are complemented by the Nonelectronic Parts Reliability Databook cited in Table 5-2.

A cooperative data exchange among United States government and industry participants is operated by the Government–Industry Data Exchange Program (GIDEP) [6]. One of the data banks maintained is the reliability–maintainability data interchange that contains information on the mode and rate of failure and the replacement rate on parts, components, and assemblies; another data interchange is devoted to failure experience *alert* reports on potential problems relating to equipment and its safety. Many of these data, however, are not processed into a more useful form giving failure rates.

A cooperative data exchange for nuclear plant systems and components is the NPRDS, Nuclear Plant Reliability Data System [7]. This system contains both an engineering data base and a failure-reporting data bank. Virtually all utilities operating nuclear plants in the United States record engineering data on their safety systems and components, and complete reports when failures occur. This system complements (and somewhat duplicates) the Licensee Event Reports (LERs) compiled by the Nuclear Regulatory Commission; an abbreviated listing of the abnormal occur-

Table 5-1 *Primary Sources of Data for the Reactor Safety Study*[a]

Reference	Office or originator	Contents
Reactor Incident File (annually updated)	Office of Operations Evaluation (OOE) of Regulatory Operations (RO), Nuclear Regulatory Commission, Bethesda, MD	Unusual occurrences at nuclear facilities and reportable abnormal occurrences observed
EEI Availability Report (Component Failure Data)	Edison Electric Institute (EEI), New York, NY	Many unit years of fossil and nuclear power plants component availability and outage statistics of contributing facilities
System Reliability Service	United Kingdom Atomic Energy Authority, Wigshaw Lane, Culcheth, Warrington, Lancashire, England	Failure rate assessments derived from UK and other available European sources
FARADA (Converged Failure Rate Data Handbooks)	Fleet Missile Systems Analysis and Evaluation Group Annex, NWS, Sea Beach, Corona, CA	Failure rate assessments derived from Army, Navy, Air Force, and NASA sources
LMEC (Failure Data Handbook for Nuclear Power Facilities)	Liquid Metal Engineering Center, Rockwell International, Canoga Park, CA	Failure rates derived from test and research reactor operating experiences up through 1969

[a] *From* Reactor Safety Study, Appendix III. U.S. Nuclear Regulatory Commission Rep. WASH-1400, NUREG 75/014 (October 1975).

Table 5-2 *Other Sources of Failure Data*

Reference	Available from	Contents
IEEE Std 500-1977	Institute of Electrical and Electronic Engineers New York, NY	Electrical, electronic, and sensing component reliability data for nuclear power generating stations
MIL-HDBK-217B	Naval Publications and Forms Center, 5801 Tabor Avenue Philadelphia, PA	Hazard rate data on military electronic equipment
NPRD (Nonelectronic Parts Reliability Databook)	Reliability Analysis Center Rome Air Development Center Griffiss Air Force Base, NY	Failure rate data on devices intrinsic to computer peripherals; other data not available in MIL-HDBK-217B
GIDEP (Government–Industry Data Exchange Program)	Department of the Navy GIDEP Operations Center Corona, CA	Reliability-maintainability data interchange; failure experience data interchange
NPRDS (Nuclear Power Plant Reliability Data System)	American National Standards Institute New York, NY	Failure-reporting data bank for systems and components designated ANSI Safety Classes 1 and 2 and IEEE Safety Class 1E
FEED (File of Evaluated and Event Data)	Electric Power Research Institute Palo Alto, CA	Personnel, component, and system failure rates

rences leading to a reactor shutdown and an LER is published in the journal *Nuclear Safety*.

A recently developed nuclear data base is FEED (File of Evaluated and Event Data), which has been compiled from Licensee Event Reports [8]. The FEED computer program is capable, for example, of retrieving all abnormal occurrences at a specific nuclear power plant or occurrences at a plant initiated by a particular cause of failure.

5-3 EXAMPLES OF FAILURE DATA FOR HARDWARE

Tables B-1 and B-2 of Appendix B are data for mechanical and electrical equipment failure by different modes [3]. As we discussed in Chapter 2, there are two kinds of conditional probabilities for failure: those arising for components in intermittent service, which fail upon demand, and those in continuous service, whose failure characteristics are described by hazard rates. In the tables, the hazard rates per hour λ for components required to function for a period of time are denoted by λ_0; the rates for those passive-type devices (such as pipes, wires, etc., which are normally dormant or in standby until tested or an accident occurs) are denoted by λ_s. The demand probabilities of failure $P(\bar{D})$ are denoted by Q_d.

The error or confidence factors of 3 or 10 in Tables B-1 and B-2 are used along with the computed mean values when assessing the effects of statistical uncertainties. The bounds of the assessed range are generally valid for a statistical confidence of 90%, which is to say that only 10% of the

Table 5-3 *Averaged Hazard Rate Estimates for 1972 Nuclear Operations*[a]

	λ, hr^{-1}	
Component	PWR	BWR
Pumps	1×10^{-6}	3×10^{-6}
Piping	1×10^{-9}	3×10^{-9}
Control rods[b]	1×10^{-6}	1×10^{-7}
Diesels	3×10^{-5}	3×10^{-5}
Valves	1×10^{-6}	3×10^{-6}
Instruments	3×10^{-7}	1×10^{-6}

[a] *Extracted from* Reactor Safety Study, Appendix III. U.S. Nuclear Regulatory Commission Rep. WASH-1400, NUREG 75/014 (October 1975).
[b] Failure rate per hour per rod, for failure to enter.

Table 5-4 *Averaged Demand Failure Probability Estimates for 1972 Nuclear Operations*

Component	PWR	BWR
Pumps	1×10^{-3}	3×10^{-3}
Control rods	1×10^{-3}	1×10^{-4}
Diesels	3×10^{-2}	3×10^{-2}
Valves	1×10^{-3}	3×10^{-3}

[a] *Extracted from* Reactor Safety Study, Appendix III. U.S. Nuclear Regulatory Commission Rep. WASH-1400, NUREG 75/014 (October 1975).

average failure rate estimates are expected to be outside the assessed range. The confidence bounds do not, however, account for errors in the averaging process used to obtain the failure estimate, such as lumping and combining failures of different modes. Furthermore, the estimates have been rounded to the nearest (base 10) half exponent, i.e., 1×10^{-X} or 3×10^{-X}.

Checks were made in developing the data of Appendix B to be certain that the results were consistent with previous nuclear plant operating history data. The nuclear hazard rate and demand probabilities of Tables B-1 and B-2, when combined into composite system results, gave results that did not contradict the operating history data in Tables 5-3 and 5-4 [3].

Tables 5-3 and 5-4 provide a convenient order-of-magnitude estimate of failure probabilities and serve as a quick guide to the relative ranking of different components in a nuclear reactor where failures may be expected to occur. When examining the data in Table 5-3, it should be remembered that there are less than 10^4 hr/yr and that a reactor typically is designed for a 40-yr lifetime. The reactor components in Tables 5-3 and 5-4 are subject to periodic inspection and maintenance so that the risk of an unexpected failure can be minimized.

5-4 EXAMPLES OF FAILURE DATA FOR HUMAN ERROR

Safety systems for nuclear reactor operation generally are designed to function automatically during the initial stages of an accident sequence. Only after such automatic operation would human interaction normally be required. In the event of a malfunction of the automatic systems, interaction by humans under stress would be necessary. A second type of human interaction with the system involves routine plant operation, testing, and maintenance. For both situations, the impact of human reliability needs to be considered in reliability and risk analyses.

Human reliability analyses are complicated by the fact that an extensive actuarial-type data base does not exist. This is because human responses under the influence of stress vary in different ways for emergency and normal conditions. Table B-3 of Appendix B presents an estimate of human error probabilities under normal operating conditions, i.e., assuming no undue time pressures or stresses related to accidents, and under abnormal conditions [3].

Following the Three Mile Island accident, the Nuclear Regulatory Commission has increased its requirements for the training of operators under simulated reactor accident conditions and especially under the occurrence of multiple equipment failures. As a consequence of such training, it is expected that the probability of incorrect operator decisions may be reduced. This is another reason why the numerical values in Table B-3 should be considered only as *very rough* estimates.

The impact of a high-stress situation leads to a conservative estimate of 0.2 to 0.3 as the average error rate for nuclear power plant personnel in a high-stress situation such as a loss-of-coolant accident [3]. This range of values was based in part upon data collected from Strategic Air Command aircraft crews who survived in-flight emergencies such as loss of an engine on takeoff, cabin fire, tire blowout, etc. [9]; in such cases, the average error rate was 0.16. By way of comparison, up to a third of new United States Army recruits have panicked when subjected to simulated emergencies such as the increasing proximity of falling mortar shells [10].

The error rate for a task is believed to be a nonlinear relation of the perceived stress level, as shown in Fig. 5-1. With very low stress levels, a task becomes so dull and uninteresting that personnel cannot perform at an optimal level; for such passive monitoring tasks, a value of 0.5 has been used [3]. On the other hand, when stress levels become too high, performance levels decline because of the effects of worry, fear, or other psychological factors.

For purposes of classifying human failures during data collection, a scheme has been developed in a format amenable to easy retrieval and up-dating of failure information [11]. This classification scheme has been used to classify all occurrences during operation of all United States light water reactors from June 1, 1973, to June 30, 1975 [12]. It was found that 14.2% of all human errors were due to incorrect or inadvertent manipulation of valves or other equipment, and 11.2% were administrative operating procedural errors. The systems most likely to be affected by human errors are shown in Table 5-5. The error rates for operation of Pressurized Water Reactor (PWR) and Boiling Water Reactor (BWR) plants were

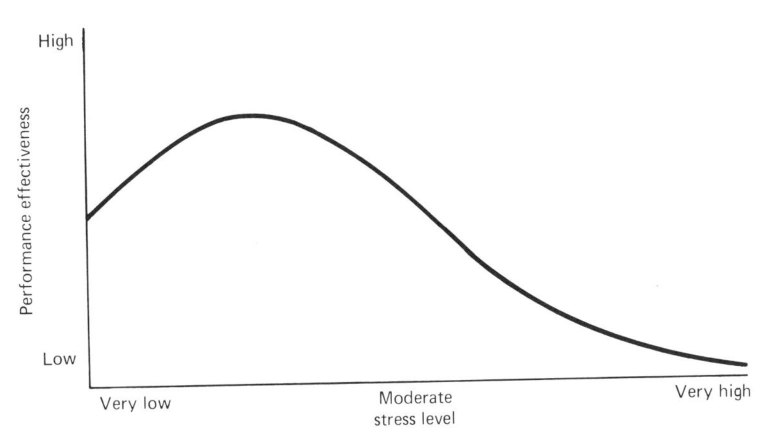

Fig. 5-1 *Hypothetical relationship between performance and stress.* [*From Reactor Safety Study, Appendix III. U.S. Nuclear Regulatory Commission Rep. WASH-1400, NUREG 75/014* (*October 1975*).]

generally comparable, with the exception of operator and maintenance errors. The operator error rate in the PWR exceeded that in the BWR by 72.2%; the maintenance error rate for the BWR, on the other hand, exceeded that in PWRs by 69%. These facts may be attributed to greater operator manipulation in PWRs and higher maintenance frequency in BWRs.

Finally, it should be mentioned that a model has now been developed to evaluate improvement in human performance acquired with operating experience [13]. Also, a statistical procedure under development [14] to evaluate major human errors during the development of a new technological system may prove useful for improving the data base for newly operating nuclear facilities.

Table 5-5 *Systems Most Likely to be Involved in Human Errors* [a]

Emergency core cooling system	14.0%
Containment atmosphere and isolation systems	7.7%
Reactor protection system	7.5%
Core	7.0%
Control system	6.7%
Liquid radwaste system	6.5%
The other 23 systems	50.6%

[a] From D. W. Joos and Z. A. Sabri, *Trans. Am. Nucl. Soc.* 26, 384 (1977).

5-5 COMMON CAUSE AND COMMON MODE FAILURES

Failure of multiple components or systems due to a single secondary event is classified as a *common cause* failure. Examples of possible common cause failures are natural dangers (such as fires, earthquakes, tornadoes, hurricanes), failures of an engineered system (e.g., a massive electrical power failure or explosions and missiles—either internally or externally initiated), and human acts (human errors or acts of violence). A phenomenon such as an extra load placed on a second pump after the first one fails is another illustration of a type of common cause failure; still another example is the rupture of a steam line that affects nearby components.

The phrase *common mode failure* describes a common cause failure that acts on a set of functionally identical (i.e., duplicate or redundant) components or systems. An example of a common mode failure is the existence of a manufacturing defect in a group of relays; because of the defect, all relays in the produced batch will thus be affected, causing all the failure rates to be higher than the average failure rate for that type of relay. Mechanisms that contribute to common mode failures can be categorized as [15]:

(a) design defects,
(b) fabrication, manufacturing, and quality control variations,
(c) test, maintenance, and repair errors,
(d) human errors,
(e) environmental variations (contamination, temperature, etc.).

The recommended measures against common mode failures are based on the introduction of diversity or periodic testing of components during the plant lifetime. Diversity is accomplished, for example [16, 17], by using different types of equipment or different manufacturers or more than one logical way to monitor the state of a system, by physically separating redundant components, and by having more than one operator.

The terms common cause and common mode frequently are used interchangeably since they are closely related and because their identification is so similar when one performs a fault tree analysis (discussed in Chapter 8). Because common cause failures entail a wide spectrum of possibilities and enter into all areas of modeling and analysis, they cannot be isolated as a separate study, but instead must be considered throughout all the modeling and quantification steps involved in any reliability or risk assessment. The main difficulty is that since events no longer remain independent, the probabilities of failure become dependent probabilities.

To assess the maximum impact of common cause failures, the simplest approach is to use a bounding technique. For the events $A_1, A_2, \cdots, A_N$, for example, it is known from Eq. (2-9) that $P(A_1A_2 \cdots A_N)$ has a *single failure upper bound* of

$$P(A_1A_2 \cdots A_N) \leq \min [P(A_1), P(A_2), \cdots P(A_N)]. \tag{5-1}$$

In a similar way, the general equation for the *double failure* bound is

$$P(A_1A_2 \cdots A_N) \leq \min[\text{Probabilities of all double combinations}]. \tag{5-2}$$

In the Reactor Safety Study [15], for example, single failure bounds were computed for combinations of failures that were judged to be possible common mode failure candidates. These bounds were then compared to the system to determine its sensitivity to possible common mode contributions. If the bounds did have impact, further investigation was performed and a more detailed analysis of the system was undertaken, or, in a number of cases, the bounds were also incorporated as part of the system result.

The way the double bounds were determined in the Study was that the noncommon cause mechanism was treated as one mechanism, c_0, termed the design environment. If the common causes are denoted as c_m, $m = 1$ to M, then by the decomposition rule of Eq. (2-19),

$$P(A_1A_2) = P(A_1A_2|c_0)P(c_0) + \sum_{m=1}^{M} P(A_1A_2|c_m)P(c_m). \tag{5-3}$$

The probabilities $P(c_m)$ must be obtained from examination of quality control processes, testing, etc. To a reasonable approximation the design environment normally exists, so

$$P(c_0) \approx 1. \tag{5-4}$$

If the design environment is also one in which events A_1 and A_2 occur independently, then

$$P(A_1A_2) = P(A_1)P(A_2) + \sum_{m=1}^{M} P(A_1A_2|c_m)P(c_m). \tag{5-5}$$

Thus the impact of common cause mechanisms can be accounted for by adding to the normal probability the contribution for all common causes.

Elimination of common mode failures and improvement of designs against common cause failures are a major objective of much of the reliability and risk analyses in the nuclear power industry today [18].

REFERENCES

1. J. B. Fussell and J. S. Arendt, System reliability engineering methodology: A discussion of the state of the art, *Nucl. Safety* **20,** 541 (1979).
2. J. F. Ablitt and A. R. Eames, Data collection for quantitative safety analysis, *Nucl. Eng. Design* **13,** 230 (1970).
3. Reactor Safety Study—An Assessment of Accident Risks in U.S. Commercial Nuclear Power Plants, Appendix III. U.S. Nuclear Regulatory Commission Rep. WASH-1400, NUREG 75/014 (October 1975).
4. IEEE Guide to the Collection and Presentation of Electrical, Electronic, and Sensing Component Reliability Data for Nuclear Power Generating Stations, IEEE Std 500-1977. Institute of Electrical and Electronic Engineers (June 30, 1977).
5. MIL-HDBK-217B, Military Standardization Handbook—Reliability Prediction of Electronic Equipment. Defense Documentation Center, Alexandria, Virginia (1974 and updates).
6. Program Description and Participation Requirements. Government–Industry Data Exchange Program, Dept. of the Navy, GIDEP Operations Center, Corona, California.
7. Reporting Procedures Manual for the Nuclear Plant Reliability Data System. Southwest Research, San Antonio, Texas.
8. R. C. Erdmann *et al.,* Probabilistic Safety Analysis. Electric Power Research Institute Rep. EPRI NP-424 (April 1977).
9. W. W. Ronan, Training for Emergency Procedures in Multiengine Aircraft, AIR-153-53-FR-44. American Institutes for Research, Pittsburgh, Pennsylvania, 1953.
10. M. M. Berkun, Performance decrement under psychological stress, *Human Factors* **6,** 21 (1964).
11. A. A. Husseiny, Z. A. Sabri, and R. A. Danofsky, Quantification of operation experience and incident reports for reliability analysis and improvement of task performance, *Trans. Am. Nucl. Soc.* **23,** 484 (1976).
12. D. W. Joos and Z. A. Sabri, Evaluation of human errors in commercial light-water reactors, *Trans. Am. Nucl. Soc.* **26,** 384 (1977).
13. Z. A. Sabri and A. A. Husseiny, Analysis of plant operation experience for performance improvement, *Trans. Am. Nucl. Soc.* **30,** 666 (1978).
14. G. Campbell and K. O. Ott, Statistical evaluation of major human errors during the development of new technological systems, *Nucl. Sci. Eng.* **71,** 267 (1979).
15. Reactor Safety Study—An Assessment of Accident Risks in U.S. Commercial Nuclear Power Plants, Appendix IV. U.S. Nuclear Regulatory Commission Rep. WASH-1400, NUREG 75/014 (October 1975).
16. G. E. Apostolakis, The effect of a certain class of potential common mode failures on the reliability of redundant systems, *Nucl. Eng. Design* **36,** 123 (1976).
17. M. E. Jolly and J. Wreathall, Common-mode failures in reactor safety systems, *Nucl. Safety* **18,** 624 (1977).
18. E. W. Hagen, Common-mode/common cause failure: a review, *Ann. Nucl. Energy* **7,** 509 (1980).

CHAPTER 6

Reliability of Simple Systems

Chapters 3 through 5 pertain to the reliability of a single component or unit of a system. We now will analyze the reliability of a system comprised of a set of components or units i, $i = 1$ to I, each of which has reliability $R_i(t) = 1 - F_i(t)$. The cumulative failure probability $F_i(t)$ can be given by one of the distributions considered in Chapter 3. When failures are random so that the exponential distribution is applicable, then $R_i(t) = e^{-\lambda_i t}$; this simple distribution frequently will be used for purposes of illustration.

To portray the manner in which a system functions, the units are connected together in a *reliability block diagram.* Such a diagram can be viewed, for example, as explaining how a signal called "system operation" would be successfully transmitted from an input to the output of the system.

6-1 SYSTEM RELIABILITY FOR SERIES AND ACTIVE-PARALLEL UNITS

Consider first the simplest of systems consisting of two independent units that operate in series or in *active-parallel,* as shown in Figs. 6-1(a) and 6-1(b), respectively. (Active-parallel means that no unit is held in standby waiting for other units in the parallel configuration to fail.) Although the units may be identical, they need not be, and so they are distinguished by the subscripts 1 and 2.

For the two units in series, both #1 AND #2 must operate for the system to function. Because the reliabilities of each unit, $R_1(t)$ and $R_2(t)$, are actually probabilities, the product rule for probabilities of Eq. (2-4) gives the reliability for the system $R_{sys}(t)$ as

$$R_{sys}(t) = R_1(t)\, R_2(t). \tag{6-1}$$

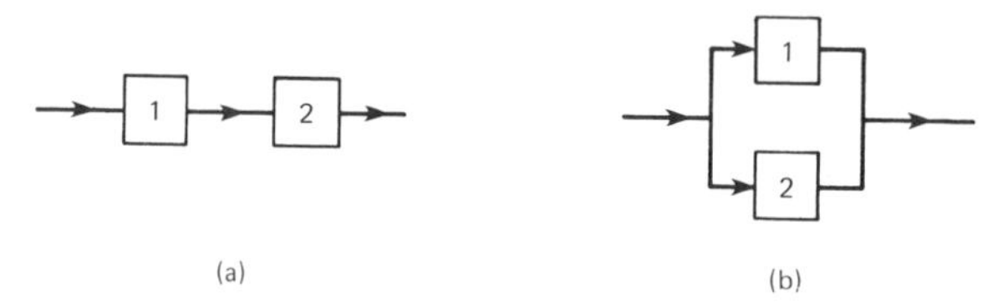

Fig. 6-1 *Reliability block diagram for (a) two units in series and (b) two units in active-parallel.*

For the two units in active-parallel, either #1 OR #2 must operate for the system to function. Hence from Eq. (2-14),

$$R_{sys}(t) = R_1(t) + R_2(t) - R_1(t)\, R_2(t). \tag{6-2}$$

For the case of random failures, the last two equations become

$$R_{sys}(t) = \exp[-(\lambda_1 + \lambda_2)t]. \tag{6-3}$$

$$R_{sys}(t) = \exp(-\lambda_1 t) + \exp(-\lambda_2 t) - \exp[-(\lambda_1 + \lambda_2)t]. \tag{6-4}$$

The reliability of two units in active-parallel is larger then that for the more reliable of the two, while the reliability for two units in series is less than that for the less reliable of the two. In Fig. 6-2, plots of system reliability versus dimensionless time from Eqs. (6-3) and (6-4) are shown for $\lambda_1 = \lambda_2 = \lambda$ and from Eqs. (6-1) and (6-2) for $\lambda_1 = \lambda_2 = kt$.

Generalizations to a system of N independent units, all in series or in active-parallel, may be made with Eqs. (2-6) and (2-16), respectively, and give [1–4]

$$R_{sys}(t) = \prod_{n=1}^{N} R_n(t)$$

$$= \exp\left[-\int_0^t \sum_{n=1}^{N} \lambda_n(\tau) d\tau\right] \qquad \text{(series)} \tag{6-5}$$

$$1 - R_{sys}(t) = \prod_{n=1}^{N} [1 - R_n(t)] \qquad \text{(active-parallel).} \tag{6-6}$$

Evaluation of integrals of Eqs. (6-5) and (6-6) for the mean time to failure for N units in a series or active-parallel system is simpler if all units fail randomly, in which case

$$\text{MTTF} = \int_0^\infty R_{sys}(t)\, dt = \int_0^\infty \exp\left(-t \sum_{n=1}^{N} \lambda_n\right) dt$$

$$= \left(\sum_{n=1}^{N} \lambda_n\right)^{-1} \qquad \text{(series)} \tag{6-7}$$

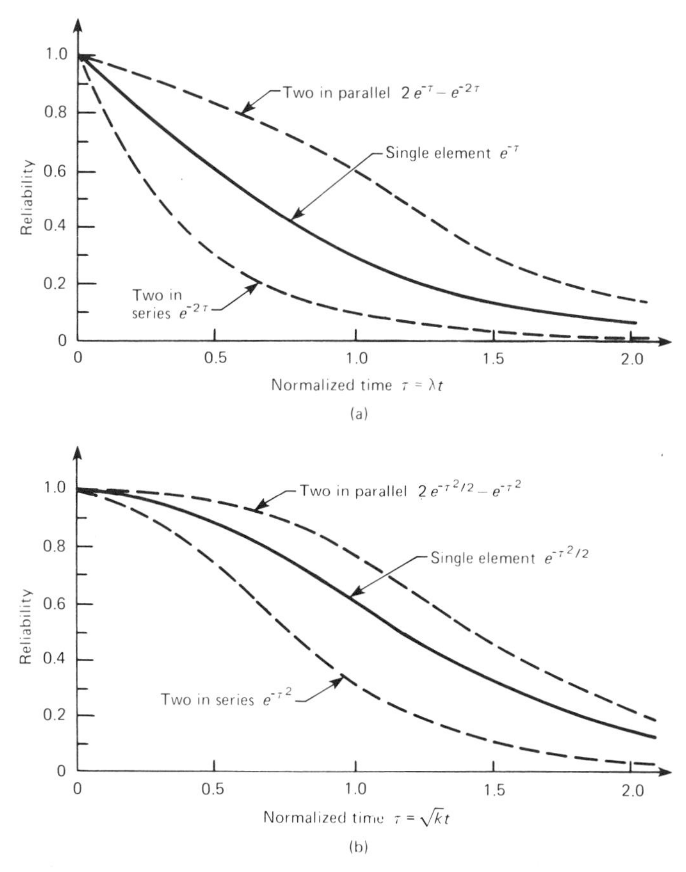

Fig. 6-2 *Comparison of system reliability functions for (a) constant hazard-rate units and (b) linearly increasing hazard-rate units. (From M. L. Shooman, "Probabilistic Reliability: An Engineering Approach." McGraw-Hill, New York, 1968.)*

$$\text{MTTF} = \sum_{n=1}^{N} \frac{1}{\lambda_n} - \sum_{n=1}^{N-1} \sum_{m=n+1}^{N} \frac{1}{\lambda_n + \lambda_m} + \cdots + (-1)^{N-1} \left(\sum_{n=1}^{N} \lambda_n \right)^{-1}$$

(active-parallel). (6-8)

If all N units are identical, then these two equations may be simplified to

$$\text{MTTF} = (N\lambda)^{-1} \qquad \text{(series)} \tag{6-9}$$

$$\text{MTTF} = \sum_{n=1}^{N} (n\lambda)^{-1} \qquad \text{(active-parallel).} \tag{6-10}$$

Thus

$$\lambda(\mathrm{MTTF})_{\mathrm{ser}} = 1/N < \sum_{n=1}^{N} (1/n) = \lambda(\mathrm{MTTF})_{\mathrm{par}}. \tag{6-11}$$

Example 6-1 Calculate the reliability and the MTTF of a system of 2n identical units that operates according to the following reliability block diagram:

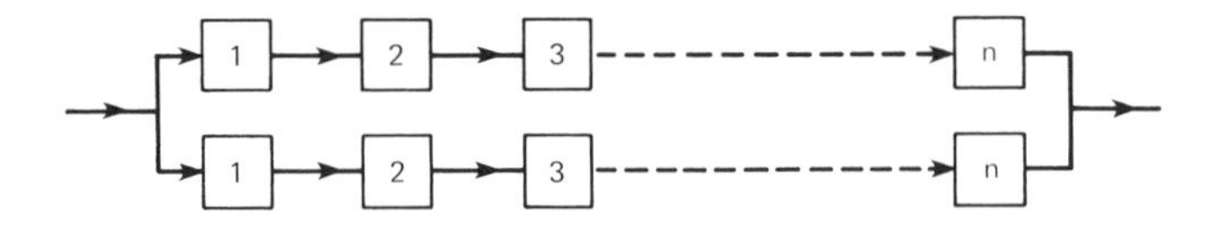

Each unit is identical and fails randomly with a hazard rate λ.

From Eq. (6-5) the reliability of the upper link $R_u(t)$ in the block diagram is $[R(t)]^n$, and similarly for the lower link, so the reliability of the system follows from Eq. (6-2) as

$$\begin{aligned} R_{\mathrm{sys}}(t) &= [R(t)]^n + [R(t)]^n - [R(t)]^{2n} \\ &= e^{-n\lambda t}(2 - e^{-n\lambda t}). \end{aligned}$$

The MTTF is

$$\begin{aligned} \mathrm{MTTF} &= \int_0^\infty R_{\mathrm{sys}}(t)\, dt \\ &= (3/2)(n\lambda)^{-1}. \end{aligned}$$

Notice for 2n = N that this MTTF is 3 times that of Eq. (6-9) and less than that of Eq. (6-10) when $N > 2$. ◇

Another class of simple systems consists of N identical units in active-parallel when only M units are needed to make the system function. This is sometimes referred to as an "M-out-of-N system" and is calculated with the binomial distribution of probability theory. For the reliability of any unit given as $R(t)$, then

$$R_{\mathrm{sys}}(t) = \sum_{n=M}^{N} \frac{N!}{n!(N-n)!} [R(t)]^n [1 - R(t)]^{N-n}. \tag{6-12}$$

As long as $M > 1$, $R_{\mathrm{sys}}(t)$ in Eq. (6-12) is smaller than that for the corresponding active-parallel system given by Eq. (6-6), which is a "1-out-of-N system" if the units are identical.

In the event that the N identical units all fail randomly, then the MTTF corresponding to Eq. (6-12) is

$$\lambda(\mathrm{MTTF}) = \sum_{n=M}^{N} (1/n). \tag{6-13}$$

Thus the MTTF for the M-out-of-N system also lies between that for the series and for the active-parallel system with N units, just as in Example 6-1.

The importance of Eqs. (6-5), (6-6), and (6-12) in the evaluation of the reliability of simple systems, made up of subsystems of units either in series or parallel, is illustrated in the next example.

Example 6-2 A system consists of seven units connected as shown in the following reliability block diagram. Units 1 to 4 all are different (with 2, 3, and 4 in active-parallel), and three identical units of type 5 constitute a 2-out-of-3 system. If R_i, i = 1 to 5, denotes the reliability of each unit as a function of time, calculate the reliability of the system.

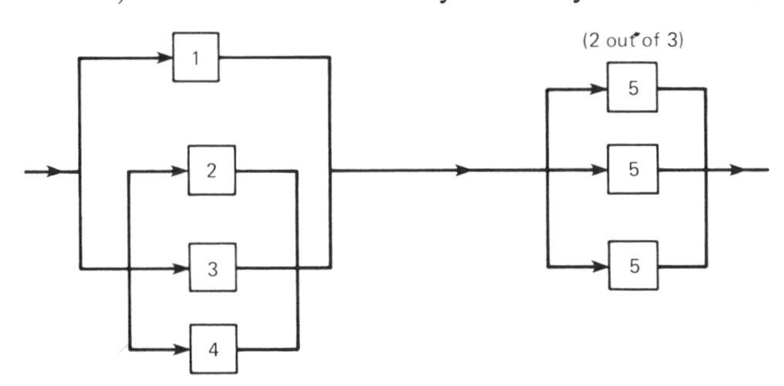

Units 2, 3, and 4 may be replaced by an equivalent unit having reliability R_{234},

$$R_{234} = R_2 + R_3 + R_4 - R_2R_3 - R_2R_4 - R_3R_4 + R_2R_3R_4.$$

Units R_1 and R_{234} may be combined to give an equivalent unit having reliability R_{1234},

$$R_{1234} = R_1 + R_{234} - R_1R_{234}.$$

The 2-out-of-3 subsystem may be replaced by an equivalent unit having reliability denoted by $R_{(55)5}$, calculated from Eq. (6-12) as

$$R_{(55)5} = 3\,R_5^2(1 - R_5) + R_5^3$$
$$= 3\,R_5^2 - 2\,R_5^3.$$

Finally, the reliability of the system R_{sys} is

$$R_{\mathrm{sys}} = R_{1234}R_{(55)5}. \quad \diamond$$

6-2 SYSTEM RELIABILITY FOR SEQUENTIALLY OPERATING UNITS

Another class of simple systems is that of parallel systems that are load-sharing or sequential in operation in such a way that only one unit of the system is in operation at a time. In order to minimize the complexity of the equations, only two independent units will be considered. A further assumption is that the second unit, initially in standby, will never fail in the standby mode of operation. Furthermore the switching from unit #1 to unit #2 will be assumed instantaneous and not subject to failure. The system may be depicted as in Fig. 6-3.

The system reliability can be calculated in either of two ways. One approach is to use the failure probability density $f_{12}(t)$ obtained for the convoluted distribution model, as in Eq. (3-47), and to calculate the reliability

$$R_{\text{sys}}(t) = \int_t^\infty f_{12}(t')dt'. \tag{6-14}$$

The second method, the "compound events" approach, gives the system reliability as

$$R_{\text{sys}}(t) = R_1(t) + \int_0^t f_1(t')\, R_2(t - t')dt'. \tag{6-15}$$

Here the first term on the right-hand side is the reliability of the first unit, while the second term accounts for the reliability of the second unit. To understand why the second term appears so complicated, it is necessary to remember that it is not known precisely when the first unit will fail; thus all possible failure times must be accounted for by integrating t' between 0 and t. Also, the second unit is not needed until the first one has failed with probability $f_1(t')dt'$. Of course, the second unit need operate only for the time after the failure at t' until t, so the reliability for the second unit is given by $R_2(t - t')$.

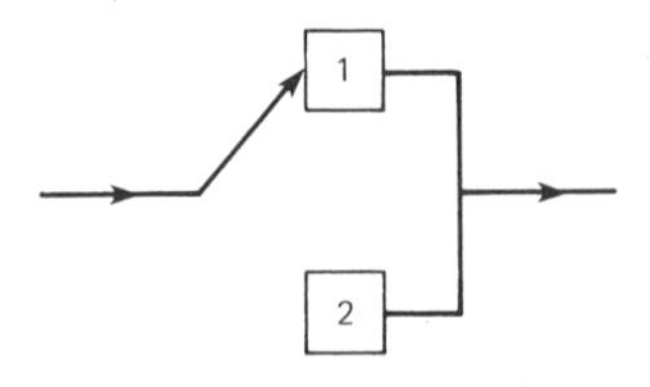

Fig. 6-3 *Reliability block diagram for two units with unit #2 in standby.*

For random failures, Eqs. (6-14) and (6-15) give

$$R_{\text{sys}}(t) = \exp(-\lambda_1 t) + \frac{\lambda_1}{\lambda_1 - \lambda_2}[\exp(-\lambda_2 t) - \exp(-\lambda_1 t)]. \tag{6-16}$$

The special case of the last equation with $\lambda_1 = \lambda_2 = \lambda$ is

$$R_{\text{sys}}(t) = e^{-\lambda t}(1 + \lambda t) \tag{6-17}$$

and follows from Eq. (6-16) with the use of the series approximation $\exp[(\lambda_2 - \lambda_1)t] \approx 1 + (\lambda_2 - \lambda_1)t$ followed by a limiting procedure. The two terms of Eq. (6-17) are those arising from the cumulative Poisson distribution of Eq. (3-18).

Equation (6-14) is not applicable if unit #2 can fail while in standby, but Eq. (6-15) can be generalized. For the reliability of this unit, the symbol $R_2^*(t)$ is used since the hazard rate $\lambda_2^*(t)$ in the standby mode is different from the $\lambda_2(t)$ in active operation (or otherwise the unit could be considered to be in active-parallel operation). If perfect switching from unit #1 to #2 is still assumed, then the system reliability is

$$R_{\text{sys}}(t) = R_1(t) + \int_0^t f_1(t')\, R_2^*(t')\, R_2(t - t')\,dt'. \tag{6-18}$$

Notice that here the second term on the right-hand side contains the reliability factor $R_2^*(t')$, which was not present before. In the case that the switching is not perfect, but is instantaneous with a reliability given by the constant R_{sw}, then that second term must be multiplied by R_{sw}.

The generalization of Eq. (6-18) to the case of three independent units operated sequentially, with perfect switching, is

$$R_{\text{sys}}(t) = R_1(t) + \int_0^t f_1(t')\, R_2^*(t')\, R_2(t - t')dt' + \int_0^t f_{12}(t')\, R_3^*(t')\, R_3(t - t')dt', \tag{6-19}$$

where now $f_{12}(t) = -dR_{12}(t)/dt$ and $R_{12}(t)$ is the sum of the first two terms of this equation. Generalizations of the last equation can be derived in the same manner.

Equation (6-18) for random failures of both units, when the second unit is in standby, is

$$R_{\text{sys}}(t) = \exp(-\lambda_1 t) + \frac{\lambda_1}{\lambda_1 + \lambda_2^* - \lambda_2}\{\exp(-\lambda_2 t) - \exp[-(\lambda_1 + \lambda_2^*)t]\}. \tag{6-20}$$

This equation obviously reduces to Eq. (6-16) when $\lambda_2^* = 0$, while if $\lambda_2^* = \lambda_2$, then the reliability is just that of Eq. (6-4) for two units in active-parallel. For the special case of $\lambda_1 = \lambda_2 = \lambda$, Eq. (6-20) has the form

$$R_{\text{sys}}(t) = e^{-\lambda t}\left[1 + \frac{\lambda}{\lambda^*}(1 - e^{-\lambda^* t})\right]. \tag{6-21}$$

The purpose of standby units is to increase the reliability and the MTTF over that which would be obtained without the standby units. For example, if there are N identical independent units, initially with one operating and the remaining in standby waiting to be operated sequentially, and if there are no failures during standby or with the switching mechanism, then

$$\lambda(\text{MTTF}) = N. \tag{6-22}$$

Comparison of Eq. (6-22) with the results in Eqs. (6-11) and (6-13) demonstrates the effect of the standby units upon the MTTF.

Example 6-3 A reactor coolant pump has an identical pump in standby that can be successfully valved into operation 99% of the time. The reliability of a pump when in operation over a year's time is 0.8, and in standby over two year's time is 0.95. What is the reliability of the pump system over a six-month time?

We use a modified form of Eq. (6-21) that accounts for the reliability of the "switch" of $R_{\text{sw}} = 0.99$,

$$R_{\text{sys}}(t) = e^{-\lambda t}\left[1 + R_{\text{sw}}\frac{\lambda}{\lambda^*}(1 - e^{-\lambda^* t})\right].$$

Since no other information is given, we assume random failures and calculate the hazard rate λ of either pump in operation as

$$\lambda = -\frac{\ln R(t)}{t} = -\frac{\ln 0.8}{1 \text{ yr}} = 0.233/\text{yr}.$$

Similarly, in standby, the hazard rate is

$$\lambda^* = -\frac{\ln R^*(t)}{t} = -\frac{\ln 0.95}{2 \text{ yr}} = 0.0256/\text{yr}.$$

Substitution of these results into the equation for $t = 0.5$ yr gives R_{sys} (0.5 yr) $= 0.993$, which exceeds the reliability for a single pump by the factor of 1.11. ◇

6-3 SYSTEM RELIABILITY AS DERIVED BY THE DECOMPOSITION METHOD

The reliability of some systems is improved by connecting components in such a way that they are in more than one subsystem. For example, consider system A shown in Fig. 6-4 in which all components act indepen-

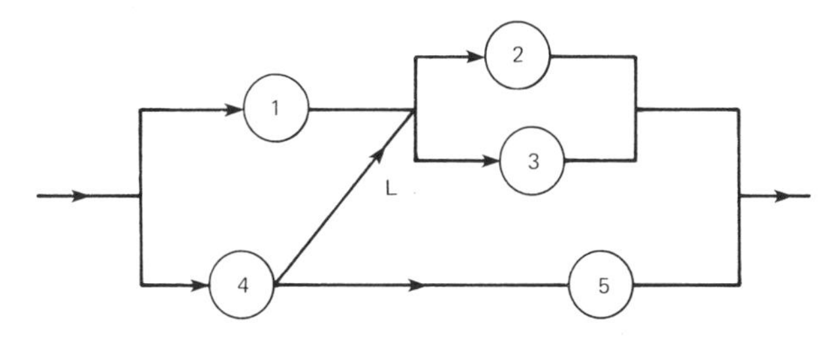

Fig. 6-4 *Reliability block diagram for cross-link system* A.

dently and have a reliability R_i. The two branches are connected by link L in such a way that a "signal" flowing from an input to output may flow along any of the paths connecting components 1–2, 1–3, 4–2, 4–3, or 4–5.

If the link L in Fig. 6-4 were *not* present, then the reliability of the system could be obtained by considering first the upper paths connecting components 1–2 and 1–3. Using Eq. (6-4) to treat units 2 and 3 in parallel, followed by use of Eq. (6-1) to treat unit 1 in series with the combination, gives the reliability of the upper portion R_u as

$$R_u = R_1 (R_2 + R_3 - R_2R_3).$$

The reliability of the path through components 4 and 5 is just R_4R_5, so the system reliability R_{sys} is obtained for R_u and R_4R_5 in parallel. The result for system A *without* the cross-link L is

$$R_{sys} = R_4R_5 + R_1 (R_2 + R_3 - R_2R_3)(1 - R_4R_5). \qquad (6\text{-}23)$$

If the components are all identical with $R_i = e^{-\lambda t}$, then the last equation gives

$$R_{sys}(t) = 3e^{-2\lambda t} - e^{-3\lambda t} - 2e^{-4\lambda t} + e^{-5\lambda t}, \qquad (6\text{-}24)$$

and integration over all time yields

$$\text{MTTF} = 13/15\lambda. \qquad (6\text{-}25)$$

If the link L in Fig. 6-4 *is* present, Eqs. (6-23) through (6-25) are no longer valid. Then the system reliability may be analyzed by means of the decomposition method, which is nothing more than successive application of a conditional probability theorem. This method is based upon the selection of a "keystone" component K at either one end or the other of the cross-link [2]. The reliability of the system may be broken down into contributions, by means of Eq. (2-19), for when K works and does not work; the *decomposition equation* is

$$R_{sys} = R_K R(\text{sys}|K) + R_{\bar{K}} R(\text{sys}|\bar{K}), \qquad (6\text{-}26)$$

where $\bar{K}$ denotes "failed K." Here $R_{\bar{K}} = 1 - R_K$.

For Fig. 6-4, we pick component K to be component 4. Then $R(\text{sys}|K)$ is calculated for the parallel combination of components 2, 3, and 5 since the signal can always bypass component 1. The reliability $R(\text{sys}|\bar{4})$ is just R_u from when cross-link L was not present, so the decomposition equation gives the reliability of system A as

$$R_{\text{sys}} = R_4[1 - (1 - R_2)(1 - R_3)(1 - R_5)] + (1 - R_4)[R_1(R_2 + R_3 - R_2R_3)]. \tag{6-27}$$

In the case that all components are identical with constant hazard rate λ, then this result becomes

$$R_{\text{sys}}(t) = 5e^{-2\lambda t} - 6e^{-3\lambda t} + 2e^{-4\lambda t}, \tag{6-28}$$

and the integral over all time gives

$$\text{MTTF} = 1/\lambda. \tag{6-29}$$

The improvement of the reliability due to the cross-link L is conveniently illustrated by comparing Eqs. (6-25) and (6-29).

Had we selected component 3 in system A to be the keystone component K, then

$$R(\text{sys}|3) = R_1 + R_4 - R_1R_4,$$

since components 2 and 5 can always be bypassed. A complication arises in the analysis, however, because if component 3 does not function, we are left with the reliability block diagram of Fig. 6-5, and we still have not removed the coupling effects arising from cross-link L. (It may be noted that the same complication would have arisen had we selected component 2 instead of 3 to be the keystone component.)

To analyze the system in Fig. 6-5, we apply the decomposition method a *second* time to calculate $R(\text{sys}|\bar{3})$. For example, if we select component 2 as the second keystone component, then we need to evaluate

$$R_{\text{sys}} = R_3\,R(\text{sys}|3) + R_{\bar{3}}\,[R_2\,R(\text{sys}|2\bar{3}) + R_{\bar{2}}\,R(\text{sys}|\bar{2}\bar{3})]. \tag{6-30}$$

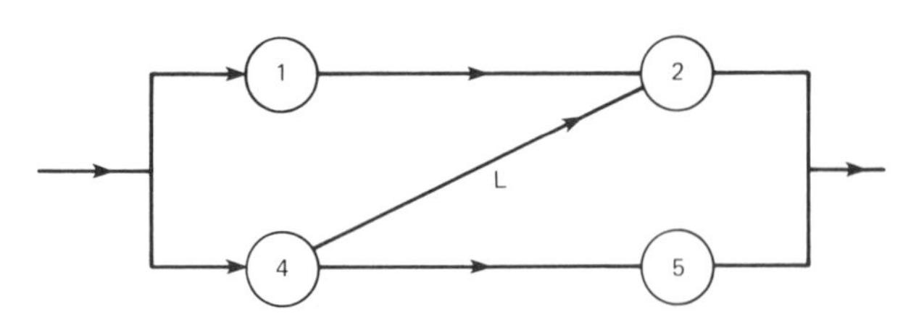

Fig. 6-5 *Reliability block diagram for cross-link system* A *with component 3 failed.*

When component 2 works, component 5 can be bypassed so

$$R(\text{sys}|2\bar{3}) = R_1 + R_4 - R_1R_4,$$

while $R(\text{sys}|\bar{2}\bar{3}) = R_4R_5$. Substitution of these results into Eq. (6-30) reproduces Eq. (6-27).

This example for system A illustrates that in the decomposition method the final answer for the reliability does not depend upon which component at the end of a link is selected as the keystone component. It also shows that generally one keystone component may be easier to use than another. Furthermore, it shows how the decomposition method can be applied repeatedly.

It is important to note that the endpoints of the cross-link in the reliability block diagram convey meaning about how the system functions. For example, in Fig. 6-4 the link L was connected between component 4 and the units 2 and 3 in parallel. Now consider the slightly different system B in which the cross-link L connects the lower and upper branches as shown in Fig. 6-6.

For system B, it is easiest to pick component 3 as the keystone component. Then $R(\text{sys}|3)$ is calculated for the parallel combination of components 1 and 4, since component 5 can be bypassed, while $R(\text{sys}|\bar{3})$ is for the parallel combination of components 1 and 2 in series and components 4 and 5 in series. Thus the reliability of system B is

$$\begin{aligned} R_{\text{sys}} &= R_3\,(R_1 + R_4 - R_1R_4) \\ &\quad + (1 - R_3)(R_1R_2 + R_4R_5 - R_1R_2R_4R_5). \end{aligned} \tag{6-31}$$

In the case that all components are identical with constant hazard rate λ, then this result becomes

$$R_{\text{sys}}(t) = 4e^{-2\lambda t} - 3e^{-3\lambda t} - e^{-4\lambda t} + e^{-5\lambda t}, \tag{6-32}$$

and the integral over all time gives

$$\text{MTTF} = 19/20\lambda. \tag{6-33}$$

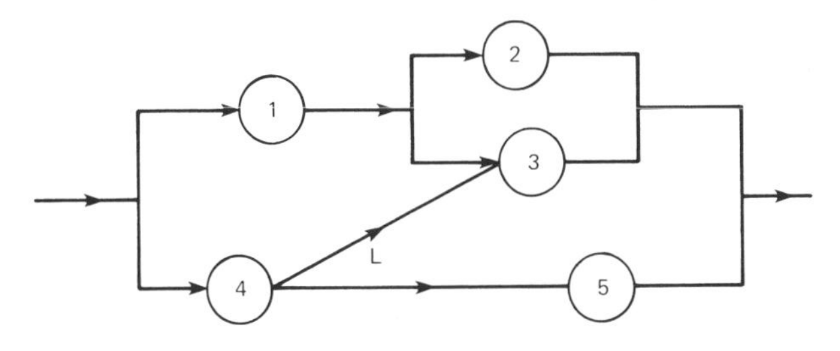

Fig. 6-6 *Reliability block diagram for cross-link system* B.

As expected, the cross-link L for system B is not as effective as that for system A, but it provides an improvement in the reliability over the case with no cross-link.

Consider now another system C shown in Fig. 6-7. If we pick component 4 as the keystone component, the result for $R(\text{sys}|4)$ is just that for the parallel combination of components 2, 3, and 5, but now $R(\text{sys}|\bar{4})$ must be calculated by selecting a second keystone component. With the use of

$$R(\text{sys}|\bar{4}) = R_1 R(\text{sys}|1\bar{4}) + R_{\bar{1}}\, R(\text{sys}|\bar{1}\bar{4})$$

and the observations that $R(\text{sys}|\bar{1}\bar{4}) = 0$ and $R(\text{sys}|1\bar{4}) = R(\text{sys}|4)$, it follows that the reliability for system C is

$$R_{\text{sys}} = (R_1 + R_4 - R_1 R_4)\,[1 - (1 - R_2)(1 - R_3)(1 - R_5)]. \quad (6\text{-}34)$$

In the case that all components are identical with constant hazard rate λ, then this result becomes

$$R_{\text{sys}}(t) = 6e^{-2\lambda t} - 9e^{-3\lambda t} + 5e^{-4\lambda t} - e^{-5\lambda t}, \quad (6\text{-}35)$$

and the integral over all time gives

$$\text{MTTF} = 21/20\lambda. \quad (6\text{-}36)$$

Thus the reliability of system C is slightly better than that for system A.

It is worth noting that system C could have been analyzed without the use of the decomposition method, since the cross-link L in Fig. 6-7 can be shortened to a single point. Then following the procedures in Section 6-1, system C is treated as a series combination of components 1 and 4 in parallel and components 2, 3, and 5 in parallel. This reasoning helps explain the result in Eq. (6-34). Thus the example of system C illustrates that sometimes the analysis is simpler without the decomposition method.

Any system containing more than one cross-link can be analyzed by a successive use of the decomposition equation (6-26), as illustrated in the next example.

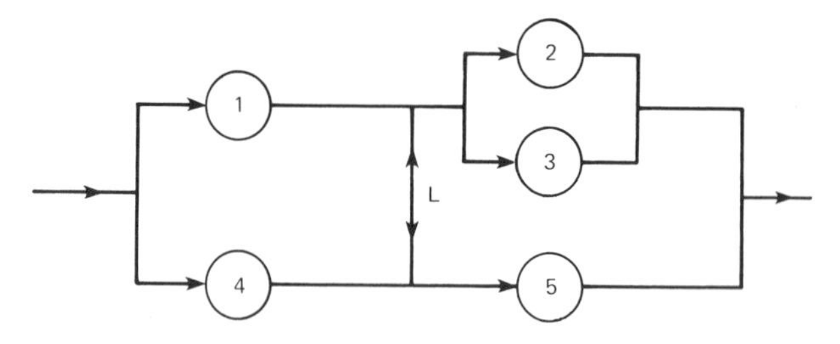

Fig. 6-7 *Reliability block diagram for cross-link system* C.

Example 6-4 A system consists of six identical units with constant hazard rate λ that are connected as shown in the following reliability block diagram:

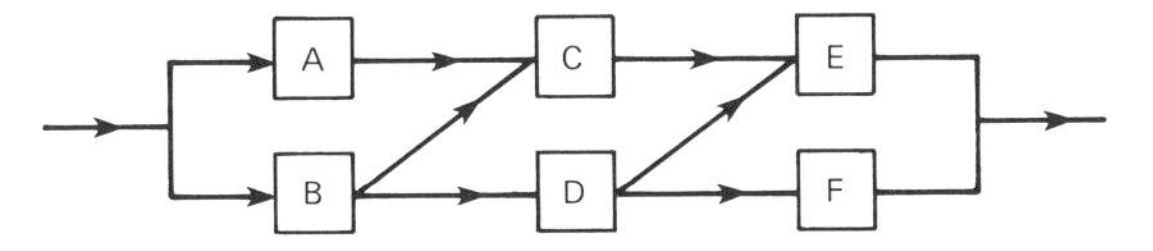

Calculate the reliability of the system and the MTTF.

If component B is selected as the first keystone component, then $R(\text{sys}|\bar{B}) = R_A R_C R_E$, but $R(\text{sys}|B)$ must be calculated by means of a second keystone component such as D. Since $R(\text{sys}|DB) = R_E + R_F - R_E R_F$ and $R(\text{sys}|\bar{D}B) = R_C R_E$,

$$R_{\text{sys}} = R_B R_D (R_E + R_F - R_E R_F) + R_B (1 - R_D) R_C R_E + (1 - R_B) R_A R_C R_E.$$

After the substitution of $e^{-\lambda t}$ for every R, we see that

$$R_{\text{sys}}(t) = 4e^{-3\lambda t} - 3e^{-4\lambda t},$$

and integration over all time gives

MTTF $= 7/12\lambda$. ◇

6-4 RELIABILITY OBTAINED USING A SIGNAL FLOW GRAPH

Signal flow graphs of various elementary systems have been extensively developed [2]. The signal flow graph technique itself is not capable of providing more information than a procedure that uses techniques similar to those presented in Sections 6-1 and 6-2. However, the use of graphs does have some analogy to circuit theories encountered in other engineering disciplines, and, for that reason, some reliability analysts prefer it for simple systems. The signal flow graph method will be introduced primarily so that it can be used in Section 6-5 to illustrate graphically the idea of "cut sets."

The graph for a signal flow analysis of a system consists of nodes $x_1, x_2, \cdots x_n$ and a collection of branches joining the nodes together. There are two nodes representing the input and output of a component i; a branch between the two nodes represents the component and has a direction indicating the flow of the signal. The "weight" of the branch, A_i, represents the event of transmitting a signal through component i. A path

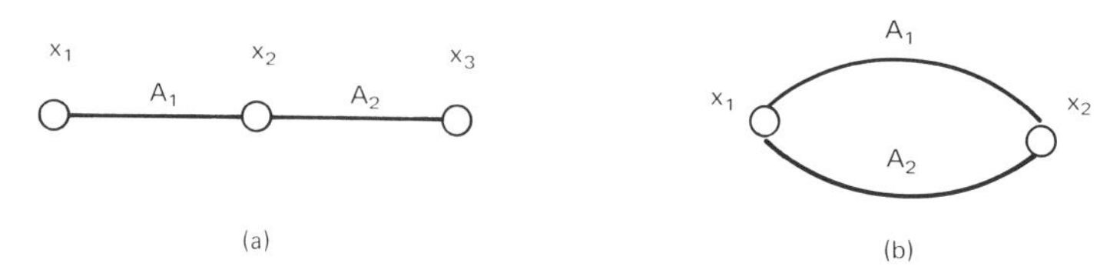

Fig. 6-8 *Signal flow graph for (a) two units in series and (b) two units in active-parallel.*

is a sequence of branches in which the output node of one branch is the input node of the next branch.

Signal flow graphs can be considered to be the analog of reliability block diagrams. For example, the signal flow graph of Fig. 6-8(a) corresponds to the block diagram of Fig. 6-1(a), while Fig. 6-8(b) corresponds to Fig. 6-1(b); the line between nodes x_1 and x_2 labeled A_1 indicates the event of transmitting a signal from x_1 to x_2.

The reliability for two units in series is written in the notation of Fig. 6-8(a) as the probability of transmission from x_1 to x_3,

$$P(x_3/x_1) = P(A_1A_2) = R_1R_2\,, \tag{6-37}$$

and follows by replacing the path between the two nodes x_1 and x_3 by a single branch equal to the AND operation of both branches in the path. This equation is the direct analog of Eq. (6-1) and can be extended as in Eq. (6-5) to the case of systems with more than two components.

The reliability of two units in active-parallel is written in the notation of Fig. 6-8(b) as

$$P(x_2/x_1) = P(A_1 + A_2) = R_1 + R_2 - R_1R_2 \tag{6-38}$$

and corresponds to the OR operation of both branches between x_1 and x_2. This equation is just a restatement of Eq. (6-2).

Using extensions of the AND and OR rules of Eqs. (6-37) and (6-38), we can analyze more complicated systems by using signal flow graphs in much the same way as in Section 6-1.

Example 6-5 Obtain the reliability for the system whose graph is

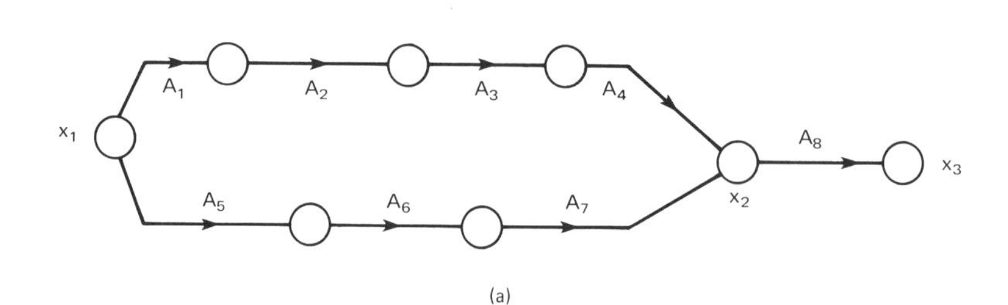

The first step is to use the AND operation to obtain the following reduced signal flow graph:

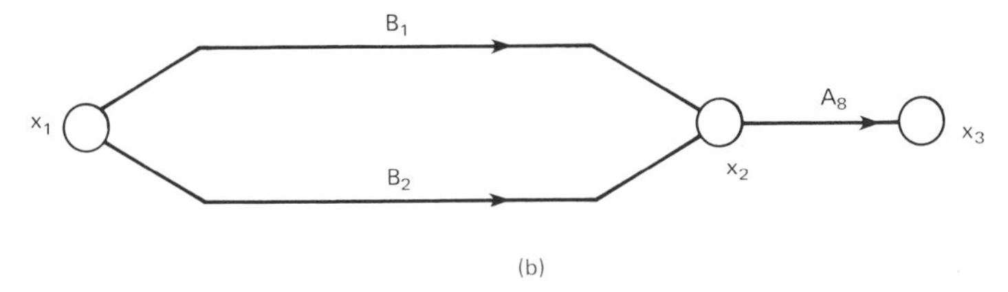

(b)

Here the product rule gives $B_1 = A_1A_2A_3A_4$ and $B_2 = A_5A_6A_7$. Then the OR operation gives the subsystem event as $(B_1 + B_2)$, and finally another AND operation gives

$$x_3/x_1 = A_8(B_1 + B_2).$$

The probability of transmission from x_1 to x_3 is

$$\begin{aligned} P(x_3/x_1) &= P[A_8(A_1A_2A_3A_4 + A_5A_6A_7)] \\ &= R_8[1 - (1 - R_1R_2R_3R_4)(1 - R_5R_6R_7)]. \quad \diamond \end{aligned}$$

6-5 CUT-SET METHOD FOR DETERMINING RELIABILITY

A *cut set* of a system is defined as a set of system events that, if they all occur, will cause system failure. Although the concept of a cut set is general, such a set can be illustrated by means of a signal flow graph; then a cut set is any cut that severs the set of all branches between the input and output nodes. A *minimal cut set* of a system is a cut set containing system events that are not a subset of the events of any other cut set [5, 6]. Another way of saying this is that the removal of any event from a minimal cut set would cause it not to be a cut set, i.e., the system would no longer fail.

Example 6-6 As an example of minimal cut sets, consider the following signal flow graph:

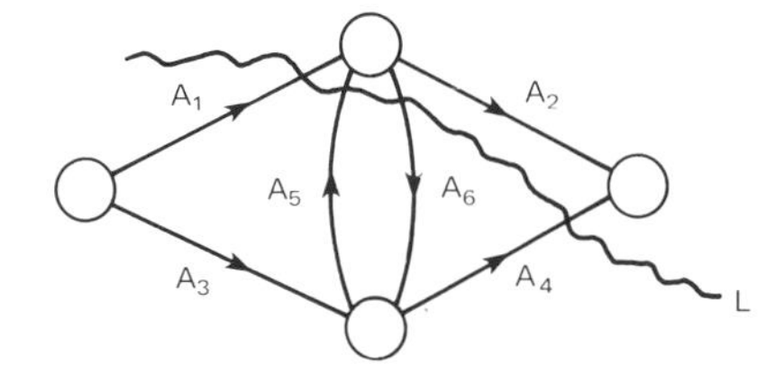

One cut set, C_1, which corresponds to the wavy line L in the figure, is the event $A_1A_5A_6A_4$. Other cut sets (denoted by C_n, $n = 2, 3$, etc.) are

$$C_2 = A_1A_3, \qquad C_3 = A_2A_4,$$
$$C_4 = A_1A_5A_3, \qquad C_5 = A_2A_6A_4, \qquad C_6 = A_3A_5A_6A_2,$$
$$C_7 = A_1A_5A_4, \qquad C_8 = A_3A_6A_2, \text{ etc.}$$

The cut sets may be obtained by cutting the graph from top to bottom (if the input and output nodes are horizontal). The cut sets C_4 and C_5 are not minimal cut sets because sets C_2 and C_3 are subsets of C_4 and C_5, respectively. Cut set C_1 is not minimal because component A_6 cannot operate if A_1 and A_5 do not; similarly, C_6 is not a minimal cut set. The cut sets C_7 and C_8 are minimal, however, as well as C_2 and C_3; there are no other minimal cut sets for this example. ◇

Now consider a general system for which all *minimal* cut sets are denoted by C_n, $n = 1$ to N. The system failure probability F_{sys} can be written as

$$F_{\text{sys}} = P(C_1 + C_2 + \cdots + C_N) \tag{6-39}$$

since a failure of the components in any one minimal cut set will lead to system failure. The F_{sys} can be bounded by use of Eq. (2-17),

$$F_{\text{sys}} \leq \sum_{n=1}^{N} P(C_n), \tag{6-40}$$

so the lower bound of the system reliability R_{sys} is

$$R_{\text{sys}} = 1 - F_{\text{sys}} \geq 1 - \sum_{n=1}^{N} P(C_n). \tag{6-41}$$

Likewise, use of Eq. (2-18) gives the upper bound of R_{sys} as

$$R_{\text{sys}} \leq 1 - \sum_{n=1}^{N} P(C_n) + \sum_{n=1}^{N-1} \sum_{m=n+1}^{N} P(C_nC_m). \tag{6-42}$$

Example 6-7 For the system of Example 6-6, calculate a minimum value of the reliability. From Eq. (6-41),

$$R_{\text{sys}} \geq 1 - [P(A_1A_3) + P(A_2A_4)$$
$$+ P(A_1A_5A_4) + P(A_3A_6A_2)].$$

If the failures of all components are independent and random, then

$$R_{\text{sys}}(t) \geq 1 - [(1 - e^{-\lambda_1 t})(1 - e^{-\lambda_3 t}) + (1 - e^{-\lambda_2 t})(1 - e^{-\lambda_4 t})$$
$$+ (1 - e^{-\lambda_1 t})(1 - e^{-\lambda_5 t})(1 - e^{-\lambda_4 t})$$
$$+ (1 + e^{-\lambda_3 t})(1 - e^{-\lambda_6 t})(1 - e^{-\lambda_2 t})]. \ ◇$$

Table 6-1 *Some Rules of Boolean Algebra for Events*[a]

Word description	Rules
1. Commutative Law	a. $XY = YX$
	b. $X + Y = Y + X$
2. Associative Law	a. $X(YZ) = (XY)Z$
	b. $X + (Y + Z) = (X + Y) + Z$
3. Idempotent Law	a. $XX = X$
	b. $X + X = X$
4. Absorption Law	a. $X(X + Y) = X$
	b. $X + XY = X$
5. Distributive Law	a. $X(Y + Z) = XY + XZ$
	b. $(X + Y)(X + Z) = X + YZ$
6. Complementation[b]	a. $X\bar{X} = \phi$
	b. $X + \bar{X} = \Omega$
	c. $\bar{\bar{X}} = X$
7. De Morgan's Theorems	a. $(\overline{XY}) = \bar{X} + \bar{Y}$
	b. $\overline{X + Y} = \bar{X}\bar{Y}$
8. Unnamed relationships but frequently useful	a. $X + \bar{X}Y = X + Y$
	b. $\bar{X}(X + \bar{Y}) = \bar{X}\bar{Y}$

[a] Selected from W. E. Vesely *et al.*, Fault Tree Handbook. U.S. Nuclear Regulatory Commission Rep. NUREG-0492 (1981).

[b] The universal event Ω is sometimes denoted by I, and the null event ϕ is sometimes denoted by 0.

To evaluate the system probability F_{sys} defined in Eq. (6-39), we first need to reduce $(C_1 + C_2 + \cdots + C_N)$ to a form involving all the events A_i. This can be done with the rules of Boolean algebra, which is the algebra for events.

A partial list of rules for Boolean algebra appears in Table 6-1. The commutative and associative laws are similar to those laws for ordinary algebra. The idempotent laws enable us to cancel out redundancies of the same event. Absorption law 4a is easily justified by observing that if event X occurs then event $(X + Y)$ also has occurred so $X(X + Y) = X$; a similar argument holds for absorption law 4b. The distributive laws 5a and 5b are very useful in fault tree analysis (Chapter 8) and may be verified by using the preceding rules. Note that De Morgan's theorems are useful when we switch from a search for the failure of a system to the successful operation of that system.

Example 6-8 Simplify the form of event H where

$$H = A_1 + A_2A_3(A_4 + A_2) + (A_2 + A_5)(A_3 + \bar{A}_2).$$

We observe the first term obviously cannot be reduced, while

$$
\begin{aligned}
A_2A_3\,(A_4 + A_2) &= A_2A_3A_4 + A_2A_3A_2 && \text{(law 5a)}\\
&= A_2A_3A_4 + A_2A_3 && \text{(law 3a)}\\
&= A_2A_3 && \text{(law 4b)}
\end{aligned}
$$

and

$$
\begin{aligned}
(A_2 + A_5)(A_3 + \bar{A}_2) &= A_2A_3 + A_2\bar{A}_2 + A_5A_3 + A_5\bar{A}_2 && \text{(law 5a)}\\
&= A_2A_3 + A_5A_3 + A_5\bar{A}_2 . && \text{(law 6a)}
\end{aligned}
$$

Combining terms gives

$$
\begin{aligned}
H &= A_1 + A_2A_3 + A_2A_3 + A_5A_3 + A_5\bar{A}_2\\
&= A_1 + A_2A_3 + A_5A_3 + A_5\bar{A}_2. \ \diamond && \text{(law 3b)}
\end{aligned}
$$

To evaluate the probability F_{sys}, defined in Eq. (6-39) as the probability of $(C_1 + C_2 + \cdots + C_N)$, it is important to note that a component A_i may be in more than one minimal cut set C_n. Thus, when calculating the system failure probability, it is necessary to avoid "double-counting" the failure probability of component A_i. This can perhaps best be illustrated by an example [6].

Example 6-9 We consider the event H, which consists of five minimal cut sets,

$$H = A + BD + BE + CD + CE,$$

and we wish to calculate the probability $P(H)$. To simplify the notation, in this example the symbol A will stand for $P(A)$, etc. From Eq. (2-15),

$$
\begin{aligned}
P(H) = {} & [A + BD + BE + CD + CE]\\
& - [ABD + ABE + ACD + ACE + BD\not{B}E + BDC\not{D}\\
& + BDCE + BECD + BEC\not{E} + CD\not{C}E]\\
& + [ABD\not{B}E + ABDC\not{D} + ABDCE + ABECD + ABEC\not{E}\\
& \quad + ACD\not{C}E + BD\not{B}EC\not{D} + BD\not{B}EC\not{E} + BDC\not{D}\not{C}E\\
& \quad + BECD\not{C}\not{E}]\\
& - [ABD\not{B}EC\not{D} + ABD\not{B}EC\not{E} + ABDC\not{D}\not{C}E + ABECD\not{C}\not{E}\\
& \quad + BD\not{B}EC\not{D}\not{C}\not{E}]\\
& + [ABD\not{B}EC\not{D}\not{C}\not{E}].
\end{aligned}
$$

This initial result shows that the initial five minimal cut sets have resulted in 31 terms in the probability expansion and that, because of lack of independence, 18 of the 31 terms contain redundant factors (denoted by slashes). (For example, had the number of minimal cut sets been ten, the complete expansion would have contained 1023 terms; if the number of cuts had been 20, then 1,048,575 terms would have been needed. The general relation is $2^n - 1$ for n minimal cut sets.)

Eliminating redundant factors and algebraically adding the probability products for identical terms, the expression for $P(H)$ is

$$\begin{aligned} P(H) = {} & [A + BD + BE + CD + CE] \\ & - [ABD + ABE + ACD + ACE + BCD + BCE \\ & \quad + BDE + CDE + 2\,BCDE] \\ & + [ABCD + ABCE + ABDE + ACDE + 4\,BCDE \\ & \quad + 2\,ABCDE] \\ & - [BCDE + 4\,ABCDE] \\ & + [ABCDE]. \quad \diamond \end{aligned}$$

As can be seen in the example, the careful calculation of the system failure probability F_{sys} could be extremely tedious, when there are a lot of different events A_i, if the analysis had to be performed by hand. Fortunately, computer programs exist for this purpose; these programs will be discussed in Chapter 10 after fault tree analysis is introduced in Chapter 8. For now it is sufficient to know that one of the ways to evaluate a fault tree numerically is to first determine the minimal cut sets and to then calculate the system failure probability from those cut sets. If the system has several hundred components, as they sometimes do, then computer assistance is mandatory unless it is only desired to bound F_{sys} by using Eq. (6-40), as illustrated in the next example.

Example 6-10 We wish to illustrate the accuracy of a failure probability bound such as Eq. (6-40) for the system in Example 6-9. We will assume the failure probabilities in that example to be

$$P(A) = 0.01$$

$$P(B) = P(C) = P(D) = P(E) = 0.1.$$

Then using the final result for $P(H)$ in Example 6-9 gives

$$\begin{aligned} P(H) &= (0.05) - (0.0046) + (0.000442) - (0.000104) + (0.000001) \\ &= 0.045739. \end{aligned}$$

If approximate answers had been obtained by truncating the series expansion as in Eqs. (6-41) and (6-42), etc., the results would have been as follows [6]:

Terms included	Failure probability	Error (%)
1	0.05	+9.3
1 and 2	0.0454	−0.74
1, 2, and 3	0.045842	+0.22
1, 2, 3, and 4	0.045738	−0.0022

If the probabilities of failure of each component had been one order of magnitude smaller, then the exact result would have been

$$P(H) = 0.00139561399$$

and taking only the first term would have given 0.0014, for an error of 0.31%. This illustrates the fact that the error bounds in Eqs. (6-41) and (6-42) are closer together when the failure probabilities of the components are smaller. ◇

6-6 SYSTEMS WITH COMMON CAUSE FAILURES

Potential common cause or common mode failures are a major concern for the reliability analyst. Here we look at only one of several possible analytical methods for approximately incorporating a certain class of potential common cause failures into a reliability analysis [7].

If it can be assumed that components of a system fail when subjected to an event that imposes an abnormal stress, i.e., a fatal shock, and that the occurrence of the shocks is governed by the Poisson process, then Eq. (3-17) gives

$$P_c(r, t) = \frac{e^{-\lambda_c t}(\lambda_c t)^r}{r!}, \tag{6-43}$$

where $P_c(r, t)$ is the probability that exactly r shocks of the common cause type occur in the interval $(0, t)$. The effect of potential common mode failures can be studied, qualitatively at least, by assuming that all components can be simultaneously destroyed with a single shock of the common mode type (so $r = 1$). Then the probability of failure of the system due to common cause failures is governed by the exponential distribution, and the reliability of the system against such failures, $R_c(t)$, is

$$R_c(t) = e^{-\lambda_c t}. \tag{6-44}$$

The overall system reliability $R_{sys}(t)$ is then

$$R_{sys}(t) = e^{-\lambda_c t} R_{sys,\bar{c}}(t), \tag{6-45}$$

if $R_{sys,\bar{c}}(t)$ denotes the system reliability with no common cause failures.

The main conclusion from Eq. (6-45) is that common cause failures cause a decrease in system reliability. In the special case of a series system of N units, for example, λ_c must satisfy the equation $\lambda_c \ll N\lambda$ in order for the contribution from common cause failures to be negligible; similar relationships are available for M-out-of-N systems [7].

Even if the assumptions leading to Eq. (6-45) are valid, the hazard rate for the system due to common cause failures, λ_c, must be estimated. This determination should be done by using any available data, but normally it must be estimated with "engineering judgment."

EXERCISES

6-1 What is the reliability of the system shown if the reliabilities of the components are R_1, R_2, and R_3?

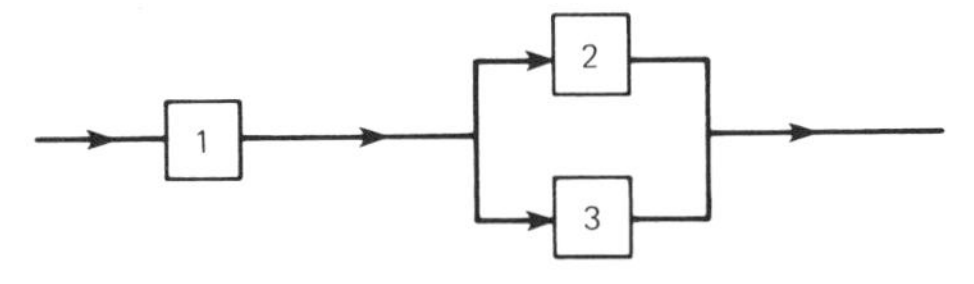

6-2 What is the reliability of the system shown of $2n$ identical components if the reliability of each individual unit is R?

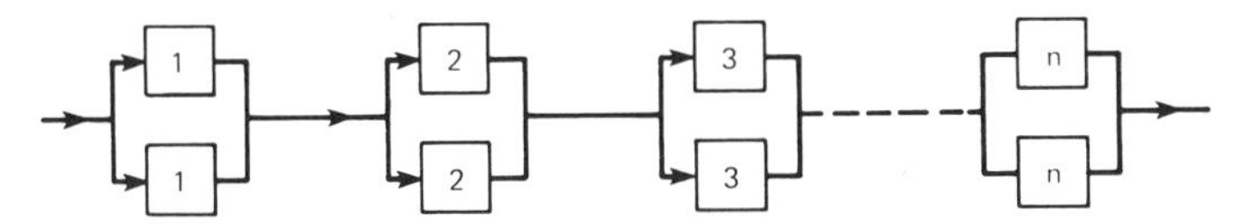

6-3 Prove the system in Exercise 6-2 is more reliable, for $n > 1$ and $R < 1$, than that in Example 6-1 for which the corresponding system reliability is $R^n(2 - R^n)$.

6-4 It is desired to increase the redundancy of the system shown, consisting of three identical components, with reliability R, by duplicating (a) each component in parallel, or (b) the whole system in parallel. Calculate the reliability of the new systems a and b and prove that the reliability of system a is greater than that of system b if $R < 1$.

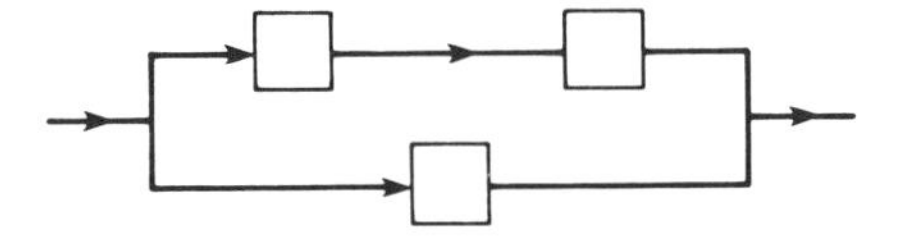

6-5 Three elements of a system, A, B, and C, have constant probabilities of success of 0.9, 0.8, and 0.7, respectively, at any time. What is the probability of success for the system if (a) all elements must be functioning? (b) any two of the units must be functioning [1]?

6-6 A reactor has four identical, independent coolant loops. Each loop consists of three subsystems: the pump S_1, the electrical power S_2 to the pump, and the piping S_3 that circulates the coolant. Over the mission time for operation, the probability of a pump failure is 0.2, of electrical power failure is 0.05, and of pipe failure is 0.001. Assume there is no repair during the operating period of interest.

(a) A coolant loop is known to have failed. If there is an electrical power failure the loop will certainly fail, but only 60% of the pump failures cause a loop failure, and only 40% of the pipe failures cause a loop failure. Calculate the probabilities that each subsystem failed, given that a loop has failed.

(b) If all three subsystems function independently, and if all must function for a loop to operate, calculate the probability of a loop failure over the mission time of operation.

(c) If each of the loops operates independently from the others, calculate the probability that at least three of the four will function for the entire mission time.

6-7 An aircraft has four engines but may land using only two engines. (a) Assuming that all four engines have a reliability of $R = 0.95$ of completing a mission, calculate the reliability of the four-engine system for a mission. (b) Calculate the reliability if the airplane must have at least one active engine on each wing.

6-8 A reactor has three identical coolant loops, each with two identical pumps connected in parallel. The reliability of each pump over the life of the plant is 0.6. At least one pump must operate for a loop to be functional, and at least two loops must operate for the coolant system to be functional. Calculate the reliability of the coolant system over the life of the plant.

6-9 A reactor has three identical coolant loops, each with four identical pumps connected in parallel. The reliability of each pump over the life of the plant is 0.4. At least two pumps must operate for a loop to be functional, and at least two loops must operate for the coolant system to be functional. Calculate the reliability of the coolant system over the life of the plant.

6-10 Derive the time-dependent reliability and calculate the MTTF of the system shown if it consists entirely of units having a constant hazard rate of 10^{-6} per hour.

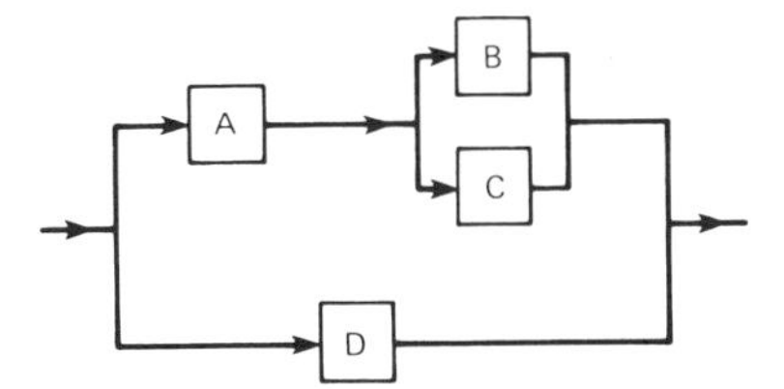

6-11 Three identical units fail randomly with rate λ. If the units for the systems are (1) in active-parallel, (2) in active-parallel, but two of the three are required for successful operation, (3) with one active and two in standby, not subject to standby or switching failures, then: (a) derive the equation for the reliability of each system, and (b) calculate the MTTF for each system.

6-12 A main coolant pump in a light water reactor has a hazard rate of 10^{-6}/hr. A (nonidentical) backup unit that automatically starts upon failure of the first unit possibly can be used. The hazard rate for the backup pump is 10^{-5}/hr during operation and 10^{-7}/hr during standby. Calculate the reliability of the system after 2000 hr (a) with the backup unit, and (b) without the backup unit.

6-13 You are given two independent units, with unit #2 in standby. The units fail randomly with hazard rates λ_1 and λ_2, and there are no failures during standby or switching. (a) Calculate the reliability for the system, (b) Calculate the MTTF, and (c) Compare your result for the MTTF from part (b) to those obtained if the two units would have been connected either in series or in parallel. For this comparison you may assume $\lambda_1 = \lambda_2 = \lambda$.

6-14 A system consists of a main unit (constant hazard rate, λ_1), a first standby unit (constant hazard rate, λ_2), and a second standby unit (constant hazard rate, λ_3). Assuming that idle standby units undergo no failures, that switching will take place when required, and that there is no repair of failed units, calculate the reliability of the system.

6-15 A system with four sequentially operated units is operated with no repair and no switch failures. The idle units do not fail in standby. The four units fail randomly with constant hazard rates λ_i, $i = 1$ to 4. (a) Derive the equation for the reliability of the system and reduce the result to a highly symmetric form involving a series of $\exp(-\lambda_i t)$ factors. (b) Calculate the MTTF for the system and reduce it to the simplest form.

6-16 A reactor has two identical coolant loops, each with two identical pumps connected in active-parallel. At least one pump must operate for a loop to remain functional. The second loop is initially in standby and is successfully switched into operation 98% of the time. All pumps fail randomly when in active operation with a hazard rate λ and in standby with a rate λ^*. Determine the system reliability as a function of time.

6-17 Use the decomposition method of Section 6-3 to verify Eq. (6-2) for two units in active-parallel.

6-18 What is the reliability of the system shown if the reliability of the components is R_A, R_B, and R_C?

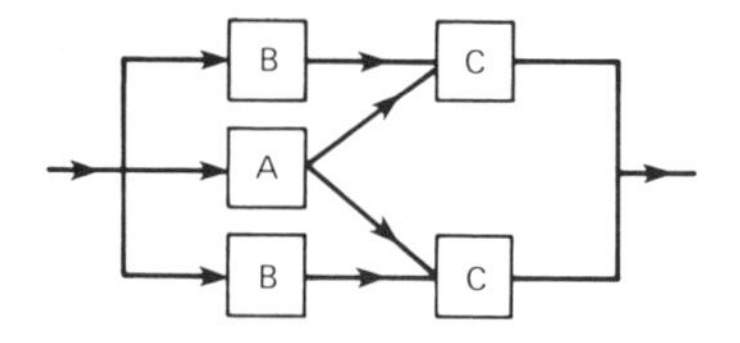

6-19 What is the reliability of the system shown if the reliability of the components is R_A, R_B, R_C, and R_D?

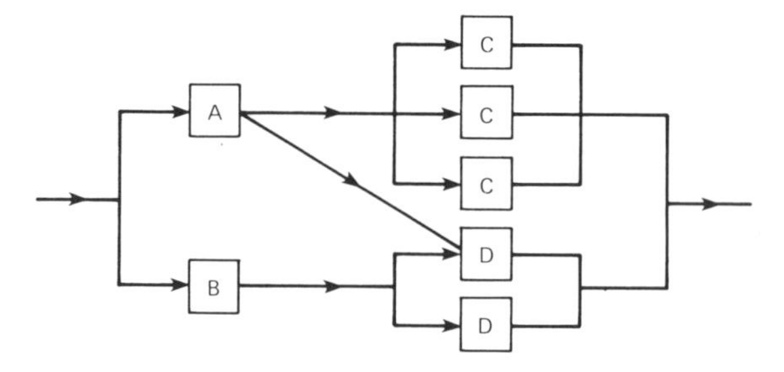

6-20 What is the reliability of the system shown if the reliability of each component is R?

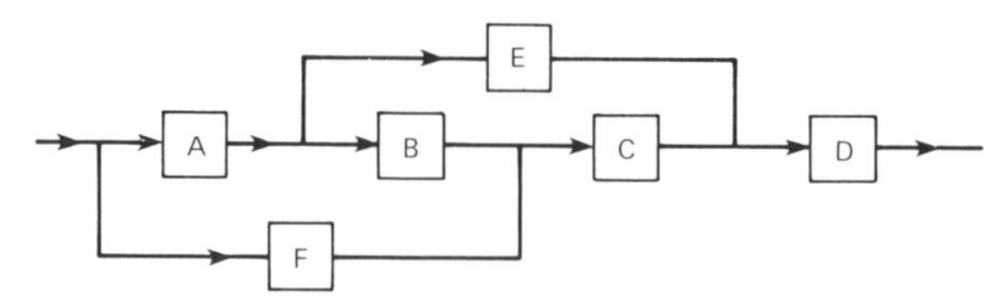

6-21 What is the reliability of the system shown if the reliability of each component is R?

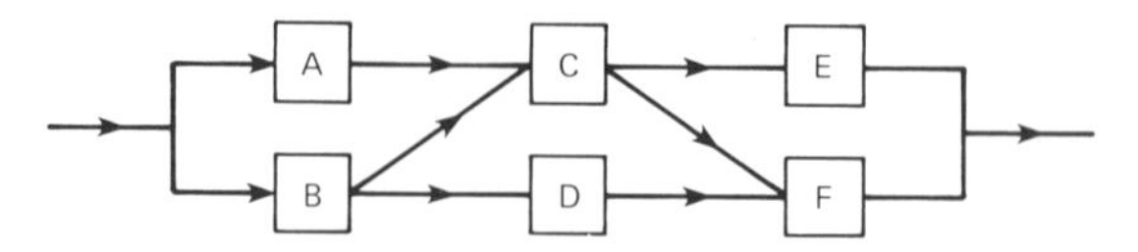

6-22 What is the reliability of the system shown if the reliability of each component is R?

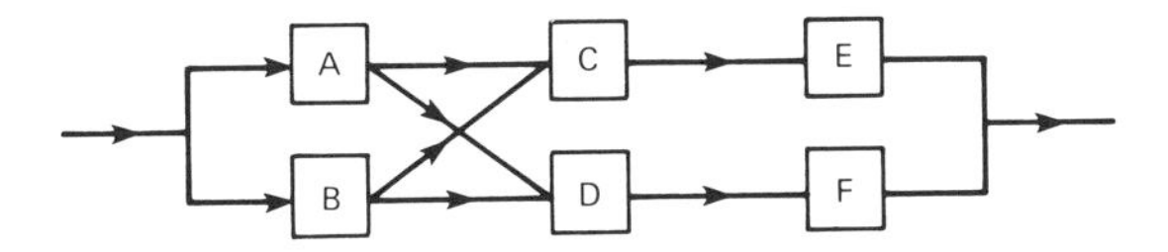

6-23 What is the reliability of the system shown if the reliability of each component is R?

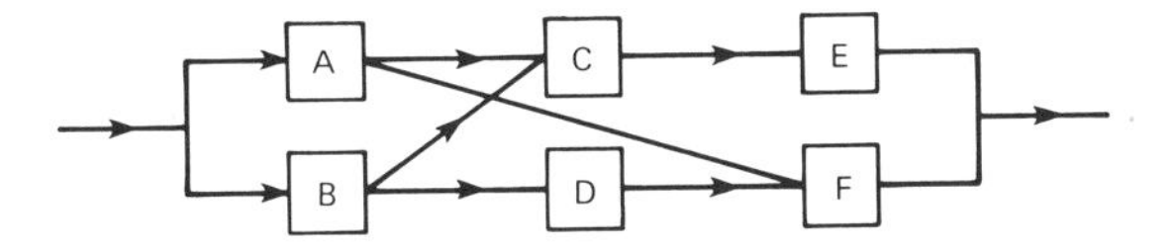

6-24 What is the reliability of the system shown if the reliability of each component is R?

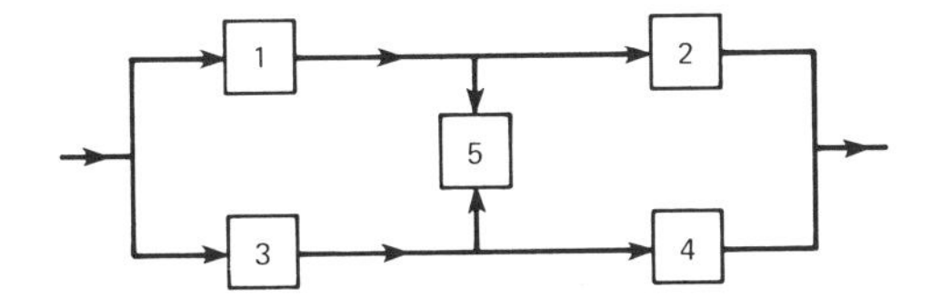

6-25 What is the reliability of the system shown if the reliability of each component is R_i, $i = 1$ to 7?

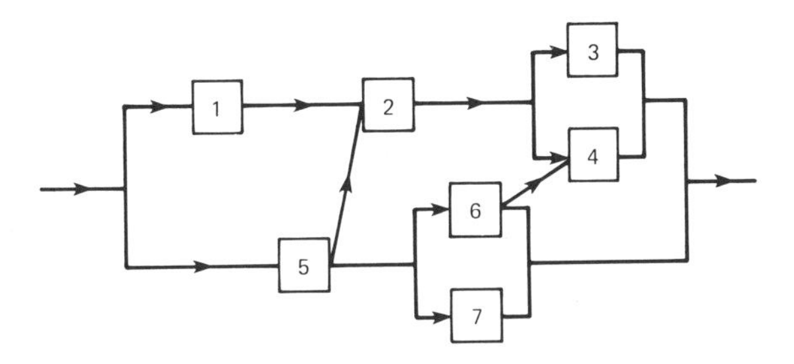

6-26 Consider the piping system shown which pumps coolant from point A to point B and (a) construct a signal flow graph. (b) Obtain the minimal cut sets. (c) Reduce the graph by using the addition and product rules. (d) Obtain the reliability in terms of the reliabilities of each component.

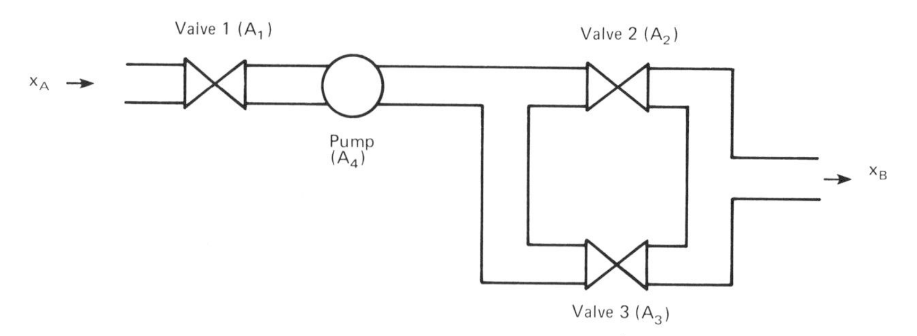

6-27 Construct a signal flow graph and obtain the minimal cut sets for the system of Exercise (a) 6-10, (b) 6-20, and (c) 6-22.

6-28 Show that [5]

$$(A + B)(A + C)(D + B)(D + C) = BC + AD.$$

6-29 Show that [5]

$$\overline{AB + A\bar{B} + \bar{A}\bar{B}} = \bar{A}B$$

6-30 Show that [5]

$$(\overline{\bar{A}B\bar{C}})(\overline{A\bar{B}\bar{C}}) = C + \bar{A}\bar{B} + AB.$$

6-31 Show that [5]

$$(AB + CDE)[AB + (\bar{C} + \bar{D} + \bar{E})] = AB.$$

REFERENCES

1. A. J. Bourne and A. E. Green, "Reliability Technology." Wiley (Interscience), New York, 1972.
2. M. L. Shooman, "Probabilistic Reliability: An Engineering Approach." McGraw-Hill, New York, 1968.
3. C. O. Smith, "Introduction to Reliability in Design." McGraw-Hill, New York, 1976.
4. B. L. Amstadter, "Reliability Mathematics." McGraw-Hill, New York, 1971.
5. W. E. Vesely, F. F. Goldberg, N. H. Roberts and D. F. Haasl, Fault Tree Handbook. U.S. Nuclear Regulatory Commission Rep. NUREG-0492 (1981).
6. Reliability Manual for Liquid Metal Fast Breeder Reactors (LMFBR) Safety Programs. General Electric Company Rep. SRD-74-113 (1974).
7. G. E. Apostolakis, The effect of a certain class of potential common mode failures on the reliability of redundant systems, *Nucl. Eng. Design* **36,** 123 (1976).

Reliability and Availability of Systems with Repair

7-1 RELIABILITY, AVAILABILITY, AND MAINTAINABILITY

Reliability $R(t)$ is the probability that a system performs a specified function or mission under given conditions *for* a prescribed time. In the simple systems analyzed in Chapter 6, there was no repair possible so a system was in one of only two possible states: either it was operating, or it was failed. More generally, the concept of reliability can be used for systems that are not repaired once the system fails, but can function while repairs of redundant subsystems are being completed.

The usual systems considered by a risk analyst can be repaired even after system failure, so the system is either operating or under repair. Hence knowledge of system reliability is not so much of interest as the availability of the system at the time an accident could occur. In this chapter, we wish to examine some of the more elementary ways of analyzing the availability of a system. As pointed out at the end of Section 2-4, the *instantaneous availability* $A(t)$ is defined as the probability of a system performing a specified function or mission under given conditions *at* a prescribed time. For this reason, $A(t)$ is sometimes called *point-wise availability* or *operational readiness*.

The determination of $R(t)$ and $A(t)$ for a system with repairable subsystems is more complicated than that for $R(t)$ for a system without repair because it is necessary to distinguish which subsystem is under repair while the system is operating. Thus an analysis of a system with repair requires distinction of the different operating states of the system.

In risk studies, detailed calculations required to obtain the instantaneous availability may not be necessary. In some cases, this can be achieved by using the *steady-state availability* $A(\infty)$, defined to be the limit of $A(t)$ as $t \to \infty$, whenever it exists. This time-independent availability is the probability that the system will be operating at any random point in time.

In the study of complicated systems, it is necessary to take into account the frequency of periodic maintenance and the test downtimes and repair times of all subsystems. The availability then used is the *interval availability* or *mission availability,* which is the fraction of time (or average probability) the system is operating in a time interval $(t_2 - t_1)$,

$$A_{\text{av}}(t_2 - t_1) = (t_2 - t_1)^{-1} \int_{t_1}^{t_2} A(t)\, dt. \tag{7-1}$$

The interval availability can vary according to the length of the time interval and the time t_1 after the beginning of system operation. The *limiting interval availability,* defined as

$$A_{\text{av}}(\infty) = \lim_{t\to\infty} \frac{1}{t} \int_0^t A(t')\, dt', \tag{7-2}$$

is the fraction of time the system operates over a very long mission time.

The availability analysis of complicated systems, comprised of subsystems that are maintained, requires consideration of *maintainability* $M(t)$, which is the probability that a subsystem is restored to one of its operating states in time Δt, given that it is presently in one of its nonoperating states. Maintainability is the third component of reliability, availability, and maintainability (RAM) analysis, which is of interest to systems designers. A detailed study of maintainability is beyond the scope of this text, so we only briefly consider periodic maintenance in the next section; then we devote the rest of the chapter to an introduction to reliability and availability analysis for systems with unscheduled repair.

7-2 PERIODIC MAINTENANCE

Complex technological systems are rarely placed in operation and allowed to operate without servicing. Instead, *preventive maintenance* or overhaul is performed to restore the system to an "as good as possible" condition. In order to determine when maintenance should be performed, testing of the system sometimes is conducted, typically on a periodic basis. Maintenance differs from *repair* of a system, which is required whenever the system stops functioning as intended; in some instances, even *replacement* of the system is required. These concepts can be schematically depicted as in Fig. 7-1.

Maintenance, repair, and replacement are all actions taken to prolong the useful operating life of the system. Elementary models are treated elsewhere for making decisions about when these actions should be performed in order to minimize costs [1].

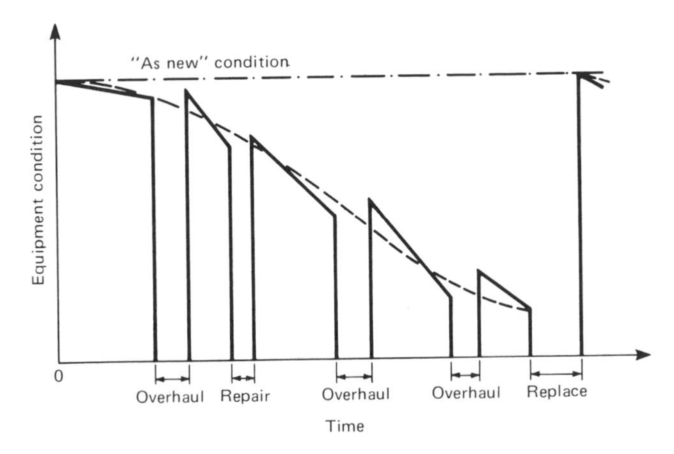

Fig. 7-1 *Effect of maintenance on equipment condition (schematic). (From A. K. S. Jardine, "Maintenance, Replacement, and Reliability." Wiley, New York, 1973.)*

The effect of periodic preventive maintenance on system failure can be examined for servicing performed every T hours, starting at time zero [2]. Then time t can be expressed in terms of the time τ in one service interval as $t = jT + \tau, j = 0, 1, 2, 3, \cdots, 0 \leq \tau < T$. *If* it can be assumed that the system is restored to the "as good as new" condition after each maintenance operation, then the probability that the system will be available at time t, $A(t)$, will appear as shown schematically in Fig. 7-2. In this case, the availability is bounded in the range $R(T) \leq A(t) \leq 1$, where $R(T)$ is the reliability of the system at time T without any servicing.

The reliability of the system with error-free periodic maintenance can be written as

$$R_{PM}(jT + \tau) = [R(T)]^j R(\tau), \qquad j = 0, 1, 2, 3 \cdots, \qquad 0 \leq \tau \leq T, \quad (7\text{-}3)$$

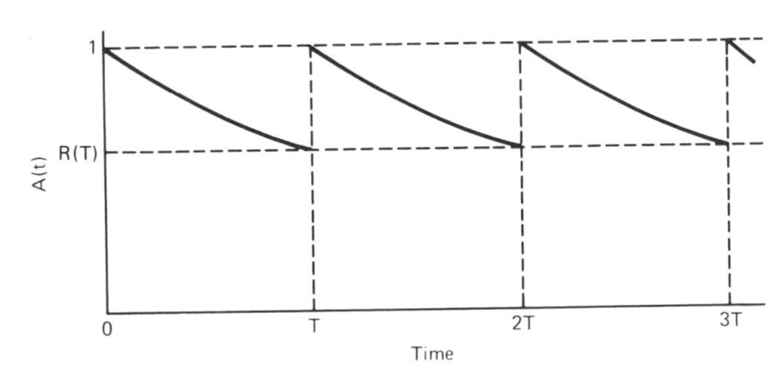

Fig. 7-2 *Effect of error-free preventive maintenance every T hours on system availability (schematic).*

since the system operation can be considered as a series of operations, each of duration T, which must transpire in order for the system to operate until time $(jT + \tau)$. To calculate the *mean time to failure* (MTTF) with periodic maintenance, we integrate Eq. (7-3) to obtain

$$\begin{aligned} \mathrm{MTTF_{PM}} &= \int_0^{\infty} R_{\mathrm{PM}}(t)\, dt \\ &= \sum_{j=0}^{\infty} \int_0^T [R(T)]^j R(\tau)\, d\tau \\ &= \frac{\int_0^T R(\tau)\, d\tau}{1 - R(T)}. \end{aligned} \tag{7-4}$$

Obviously the MTTF with error-free periodic maintenance exceeds the MTTF without maintenance.

More complicated models for periodic maintenance can be constructed but are of less interest to us here than the analysis of problems with unscheduled repairs, which are analyzed in the following sections by means of Markov models.

7-3 INTRODUCTION TO MARKOV MODELS

The analysis of reliability and availability of systems with repair becomes quite complex except for the special case of random failures [3–7]. To analyze problems with repair, it is necessary to introduce a conditional probability $\mu(t)\, dt$ that a unit which failed at time t will be repaired in time dt about t. Thus, $\mu(t)$ has units of $(\text{time})^{-1}$ and serves to describe the instantaneous rate of repair, just as $\lambda(t)$ in units of $(\text{time})^{-1}$ describes the instantaneous rate of failure. If $\mu(t) = \mu$, so that the instantaneous repair rate is constant, then this assumption on $\mu(t)$ is equivalent to assuming random repairs, which means only that the repair is *completed* at a random time after it is started, and not that the repairs are initiated at random times after the failure occurs. Another way of saying the same thing is that the *mean time to repair* (MTTR) of a unit is μ^{-1}, just as the mean time to failure (MTTF) is λ^{-1}; since repairs of devices normally take much less time than failures, usually $\lambda \ll \mu$. The *mean time between failure* (MTBF) is just the sum of MTTF and MTTR.

Although the assumptions of constant $\lambda(t)$ and $\mu(t)$ are not always desirable, both assumptions are necessary in order to avoid extensive mathematical complications, so both will be used in the remainder of this chapter. In any case [8],

> assuming constant component repair rates does not usually introduce serious limitations when unavailability or unreliability calculations are performed. The asymptotic un-

availability of a repairable component is dependent only on the mean time to repair and the mean time to failure and is nearly reached after about three mean times to repair in most cases of practical interest. In addition, system reliability characteristics, such as the expected number of failures, depend on the repair distribution through their dependence on the component unavailabilities. Therefore, assuming constant component repair rates generally only affects system results during the early transient, and even this effect is slight in most cases. . . .

A system that consists of several components, each having a constant hazard and repair rate, is a simple multiple-state system since each component in the system can be either in operation or under repair (or even in standby awaiting repair). This sort of model is generally called a Markov chain model or discrete-state, continuous-time model. One of the most important features of any Markov model is that the transition probability from state i to state j depends only on states i and j and is completely independent of all earlier states.

In this section, we shall consider a very simple Markov process for which transitions from a state with n occurrences to one with $(n + 1)$ occurrences are irreversible (i.e., there is no repair of any component) and occur with a constant rate of λ; this is just the Poisson process of Section 3-1. The purpose of this analysis is to introduce some basic mathematical tools for analyzing Markov models while examining a physical process with which we are already familiar. If we consider the possibility of only N transitions, the states of the system are:

0	no occurrences,
1	one occurrence,
⋮	⋮
N	N occurrences.

A *state transition diagram* illustrating the rates of transitions from one state to another would appear as in Fig. 7-3.

If $P_n(t)$ is defined as the probability that the system is in state n at time t, i.e., that there have been n occurrences, then for $n = 0$,

$$P_0(t + \Delta t) = (1 - \lambda \Delta t) P_0(t). \tag{7-5}$$

Here $\lambda \Delta t$ is the probability that there has been a transition in time interval Δt. To describe transitions to other states, the following equations are

Fig. 7-3 *State transition diagram for ≤N transitions between states for a Poisson process.*

needed:

$$P_n(t + \Delta t) = \lambda \,\Delta t\, P_{n-1}(t) + (1 - \lambda \,\Delta t)\, P_n(t), \qquad n = 1, 2, \cdots (N - 1)$$
$$P_N(t + \Delta t) = \lambda \,\Delta t\, P_{N-1}(t) + P_N(t). \tag{7-6}$$

Each of these equations has been written for the probability that the system will be in state n at time $t + \Delta t$, given that it was in that state at time t with unit probability. Note that Δt is assumed small enough that terms of order $(\Delta t)^2$, corresponding to two nearly simultaneous transitions, are negligible.

After division by Δt in Eqs. (7-5) and (7-6), the limit $\Delta t \to 0$ yields the equations

$$dP_0(t)/dt = -\lambda P_0(t),$$
$$dP_n(t)/dt = \lambda\, P_{n-1}(t) - \lambda\, P_n(t), \qquad n = 1, 2, \cdots (N - 1),$$
$$dP_N(t)/dt = \lambda P_{N-1}(t). \tag{7-7}$$

The initial conditions for this set of equations are

$$P_n(0) = 1, \qquad n = 0,$$
$$= 0, \qquad n \geq 1, \tag{7-8}$$

corresponding to the fact that originally the system was in state 0.

A quick check of the consistency of Eq. (7-7) follows if all are added together: the result should be

$$\frac{d}{dt} \sum_{n=0}^{N} P_n(t) = 0, \tag{7-9}$$

if there are $(N + 1)$ possible states of the system. Equation (7-9) generally must be true since, after integrating over time and remembering that initially the system is in one of the states, it follows for all times t that

$$\sum_{n=0}^{N} P_n(t) = 1. \tag{7-10}$$

This equation is a restatement of the fact that the system must be in one of the possible states $n = 0$ to N.

Using the standard theory of matrices, which is reviewed in Appendix C, Eq. (7-7) also may be written in the form

$$d\mathbf{P}(t)/dt = \mathbf{M}\, \mathbf{P}(t), \tag{7-11}$$

where $\mathbf{P}(t)$ is the *state vector* with $(N + 1)$ elements $P_n(t)$; the state vector and its initial condition are given by

$$\mathbf{P}(t) = \begin{bmatrix} P_0(t) \\ P_1(t) \\ P_2(t) \\ \vdots \\ P_N(t) \end{bmatrix}, \qquad \mathbf{P}(0) = \begin{bmatrix} 1 \\ 0 \\ 0 \\ \vdots \\ 0 \end{bmatrix}. \tag{7-12}$$

The *transition matrix* $\mathbf{M}$ has $(N + 1) \times (N + 1)$ time-independent matrix elements M_{nm} in the nth row and mth column $(n, m = 0$ to $M)$. For $i \neq j$, M_{ij} is the instantaneous rate of transition from state j into state i; M_{ii} is the corresponding rate out of state i. In Eq. (7-7), $\mathbf{M}$ is given by

$$\mathbf{M} = \begin{bmatrix} -\lambda & & & & & & \\ \lambda & -\lambda & & & & & \\ 0 & \lambda & -\lambda & & & & \\ \vdots & \vdots & \vdots & \vdots & & & \\ 0 & 0 & 0 & 0 & \cdots & \lambda & 0 \end{bmatrix}. \tag{7-13}$$

The quick check for accuracy of the transition matrix is done by summing all the matrix elements in each column to obtain zero; in other words, the determinant of $\mathbf{M}$ must vanish.

Equations (7-9) through (7-12) are general results that will be obtained for all the Markov models for repairable systems to be considered in this chapter; only the time-independent matrix elements of $\mathbf{M}$ will change for the different systems. Therefore it is worthwhile to discuss the general solution of Eqs. (7-11) and (7-12). The solution may be formally written as

$$\mathbf{P}(t) = \exp(\mathbf{M}t)\,\mathbf{P}(0), \tag{7-14}$$

where $\mathbf{P}(0)$ is given by Eq. (7-12). The matrix exponential function is a $(N + 1) \times (N + 1)$ matrix, which is defined as

$$\exp(\mathbf{M}t) = \mathbf{I} + \mathbf{M}t + (\mathbf{M}t)^2/2! + \cdots = \sum_{j=0}^{\infty} (\mathbf{M}t)^j/j!, \tag{7-15}$$

where $\mathbf{I}$ is the unit matrix with all diagonal elements equal to unity and with off-diagonal elements all zero. In cases where the summation in Eq. (7-15) can be readily performed, a general solution for $\mathbf{P}(t)$ valid for all times can be derived; otherwise, a few-term approximation to Eq. (7-15) can be used in Eq. (7-14) to obtain an approximate $\mathbf{P}(t)$. The matrix exponential method is used in the computer program ORIGEN, for example, to solve the analogous nuclear radioisotope generation and depletion equations [9].

Example 7-1 Obtain $\mathbf{P}(t)$ for the Poisson model with $(N + 1)$ possible states. The matrix $(\mathbf{M}t)$ is given by

See note for 7-2 P. 126

$$\mathbf{M}t = (\lambda t)\begin{bmatrix} -1 & & & & & & 0 \\ 1 & -1 & & & & & \\ 0 & 1 & -1 & & & & \\ 0 & 0 & 1 & -1 & & & \\ \vdots & \vdots & \vdots & \vdots & & & \\ 0 & 0 & 0 & 0 & \cdots & 1 & 0 \end{bmatrix}$$

and hence $(\mathbf{M}t)\ \mathbf{P}(0)$ simplifies to only the first column of $(\mathbf{M}t)$,

$$(\mathbf{M}t)\ \mathbf{P}(0) = (\lambda t)\begin{bmatrix} -1 \\ 1 \\ 0 \\ 0 \\ \vdots \\ 0 \end{bmatrix}.$$

Similarly, the matrix product $(\mathbf{M}t)^2$ is

$$(\mathbf{M}t)^2 = (\lambda t)^2\begin{bmatrix} 1 & & & & & & & \\ -2 & 1 & & & & & & \\ 1 & -2 & 1 & & & & & \\ 0 & 1 & -2 & 1 & & & & \\ \vdots & \vdots & \vdots & \vdots & & & & \\ 0 & 0 & 0 & 0 & \cdots & 1 & -1 & 0 \end{bmatrix}$$

and $(\mathbf{M}t)^2\ \mathbf{P}(0)$ is again the first column. The jth matrix product $(\mathbf{M}t)^j\ \mathbf{P}(0)$ is a vector whose nth element is

$$(-1)^j\ (\lambda t)^j \binom{j}{n}, \qquad n = 0 \text{ to } N,$$

where $\binom{j}{n}$ is the binomial coefficient defined in Eq. (3-5), and where $\binom{j}{n}$ is defined to vanish for $n > j$.

The sum of the vectors given by

$$\sum_{j=0}^{\infty} (\mathbf{M}t)^j \mathbf{P}(0)/j!$$

is obtained using the identity

$$\sum_{j=0}^{\infty} (-x)^j/j! = \exp(-x).$$

The final result is a state vector $\mathbf{P}(t)$ with elements

$$P_n(t) = (\lambda t)^n e^{-\lambda t}/n!, \qquad \text{n} = 0 \text{ to } N,$$

which agrees with Eq. (3-17). ◇

The matrix exponential method has computational advantages, but it is cumbersome to use if a general solution for all times is desired. For this reason, we consider a second method for solution of the coupled first-order differential equations, namely the use of Laplace transforms. For our purposes, we need only the definition

$$\mathscr{L}[f(t)] = f(s) = \int_0^{\infty} e^{-st} f(t)\, dt \tag{7-16}$$

and the properties

$$\mathscr{L}[df(t)/dt)] = sf(s) - f(0) \tag{7-17}$$

and, for constant a,

$$\mathscr{L}^{-1}[(s - a)^{-k}] = t^{k-1}e^{at}/(k - 1)!, \qquad k = 1, 2, \cdots. \tag{7-18}$$

[The change of variables from t to s thus denotes the transformed quantity provided we remember that $f(0)$ in Eq. (7-17) stands for $f(t)$ with $t = 0$.]

Equation (7-17) enables a first-order differential equation in terms of time to be converted into an algebraic equation in terms of the Laplace transform variable s, while the inverse transform Eq. (7-18) permits one to convert a particular term of an algebraic expression in s back into a function of time. The derivations of Eqs. (7-17) and (7-18), including constraints (not relevant here) on the type of $f(t)$ for which the theory may be used, are presented in standard advanced engineering mathematics texts [10, 11].

After application of Eq. (7-17) to Eq. (7-11) and collecting terms, we see that

$$(s\mathbf{I} - \mathbf{M})\mathbf{P}(s) = \mathbf{P}(0) \tag{7-19}$$

and this matrix equation can be inverted to give

$$\mathbf{P}(s) = (s\mathbf{I} - \mathbf{M})^{-1}\mathbf{P}(0). \tag{7-20}$$

Because of the simple form of $\mathbf{P}(0)$ in Eq. (7-12), only the matrix elements in the first column of $(s\mathbf{I} - \mathbf{M})^{-1}$ need be calculated,

$$P_n(s) = (s\mathbf{I} - \mathbf{M})^{-1}_{n0}, \qquad n = 0 \text{ to } N. \tag{7-21}$$

These elements are the Laplace transform of the probability that the system is in the nth state.

The matrix elements in Eq. (7-21) are calculated from the equation

$$P_n(s) = [\text{cof}(s\mathbf{I} - \mathbf{M})^T]_{n0}/\Delta, \qquad n = 0 \text{ to } N. \tag{7-22}$$

Here the transpose matrix $(s\mathbf{I} - \mathbf{M})^T$ is obtained, as discussed in Appendix C, from the matrix elements $(s\mathbf{I} - \mathbf{M})_{nm}$ with interchanged row and column indices. Then the $n0$ element of the (cofactor) $\text{cof}(s\mathbf{I} - \mathbf{M})^T$ is obtained from $(-1)^n$ times the determinant of the matrix obtained by omitting the first column (with $m = 0$) and the nth row of $(s\mathbf{I} - \mathbf{M})^T$. The Δ in Eq. (7-22) is the product

$$\Delta = (s - s_0)(s - s_1) \cdots (s - s_N), \tag{7-23}$$

where the eigenvalues s_n, $n = 0$ to N, of $\mathbf{M}$ are the roots of the characteristic equation given by the determinant

$$|s\mathbf{I} - \mathbf{M}| = 0. \tag{7-24}$$

Once the $(s\mathbf{I} - \mathbf{M})^{-1}_{n0}$ in Eq. (7-21) have been calculated, a partial fraction decomposition can be performed [10, 11]. If the eigenvalues are all different, the result will be a series of the form

$$P_n(s) = \sum_{j=0}^{N} b_{nj}(s - s_j)^{-1}, \tag{7-25}$$

where the b_{nj} are constants. Then application of Eq. (7-18) gives the probability the system is in the nth state as

$$P_n(t) = \sum_{j=0}^{N} b_{nj}\exp(s_j t), \qquad n = 0 \text{ to } N. \tag{7-26}$$

Example 7-2 Obtain $P_n(t)$ for the Poisson model with $(N + 1)$ possible states by using the Laplace transform method.

We first use Eqs. (7-13) and (7-24) to calculate the eigenvalues s_n and find

$$(s + \lambda)^{N+1} = 0,$$

so $s_n = -\lambda$, $n = 0$ to N, which means there is a degeneracy of eigenvalues. This unusual situation does not occur unless there are transitions

Page 124, Example 7-1, and page 126, Example 7-2: The discussion is not correct as stated without the important qualification that the number of states N must tend to infinity.

for which there are no repairs possible. From Eq. (7-23) it follows that $\Delta = (s + \lambda)^{N+1}$.

The calculation of the cofactors is also simplified for the Poisson process since

$$[\mathrm{cof}(s\mathbf{I} - \mathbf{M})^T]_{n0} = \lambda^n(s + \lambda)^{N-n}.$$

Hence use of Eq. (7-22) shows that

$$P_n(s) = \frac{\lambda^n(s + \lambda)^{N-n}}{s(s + \lambda)^{N+1}} = \lambda^n(s + \lambda)^{-(n+1)},$$

and from Eq. (7-18) the inverse Laplace transform gives

$$P_n(t) = (\lambda t)^n e^{-\lambda t}/n!, \qquad n = 0 \text{ to } N,$$

which agrees with Example 7-1. ◇

The Laplace transform method for a general transition matrix **M** can become tedious to apply by hand when the number of states $(N + 1)$ exceeds three or four. This is because the eigenvalues s_n are the roots of a $(N + 1)$-degree polynomial and because the cofactors in Eq. (7-22) are quite complicated. For this reason, in Section 7-4, we shall only derive the matrices **M** appropriate for reliability and availability calculations on several systems with repair; then, in Section 7-5, we shall consider a few simple systems to obtain the time-dependent reliability and availability.

7-4 MARKOV MODELS FOR SYSTEMS

7-4-1 INTRODUCTION

In this section, we shall examine several examples of systems that have subsystems that can undergo failures and repairs while the system continues to operate. The objective will be to develop the transition matrix **M**, which contains all the information about the rate of transitions between the different states of the system n, $n = 0$ to N. Then the state vector $\mathbf{P}(t)$ can be calculated numerically from the matrix exponential function of Eq. (7-14) or from the inverse Laplace transform of Eq. (7-20). The element $P_n(t)$ of $\mathbf{P}(t)$ gives the predicted probability that the system is in its nth state.

The first step in the analysis is to define accurately the different states for which the system is operating and is failed. In all cases, we shall continue to assume that the system is initially in state 0, corresponding to the state in which no units are under repair, so there are N states in which some subsystem is failed or under repair. For convenience, we shall order the system states so that for $0 \le n \le N_u$ the system is "up" and continues to function, while for $(N_u + 1) \le n \le N$, the system will not function. Thus, since the system instantaneous availability $A(t)$ is the probability that the system is in state n, for $n = 0$ OR 1 OR $\cdots N_u$, we have

$$A(t) = P(0 + 1 + 2 + \cdots + N_u). \tag{7-27}$$

Since the states are all different (i.e., mutually exclusive), it follows from Eq. (2-15) that

$$A(t) = \sum_{n=0}^{N_u} P_n(t), \tag{7-28}$$

and from Eq. (7-10) the system instantaneous *unavailability* $Q(t)$ is

$$Q(t) = 1 - A(t) = \sum_{n=N_u+1}^{N} P_n(t). \tag{7-29}$$

Usually it is easier to calculate the system availability using Eq. (7-29) since there are fewer down states than up states, i.e., $(N - N_u) < (N_u + 1)$, in which case, fewer matrix elements (7-24) and inverse Laplace transforms are required.

The equations for reliability and unreliability for a system with repair also obey Eqs. (7-28) and (7-29), respectively, since availability and reliability are calculated in much the same way. The difference is that $R(t)$ is calculated for the system in which there is a single *absorbing state,* state $(N_u + 1)$, from which no repairs can be made.

In the following sections the transition matrix for an availability analysis will be denoted as $\mathbf{M}_a$ whenever confusion could exist, while the transition matrix $\mathbf{M}_r$ is the matrix for a reliability analysis. The matrix $\mathbf{M}_r$ can be obtained from $\mathbf{M}_a$ by adding all the columns corresponding to the *failed system states* of $\mathbf{M}_a$ to get the column for the absorbing state, and then repeating the steps for the rows corresponding to the failed system states, and finally setting all repair rates in the new column to zero.

7-4-2 Two Nonidentical Units

For the first example, we construct the transition matrix $\mathbf{M}_a$ for a system that can operate with either of two units, numbered 1 and 2. Whenever both are operable, unit 1 will have priority for use and unit 2 will be in standby. The instantaneous failure rates during operation will be denoted by λ_1 and λ_2, and the failure rate of unit 2 during standby will be λ_2^*. (Of course, λ_2^* may vanish, in which case there are no failures during standby.) Examples of this system might be two pumps, one of which is an emergency pump, or an electrical generating system that has a diesel-driven backup generator.

The repair of a unit is assumed to begin instantaneously after it fails, but the repair will be completed randomly so that the instantaneous repair rates will be μ_1 and μ_2. For purposes of illustration, we will first assume that repairs can be done on only one unit at a time, and that any unit under repair will remain so until the task is completed; for simplicity, this is referred to as the "one repairman" case. (In Example 7-4, we shall examine the "two repairmen" case in which both units could be under repair simultaneously.)

In this example, as in all examples presented in this chapter, the switching from one unit to another will be assumed to be instantaneous and perfect.

The five possible states of the system for an availability analysis are:

0 unit 1 is in operation and unit 2 is in standby,
1 unit 1 is under repair and unit 2 is in operation,
2 unit 2 is under repair and unit 1 is in operation,
3 both units are down and unit 1 is under repair,
4 both units are down and unit 2 is under repair.

In this case, $N_u = 2$ and $N = 4$ for Eqs. (7-28) and (7-29). When we look for possible transitions between system states, it is convenient to set up a state transition diagram as in Fig. 7-4. Each directed line in the diagram denotes a transition between one state and another at the rate shown. As we can see, transitions with nonzero probabilities from state i to j, $i < j$, correspond to failures described by λ rates, whereas those from state i to j, $i > j$, correspond to repairs described by μ rates. Not all transitions are possible during a small time interval Δt, however, and the ones of order $(\Delta t)^2$ or higher are not shown; thus there are no factors proportional to order λ^2 or μ^2 in the equations.

To write a set of equations for the $P_n(t)$, we examine in turn each state in Fig. 7-4 and assume that at time t, with unit probability, the system is in

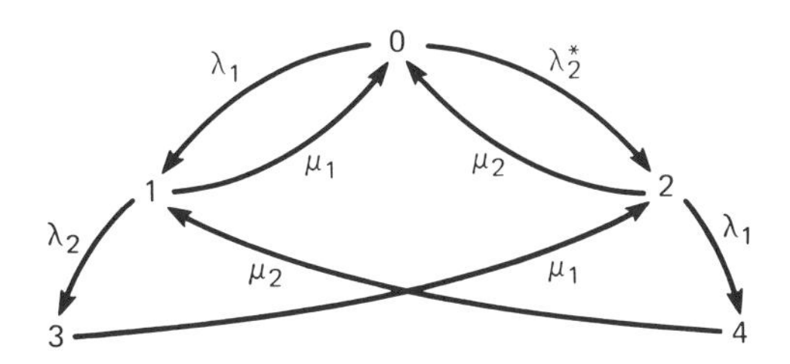

Fig. 7-4 *State transition diagram for availability analysis of two nonidentical units with priority operation and one repairman.*

the nth state, but that transitions during a small time Δt occur with conditional probabilities $\lambda\Delta t$ and $\mu\Delta t$. The equations are:

$$
\begin{aligned}
P_0(t+\Delta t) &= (1-\lambda_1\,\Delta t - \lambda_2^*\,\Delta t)P_0(t) + \mu_1\,\Delta t\,P_1(t) + \mu_2\,\Delta t\,P_2(t),\\
P_1(t+\Delta t) &= \lambda_1\,\Delta t P_0(t) + (1-\lambda_2\,\Delta t - \mu_1\,\Delta t)P_1(t) + \mu_2\,\Delta t\,P_4(t),\\
P_2(t+\Delta t) &= \lambda_2^*\,\Delta t\,P_0(t) + (1-\lambda_1\,\Delta t - \mu_2\,\Delta t)P_2(t) + \mu_1\,\Delta t\,P_3(t),\\
P_3(t+\Delta t) &= \lambda_2\,\Delta t\,P_1(t) + (1-\mu_1\,\Delta t)P_3(t),\\
P_4(t+\Delta t) &= \lambda_1\,\Delta t\,P_2(t) + (1-\mu_2\,\Delta t)P_4(t).
\end{aligned} \tag{7-30}
$$

After division by Δt, the limit $\Delta t \to 0$ yields the equations

$$
\begin{aligned}
dP_0(t)/dt &= -(\lambda_1+\lambda_2^*)P_0(t) + \mu_1 P_1(t) + \mu_2 P_2(t)\\
dP_1(t)/dt &= \lambda_1 P_0(t) - (\lambda_2+\mu_1)P_1(t) + \mu_2 P_4(t)\\
dP_2(t)/dt &= \lambda_2^* P_0(t) - (\lambda_1+\mu_2)P_2(t) + \mu_1 P_3(t)\\
dP_3(t)/dt &= \lambda_2 P_1(t) - \mu_1 P_3(t)\\
dP_4(t)/dt &= \lambda_1 P_2(t) - \mu_2 P_4(t).
\end{aligned} \tag{7-31}
$$

These equations may be rewritten in the form

$$d\mathbf{P}(t)/dt = \mathbf{M}_a\mathbf{P}(t), \tag{7-32}$$

where

$$
\mathbf{M}_a = \begin{bmatrix}
-(\lambda_1+\lambda_2^*) & \mu_1 & \mu_2 & 0 & 0\\
\lambda_1 & -(\lambda_2+\mu_1) & 0 & 0 & \mu_2\\
\lambda_2^* & 0 & -(\lambda_1+\mu_2) & \mu_1 & 0\\
0 & \lambda_2 & 0 & -\mu_1 & 0\\
0 & 0 & \lambda_1 & 0 & -\mu_2
\end{bmatrix}. \tag{7-33}
$$

Example 7-3 Two nonidentical units 1 and 2 are to operate in active-parallel in a system. Only one repairman is available who (is stubborn and) continues working on a failed unit until the repair is completed. Construct the transition matrix $\mathbf{M}_a$.

State 0 now corresponds to the condition in which both units are up and in active operation, but all other states are the same as for the case just discussed with unit 2 initially in standby. Just replace λ_2^* by λ_2 in Fig. 7-4 and Eqs. (7-30) through (7-33). ◇

Example 7-4 Two nonidentical units 1 and 2 comprise a system that operates with unit 1 in priority and unit 2 in standby whenever both are up. Everything is the same as described by Eqs. (7-30) to (7-33) except that both units can be under repair simultaneously (by two repairmen). Construct the transition matrix $\mathbf{M}_a$.

States 0 through 2 are the same as before, and $N_u = 2$, but now $N = 3$ since state 3 corresponds to the condition in which both are under repair simultaneously. Now the state transition diagram is as shown in Fig. E7-4. The transition matrix is

$$\mathbf{M}_a = \begin{bmatrix} -(\lambda_1 + \lambda_2^*) & \mu_1 & \mu_2 & 0 \\ \lambda_1 & -(\lambda_2 + \mu_1) & 0 & \mu_2 \\ \lambda_2^* & 0 & -(\lambda_1 + \mu_2) & \mu_1 \\ 0 & \lambda_2 & \lambda_1 & -(\mu_1 + \mu_2) \end{bmatrix}.$$

Notice that this matrix could have been obtained from that of Eq. (7-33) by adding the last two rows and then adding the last two columns. ◇

Example 7-5 Two nonidentical units 1 and 2 comprise a system that operates with unit 1 in priority and unit 2 in standby whenever both are up. Everything is the same as for the case discussed in Eqs. (7-30) to (7-33) except that there is no repair of the system when both units fail. Construct the reliability transition matrix $\mathbf{M}_r$ for the set of equations

$$d\mathbf{P}(t)/dt = \mathbf{M}_r \mathbf{P}(t).$$

The analysis of this system is more similar to that in Example 7-4 than that of Eq. (7-33) since there is only failure state 3 when both units are failed. The transition matrix $\mathbf{M}_r$ is obtained from that of Example 7-4 by

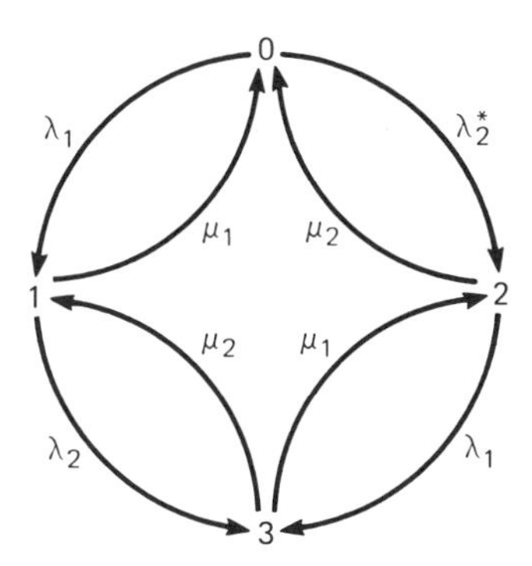

Fig. E7-4 *State transition diagram for availability analysis of two nonidentical units with priority operation and two repairmen.*

setting the repair rates in the last column to zero, i.e.,

$$\mathbf{M}_r = \begin{bmatrix} -(\lambda_1 + \lambda_2^*) & \mu_1 & \mu_2 & 0 \\ \lambda_1 & -(\lambda_2 + \mu_1) & 0 & 0 \\ \lambda_2^* & 0 & -(\lambda_1 + \mu_2) & 0 \\ 0 & \lambda_2 & \lambda_1 & 0 \end{bmatrix}. \quad \diamond$$

The last case we shall consider for two nonidentical units treats a system in which unit 1 is initially in operation with unit 2 in standby, but thereafter a unit that is operating will remain in operation until it fails. That is, after $t = 0$, unit 1 does not have priority in operation. Unit i, $i = 1$ or 2, can fail when in active operation with a MTTF of λ_i^{-1} or in standby with rate $(\lambda_i^*)^{-1}$; the MTTR of unit i is μ_i^{-1}. For increased complexity, we shall assume there is only one repairman who continues working on a failed unit until it is repaired.

The six system states for an availability analysis are:

0 unit 1 is in operation and unit 2 is in standby,
1 unit 1 is under repair and unit 2 is in operation,
2 both units are up and unit 2 is in operation,
3 unit 2 is under repair and unit 1 is in operation,
4 both units are down and unit 1 is under repair,
5 both units are down and unit 2 is under repair.

Now $N_u = 3$ and $N = 5$ for Eqs. (7-28) and (7-29). The state transition diagram is shown in Fig. 7-5.

The transition matrix is given by

$$\mathbf{M}_a = \begin{bmatrix} -(\lambda_1 + \lambda_2^*) & 0 & 0 & \mu_2 & 0 & 0 \\ \lambda_1 & -(\lambda_2 + \mu_1) & \lambda_1^* & 0 & 0 & \mu_2 \\ 0 & \mu_1 & -(\lambda_2 + \lambda_1^*) & 0 & 0 & 0 \\ \lambda_2^* & 0 & \lambda_2 & -(\lambda_1 + \mu_2) & \mu_1 & 0 \\ 0 & \lambda_2 & 0 & 0 & -\mu_1 & 0 \\ 0 & 0 & 0 & \lambda_1 & 0 & -\mu_2 \end{bmatrix}. \tag{7-34}$$

Again, as in Example 7-4, if there are two repairmen, then the last two columns are added and the last two rows are added, because states 4 and 5 are collapsed into one state to obtain the $\mathbf{M}_a$.

Several other models for two nonidentical units have been developed,

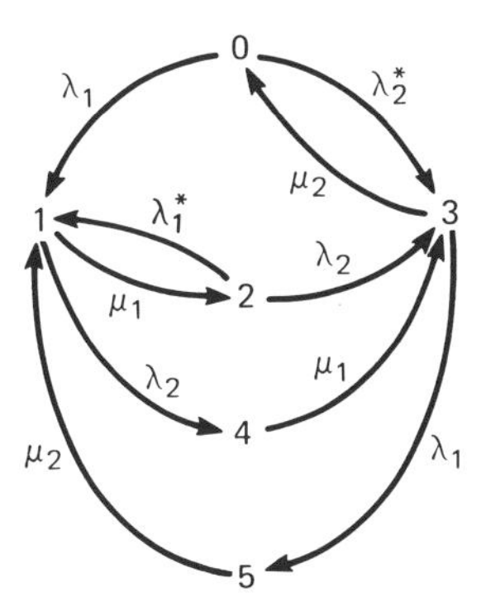

Fig. 7-5 *State transition diagram for availability analysis of two dissimilar units alternating in operation and with one repairman.*

including those for priority repair and for systems having different repair rates depending upon whether the failure was during operation or standby [12].

7-4-3 Two Identical Units

Systems with identical units are less complicated to analyze than those with nonidentical units because the number of states is reduced. Another simplification when there are only two units is that the number of states does not depend upon the number of repairmen, so only one state transition diagram need be constructed for the case with n repairmen, $n = 1$ or 2.

Consider a system of two identical units in which only one unit is required for operation. The instantaneous failure rate of either unit is λ in operation and λ^* in standby, and the repair rate of each is μ. The three states of the system for an availability analysis are:

0 one unit is in operation and the other is in standby,
1 one unit is under repair and the other is in operation,
2 both units are down and n units are under repair, $n = 1$ or 2.

For this example, $N_u = 1$ and $N = 2$ and the state transition diagram is shown in Fig. 7-6.
The transition matrix is

$$\mathbf{M}_a = \begin{bmatrix} -(\lambda + \lambda^*) & \mu & 0 \\ (\lambda + \lambda^*) & -(\lambda + \mu) & n\mu \\ 0 & \lambda & -n\mu \end{bmatrix}. \tag{7-35}$$

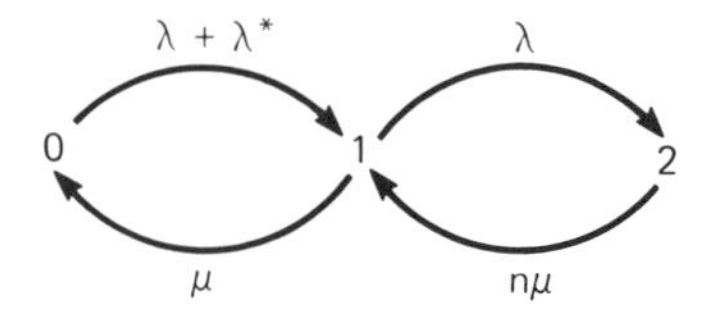

Fig. 7-6 *State transition diagram for availability analysis of two identical units with n repairmen, n = 1 or 2.*

Again, as in Example 7-3, if the two units are connected in active-parallel rather than with one in standby, then $\lambda^* = \lambda$.

Example 7-6 Two identical and independent units comprise a system that is to operate with one unit in standby whenever both are up. Each unit has an instantaneous failure rate during operation of λ and during standby of λ^*. In addition, whenever a unit is not under repair, it can fail randomly by common causes. Construct the state transition matrix to analyze the availability for the case of n repairmen, $n = 1$ or 2.

We let λ_c denote the instantaneous rate of failure of a unit due to the common causes. Since a unit in active operation can fail either by operation or by the common causes (and if we presume it does not fail for both reasons), then the probability of failure of a single unit in time Δt will be $(\lambda + \lambda_c)\Delta t$; initially when both units are up and one is operating, either the active or standby unit fails with a combined probability of $(\lambda + \lambda^*)\Delta t$, or both can fail simultaneously with probability $\lambda_c \Delta t$. Thus the state transition diagram becomes that shown in Fig. E7-6. Notice that there is now a direct transition between states 0 and 2.

This Markov model for the treatment of common cause failures is called the *Marshall–Olkin* model in which each failure, because of either independent or common causes, is characterized by an exponential distribution for the time to first failure [13]. This method also is sometimes known as the *beta factor* method in which the parameter β is interpreted as the

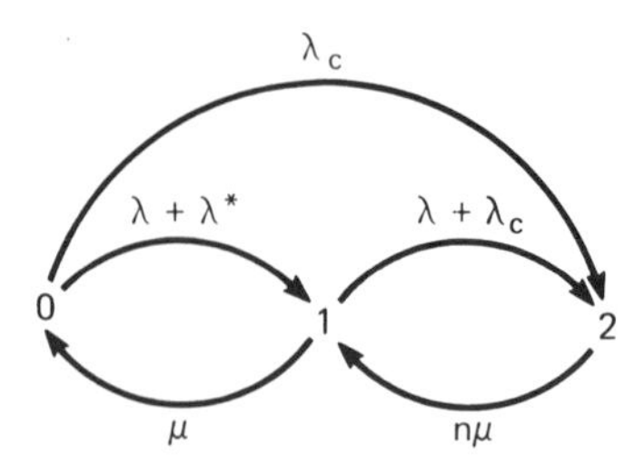

Fig. E7-6 *State transition diagram for availability analysis of two identical units with common cause failures and with n repairmen, n = 1 or 2.*

fraction of the total failure rate for a unit attributable to common cause failures [14], i.e.,

$$\beta = \frac{\lambda_c}{\lambda + \lambda_c}.$$

Estimates for the parameter β typically vary from 0.1 to 0.2 for such equipment as diesel generators or pumps that either fail to start or to run [14]. ◇

As the last example with two identical units, we consider a system that initially has one unit in standby, and presume that after repair of a unit is completed, it must be tested. Furthermore, we assume that the testing is initiated immediately after a repair is finished and is completed at a random time after it began, with a mean time for testing of τ^{-1} for each unit. There are m testing men, $m = 1$ or 2, and n repairmen, $n = 1$ or 2.

The six states of the system for an availability analysis are:

0 one unit is in operation and the other is in standby,
1 one unit is under repair and the other is in operation,
2 one unit is being tested and the other is in operation,
3 both units are failed, n are under repair and none are being tested,
4 one unit is under repair and one is being tested,
5 m units are being tested and no units are under repair.

Now $N_u = 2$ and $N = 5$ for Eqs. (7-28) and (7-29), and the state transition diagram is shown in Fig. 7-7. The transition matrix $\mathbf{M}_a$ can be ascertained from the diagram.

Many other Markov models for two units with repair have been published in the journal *IEEE Transactions on Reliability*. As examples, cases have been considered that treat cold standby units that can be supplied only after a delay in time [15] and systems that can undergo switching failures [16]. Two-unit systems subjected to intermittent use also have been analyzed [17].

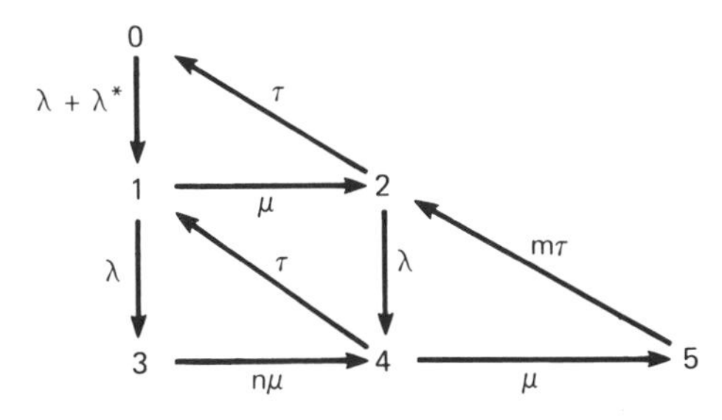

Fig. 7-7 *State transition diagram for availability analysis of two identical units with separate repair and testing operations.*

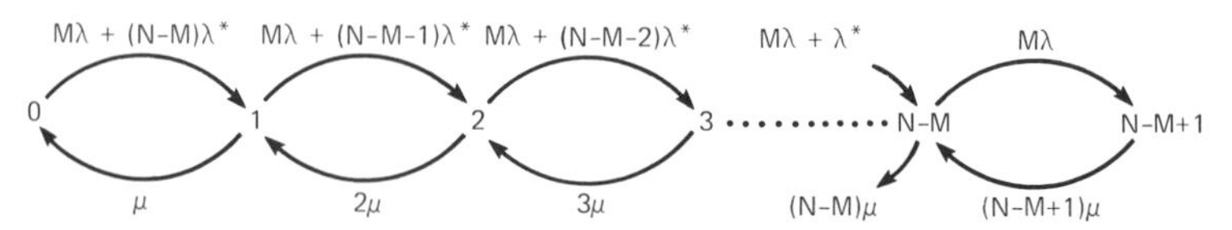

Fig. 7-8 *State transition diagram for availability analysis of an M-out-of-N system with* $(N - M + 1)$ *repairmen.*

7-4-4 M-Out-of-N Units

Consider the system of N identical units, of which M are needed at any time for the system to operate. Initially $(N - M)$ units will be in standby. All units fail randomly in operation with the constant rate λ and in standby with rate λ^*. For simplicity, we will assume there are enough repairmen, $N - M + 1$, to have each failed unit simultaneously under repair; the mean time to repair is μ^{-1}. An example of this type system might be the cooling system of a nuclear reactor in which two loops out of three are required, or a safety system in which two-out-of-four units must be activated for the system to function.

The possible system states needed for an availability analysis are:

- 0 — M units are in operation, $(N - M)$ are in standby, and no units are under repair,
- n — M units are in operation, $(N - M - n)$ units are in standby, and n units are under repair, $n = 1$ to $(N - M)$,
- $(N - M + 1)$ — the system has failed, and $(N - M + 1)$ units are under repair.

For this system, N_u and N in Eqs. (7-28) and (7-29) are given by $(N - M)$ and $(N - M + 1)$, respectively. The state transition diagram used to construct the transition matrix is shown in Fig. 7-8.

7-5 TIME-DEPENDENT AVAILABILITY AND RELIABILITY

In this section, we illustrate the use of the Laplace transform procedure of Section 7-3, as applied to a transition matrix **M** constructed as in Section 7-4, to obtain the instantaneous availability or reliability with repair. We will first consider elementary examples of availability analyses.

Example 7-7 Consider the system consisting of a single unit that can be repaired. Derive the instantaneous availability.

Either the unit is working or it is under repair, so the transition matrix for the two-state system is

$$\mathbf{M}_a = \begin{bmatrix} -\lambda & \mu \\ \lambda & -\mu \end{bmatrix}.$$

From Eq. (7-24) the eigenvalues s_0 and s_1 are calculated from the equation

$$\begin{vmatrix} s+\lambda & -\mu \\ -\lambda & s+\mu \end{vmatrix} = 0,$$

which gives $s_0 = 0$ and $s_1 = -(\lambda + \mu)$. Since the system functions only in state 0 and the number of up states equals the number of down states (one each), it is just as easy to use Eq. (7-28) for $A(t)$. Thus we need to obtain only one cofactor of $(s\mathbf{I} - \mathbf{M}_a)^T$, namely

$$[\operatorname{cof}(s\mathbf{I} - \mathbf{M}_a)^T]_{00} = (s + \mu).$$

From Eq. (7-22), we find

$$P_0(s) = \frac{s+\mu}{s(s+\lambda+\mu)},$$

so a partial fraction decomposition gives

$$P_0(s) = \frac{\mu/(\lambda+\mu)}{s} + \frac{\lambda/(\lambda+\mu)}{s+\lambda+\mu}.$$

From Eqs. (7-25) and (7-26) it follows that

$$A(t) = P_0(t) = (\lambda + \mu)^{-1}\,[\mu + \lambda e^{-(\lambda+\mu)t}].$$

By comparison, when $\mu = 0$ the reliability $R(t) = e^{-\lambda t}$ is recovered.

The steady-state availability $A(\infty)$ follows when one takes the limit as $t \to \infty$ to obtain

$$A(\infty) = \mu/(\lambda + \mu).$$

A comparison of the time-dependent availability and reliability is shown in Fig. E7-7. ◇

Example 7-7 illustrates the general rule that for an availability analysis one eigenvalue should be zero and all other s_n should be negative so that $P_n(t)$ will consist of a constant plus a sum of exponential functions that decrease with time. A second general rule is that the reliability *without* repair can be recovered from $P_0(t)$ by taking the limit as $\mu \to 0$ (which means that the MTTR, μ^{-1}, is infinite).

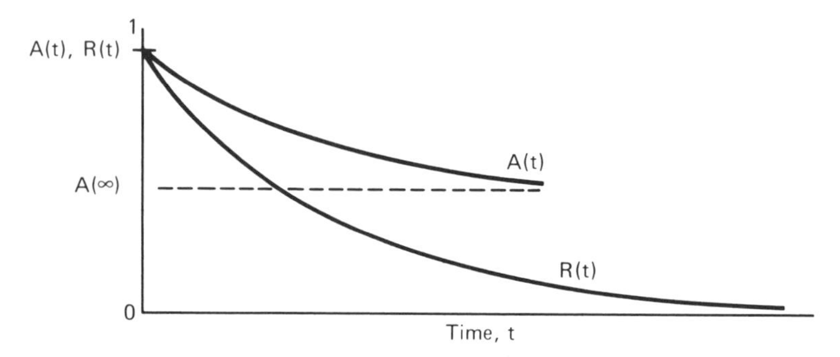

Fig. E7-7 *Time-dependent availability and reliability of a single unit (schematic).*

Example 7-8 Derive the availability of a system of two identical units in which either unit fails during operation at a rate λ, but only one unit is required to make the system function; there are no failures during standby. There is only one repairman.

This system is a special case of the first one considered in Section 7-4-3 in which $\lambda^* = 0$ and $n = 1$. From Eq. (7-35), the transition matrix is

$$\mathbf{M}_a = \begin{bmatrix} -\lambda & \mu & 0 \\ \lambda & -(\lambda + \mu) & \mu \\ 0 & \lambda & -\mu \end{bmatrix},$$

and so from Eq. (7-24) the three eigenvalues are given by $s_0 = 0$ and $s_1 = -(\lambda + \mu) - (\mu\lambda)^{1/2}$, $s_2 = -(\lambda + \mu) + (\mu\lambda)^{1/2}$. For this system, it is easiest to calculate the system availability by first calculating the unavailability from Eq. (7-29) since there are two up states and one down state; thus we need to derive only $P_2(s)$ and take the inverse Laplace transform.

From Eqs. (7-22) and (7-23),

$$P_2(s) = \frac{(-1)^2}{\Delta}\begin{vmatrix} -\lambda & 0 \\ (s + \lambda + \mu) & -\lambda \end{vmatrix} = \frac{\lambda^2}{s(s - s_1)(s - s_2)}.$$

After some algebra it may be shown that

$$P_2(s) = \sum_{j=0}^{2} b_{2j}(s - s_j)^{-1}$$

for

$$b_{20} = \lambda^2/s_1 s_2 = \lambda^2/(\mu^2 + \mu\lambda + \lambda^2)$$

$$b_{21} = \lambda^2/s_1(s_1 - s_2)$$

$$b_{22} = -\lambda^2/s_2(s_1 - s_2).$$

The inverse Laplace transform of $P_2(s)$ follows from Eq. (7-26), and so the final result is

$$A(t) = 1 - P_2(t) = A(\infty) - \frac{\lambda^2(s_2 e^{s_1 t} - s_1 e^{s_2 t})}{s_1 s_2 (s_1 - s_2)},$$

where the steady-state availability is

$$A(\infty) = \frac{\mu^2 + \mu\lambda}{\mu^2 + \mu\lambda + \lambda^2}. \quad \diamond$$

The time-dependent and steady-state availabilities from Examples 7-7 and 7-8 are shown in Table 7-1, along with those from seven other more complicated systems in which only one unit need function for the system to operate [18]. The systems consist of one, two, or three identical units that operate with redundant units either in standby, with no standby failures, or in active-parallel; the availability of a system operated in standby with standby failures is bounded by these two availabilities. The number of repairmen is variable. A numerical comparison of $A(\infty)$ for $\lambda = 0.01$ and $\mu = 0.2$, measured in the same units of $(\text{time})^{-1}$, also is provided to numerically illustrate the relative improvement provided by standby operation rather than active-parallel operation, and by multiple repairmen rather than only one repairman.

We now consider an example of a reliability analysis for a system comprised of subsystems that can be repaired during system operation.

Example 7-9 Derive the reliability of a system of two identical units in which either unit fails during operation at a rate λ, but only one unit is required to make the system function; there are no failures during standby. There is one repairman who can complete a repair with a mean time of μ^{-1}.

The system is like that of Example 7-8 except that there is no repair from failure state 2, so the transition matrix becomes

$$\mathbf{M}_r = \begin{bmatrix} -\lambda & \mu & 0 \\ \lambda & -(\lambda + \mu) & 0 \\ 0 & \lambda & 0 \end{bmatrix}.$$

From Eq. (7-24), the three eigenvalues are given by $s_0 = 0$ and

$$s_1 = -\tfrac{1}{2}[(2\lambda + \mu) + (\mu^2 + 4\lambda\mu)^{1/2}]$$

$$s_2 = -\tfrac{1}{2}[(2\lambda + \mu) - (\mu^2 + 4\lambda\mu)^{1/2}].$$

Table 7-1 *Availability of Systems Comprised of Identical Units with MTTF* $= \lambda^{-1}$ *and MTTR* $= \mu^{-1}$[a]

Conditions			
No. of identical units	Type of system	No. of repair-men	Instantaneous availability $A(t)$
1	—	1	$A(t) = \frac{\mu}{\mu + \lambda} + \frac{\lambda}{\mu + \lambda} e^{s_1 t}$
2	Standby ($\lambda^* = 0$)	1	$A(t) = \frac{\mu^2 + \mu\lambda}{\mu^2 + \mu\lambda + \lambda^2} - \frac{\lambda^2(s_2 e^{s_1 t} - s_1 e^{s_2 t})}{s_1 s_2(s_1 - s_2)}$
		2	$A(t) = \frac{2\mu^2 + 2\mu\lambda}{2\mu^2 + 2\mu\lambda + \lambda^2} - \frac{\lambda^2(s_2 e^{s_1 t} - s_1 e^{s_2 t})}{s_1 s_2(s_1 - s_2)}$
	Active-parallel	1	$A(t) = \frac{\mu^2 + 2\mu\lambda}{\mu^2 + 2\mu\lambda + 2\lambda^2} - \frac{2\lambda^2(s_2 e^{s_1 t} - s_1 e^{s_2 t})}{s_1 s_2(s_1 - s_2)}$
		2	$A(t) = \frac{\mu^2 + 2\mu\lambda}{\mu^2 + 2\mu\lambda + \lambda^2} - \frac{2\lambda^2(s_2 e^{s_1 t} - s_1 e^{s_2 t})}{s_1 s_2(s_1 - s_2)}$
3	Standby ($\lambda^* = 0$)	1	$A(t) = \frac{\mu^3 + \mu^2\lambda + \mu\lambda^2}{\mu^3 + \mu^2\lambda + \mu\lambda^2 + \lambda^3} + \frac{\lambda^3[s_2 s_3(s_2 - s_3)e^{s_1 t} - s_1 s_3(s_1 - s_3)e^{s_2 t} + s_1 s_2(s_1 - s_2)e^{s_3 t}]}{s_1 s_2 s_3(s_1 - s_2)(s_1 - s_3)(s_2 - s_3)}$
		3	$A(t) = \frac{6\mu^3 + 6\mu^2\lambda + 3\mu\lambda^2}{6\mu^3 + 6\mu^2\lambda + 3\mu\lambda^2 + \lambda^3} + \frac{\lambda^3[s_2 s_3(s_2 - s_3)e^{s_1 t} - s_1 s_3(s_1 - s_3)e^{s_2 t} + s_1 s_2(s_1 - s_2)e^{s_3 t}]}{s_1 s_2 s_3(s_1 - s_3)(s_1 - s_2)(s_2 - s_3)}$
	Active-parallel	1	$A(t) = \frac{\mu^3 + 3\mu^2\lambda + 6\mu\lambda^2}{\mu^3 + 3\mu^2\lambda + 6\mu\lambda^2 + 6\lambda^3} + \frac{6\lambda^3[s_2 s_3(s_2 - s_3)e^{s_1 t} - s_1 s_3(s_1 - s_3)e^{s_2 t} + s_1 s_2(s_1 - s_2)e^{s_3 t}]}{s_1 s_2 s_3(s_1 - s_2)(s_1 - s_3)(s_2 - s_3)}$
		3	$A(t) = \frac{\mu^3 + 3\mu^2\lambda + 3\mu\lambda^2}{(\mu + \lambda)^3} + \frac{6\lambda^3[s_2 s_3(s_2 - s_3)e^{s_1 t} - s_1 s_3(s_1 - s_3)e^{s_2 t} + s_1 s_2(s_1 - s_2)e^{s_3 t}]}{s_1 s_2 s_3(s_1 - s_2)(s_1 - s_3)(s_2 - s_3)}$

[a] (From R. H. Myers, K. L. Wong, and H. M. Gordy (eds.), "Reliability Engineering for Electronic Systems." Wiley, New York, 1964.)

To calculate the unreliability of the system, we use Eqs. (7-22) and (7-23) to obtain the same $P_2(s)$ as in Example 7-8, except that the values of s_1 and s_2 are now different so $s_1 s_2 = \lambda^2$ and $b_{20} = 1$. After use of Eq. (7-26) to invert the Laplace transform of $P_2(s)$, the result for the reliability simplifies to

$$R(t) = 1 - P_2(t) = \frac{-s_2 e^{s_1 t} + s_1 e^{s_2 t}}{s_1 - s_2}. \quad \diamond$$

Eigenvalues other than $s_0 = 0$	Steady-state availability $A(\infty)$	$A(\infty)$ for $\lambda = 0.01$, $\mu = 0.2$
$s_1 = -(\lambda + \mu)$	$\dfrac{\mu}{\mu + \lambda}$	0.95
$s_1 = -(\lambda + \mu) - \sqrt{\mu\lambda}$ $s_2 = -(\lambda + \mu) + \sqrt{\mu\lambda}$	$\dfrac{\mu^2 + \mu\lambda}{\mu^2 + \mu\lambda + \lambda^2}$	0.998
$s_1 = -\frac{1}{2}[(2\lambda + 3\mu) + \sqrt{\mu^2 + 4\mu\lambda}]$ $s_2 = -\frac{1}{2}[(2\lambda + 3\mu) - \sqrt{\mu^2 + 4\mu\lambda}]$	$\dfrac{2\mu^2 + 2\mu\lambda}{2\mu^2 + 2\mu\lambda + \lambda^2}$	0.999
$s_1 = -\frac{1}{2}[(3\lambda + 2\mu) + \sqrt{\lambda^2 + 4\mu\lambda}]$ $s_2 = -\frac{1}{2}[(3\lambda + 2\mu) - \sqrt{\lambda^2 + 4\mu\lambda}]$	$\dfrac{\mu^2 + 2\mu\lambda}{\mu^2 + 2\mu\lambda + 2\lambda^2}$	0.996
$s_1 = -2(\mu + \lambda)$ $s_2 = -(\mu + \lambda)$	$\dfrac{\mu^2 + 2\mu\lambda}{\mu^2 + 2\mu\lambda + \lambda^2}$	0.998
s_1, s_2, and s_3 correspond to the three roots of $s^3 + s^2(3\lambda + 3\mu) + s(3\lambda^2 + 4\mu\lambda + 3\mu^2) + (\lambda^3 + \mu\lambda^2 + \lambda\mu^2 + \mu^3)$	$\dfrac{\mu^3 + \mu^2\lambda + \lambda^2\mu}{\mu^3 + \mu^2\lambda + \lambda^2\mu + \lambda^3}$	0.9999
s_1, s_2, and s_3 correspond to the three roots of $s^3 + s^2(3\lambda + 6\mu) + s(3\lambda^2 + 9\mu\lambda + 11\mu^2) + (\lambda^3 + 3\mu\lambda^2 + 6\mu^2\lambda + 6\mu^3)$	$\dfrac{6\mu^3 + 6\mu^2\lambda + 3\mu\lambda^2}{6\mu^3 + 6\mu^2\lambda + 3\mu\lambda^2 + \lambda^2}$	0.99998
s_1, s_2, and s_3 correspond to the three roots of $s^3 + s^2(6\lambda + 3\mu) + s(11\lambda^2 + 9\mu\lambda + 3\mu^2) + (6\lambda^3 + 6\mu\lambda^2 + 3\mu^2\lambda + \mu^3)$	$\dfrac{\mu^3 + 3\mu^2\lambda + 6\mu\lambda^2}{\mu^3 + 3\mu^2\lambda + 6\mu\lambda^2 + 6\lambda^3}$	0.9993
s_1, s_2, and s_3 correspond to the three roots of $s^3 + s^2(6\lambda + 6\mu) + s[11(\mu + \lambda)^2] + 6(\mu + \lambda)^3$	$\dfrac{\mu^3 + 3\mu^2\lambda + 3\mu\lambda^2}{\mu^3 + 3\mu^2\lambda + 3\mu\lambda^2 + \lambda^3}$	0.9999

The reliability of systems consisting of one, two, or three identical units is given in Table 7-2; only one unit need function for the system to operate. The systems operate with redundant units either in standby, with no standby failures, or in active-parallel, and the number of repairmen is variable. Also given is the approximate result for $R(t)$ when $\lambda \leq 10\mu$. A comparison of the corresponding systems in Table 7-1 and 7-2 shows that the reliability result for a system with repair is slightly less complicated than that for availability; this is because coefficient b_{no}, $n = N_u + 1$, of

Table 7-2 *Reliability of Systems Comprised of Identical Units with MTTF* $= \lambda^{-1}$ *and MTTR* $= \mu^{-1}$[a]

Conditions			
No. of identical units	Type of system	No. of repairmen	Reliability, $R(t)$
1	—	—	$R(t) = e^{s_1 t}$
2	Standby ($\lambda^* = 0$)	1	$R(t) = \dfrac{-s_2 e^{s_1 t} + s_1 e^{s_2 t}}{s_1 - s_2}$
	Active-parallel	1	$R(t) = \dfrac{-s_2 e^{s_1 t} + s_1 e^{s_2 t}}{s_1 - s_2}$
3	Standby ($\lambda^* = 0$)	1	$R(t) = \dfrac{-s_2 s_3(s_2 - s_3)e^{s_1 t} + s_1 s_3(s_1 - s_3)e^{s_2 t} - s_1 s_2(s_1 - s_2)e^{s_3 t}}{(s_1 - s_2)(s_3 - s_2)(s_1 - s_3)}$
		2	$R(t) = \dfrac{-s_2 s_3(s_2 - s_3)e^{s_1 t} + s_1 s_3(s_1 - s_3)e^{s_2 t} - s_1 s_2(s_1 - s_2)e^{s_3 t}}{(s_1 - s_2)(s_3 - s_2)(s_1 - s_3)}$
	Active-parallel	1	$R(t) = \dfrac{-s_2 s_3(s_2 - s_3)e^{s_1 t} + s_1 s_3(s_1 - s_3)e^{s_2 t} - s_1 s_2(s_1 - s_2)e^{s_3 t}}{(s_1 - s_2)(s_3 - s_2)(s_1 - s_3)}$
		2	$R(t) = \dfrac{-s_2 s_3(s_2 - s_3)e^{s_1 t} + s_1 s_3(s_1 - s_3)e^{s_2 t} - s_1 s_2(s_1 - s_2)e^{s_3 t}}{(s_1 - s_2)(s_3 - s_2)(s_1 - s_3)}$

[a] Extracted from R. H. Myers, K. L. Wong, and H. M. Gordy (eds.), "Reliability Engineering for Electronic Systems." Wiley, New York, 1964.

partial fraction expansion (7-25), corresponding to $s_0 = 0$, is unity so that $R(\infty) = 0$.

Calculation of the general solution for time-dependent availability or reliability can be quite complicated, even when the number of states is only moderately large, and sometimes is not even needed. In the next section, for example, we obtain the time-dependent availability and reliability for the case when failures only rarely occur without solving the set of differential equations with Laplace transforms. Then in Sections 7-7 and 7-8, we obtain the steady-state availability and the mean time to failure by purely algebraic means.

7-6 TIME-DEPENDENT UNAVAILABILITY FOR RARE FAILURES

The rare-event approximation corresponds to the situation in which the failure rates of all units during active operation (λ) and standby (λ^*) are small so that failures rarely occur. Such an approximation is quite good for nuclear power safety systems that are regularly inspected and maintained.

Eigenvalues other than $s_0 = 0$	Approximate reliability models for $\lambda \le 10\mu$
$s_1 = -\lambda$	$R(t) = \exp - \lambda t$
$s_1 = -\frac{1}{2}[(2\lambda + \mu) + \sqrt{\mu^2 + 4\mu\lambda}]$ $s_2 = -\frac{1}{2}[(2\lambda + \mu) - \sqrt{\mu^2 + 4\mu\lambda}]$	$R(t) = \exp - \frac{1}{2}[(2\lambda + \mu) - \sqrt{\mu^2 + 4\mu\lambda}]t$
$s_1 = -\frac{1}{2}[(3\lambda + \mu) + \sqrt{\mu^2 + 6\mu\lambda + \lambda^2}]$ $s_2 = -\frac{1}{2}[(3\lambda + \mu) - \sqrt{\mu^2 + 6\mu\lambda + \lambda^2}]$	$R(t) = \exp - \frac{1}{2}[(3\lambda + \mu) - \sqrt{\mu^3 + 6\mu\lambda + \lambda^3}]t$
s_1, s_2, and s_3 correspond to the three roots of $s^3 + s^2(3\lambda + 2\mu) + s(3\lambda^2 + 2\mu\lambda + \mu^2) + \lambda^3$	$R(t) = \exp - \left(\frac{\lambda^3}{3\lambda^2 + 2\lambda\mu + \mu^2}\right) t$
s_1, s_2, and s_3 correspond to the three roots of $s^3 + s^2(3\lambda + 3\mu) + s(3\lambda^2 + 3\mu\lambda + 2\mu^2) + \lambda^3$	$R(t) = \exp - \left(\frac{\lambda^3}{3\lambda^2 + 3\lambda\mu + 2\mu^2}\right) t$
s_1, s_2, and s_3 correspond to the three roots of $s^3 + s^2(6\lambda + 2\mu) + s(11\lambda^2 + 4\mu\lambda + \mu^2) + 6\lambda^3$	$R(t) = \exp - \left(\frac{6\lambda^3}{11\lambda^2 + 4\mu\lambda + \mu^2}\right) t$
s_1, s_2, and s_3 correspond to the three roots of $s^3 + s^2(6\lambda + 3\mu) + s(11\lambda^2 + 7\mu\lambda + 2\mu^2) + 6\lambda^3$	$R(t) = \exp - \left(\frac{6\lambda^3}{11\lambda^2 + 7\mu\lambda + 2\mu^2}\right) t$

From the definition of unavailability $\bar{A}(t)$ in Eq. (7-29), it follows that we need to approximate $P_n(t)$ for small λ for the states $n = (N_u + 1)$ to N in which the system has failed. This is most easily done by using Eqs. (7-12), (7-14), and (7-15) to obtain

$$P_n(t) \approx [\mathbf{I} + \mathbf{M}_a t + (\mathbf{M}_a t)^2/2! + \cdots]_{n0}, \qquad (7\text{-}36)$$

where only as many terms are taken as necessary to obtain a nonzero contribution to $P_n(t)$.

The system unavailability can be obtained by summing the results for $n = (N_u + 1)$ to N and rounding to the lowest order power in λ. This approach does not require that the general solution for $Q(t)$ first be obtained by use of Laplace transforms.

Example 7-10 Derive the approximate unavailability for a system of two identical units in which each unit rarely fails during operation. Only one unit is needed for system operation. There are n repairmen, $n = 1$ or 2.

This physical situation was analyzed in Section 7-4-3 and $\mathbf{M}_a$ is given in Eq. (7-35). Use of that $\mathbf{M}_a$ in Eq. (7-36) for the only failure state 2 gives

$$Q(t) = P_2(t) \approx \lambda(\lambda + \lambda^*)t^2/2.$$

When there are no failures during standby ($\lambda^* = 0$), this result reduces to $(\lambda t)^2/2$, which also can be obtained from the general time-dependent unavailability from Example 7-8 if $e^{s_1 t}$ and $e^{s_2 t}$ are expanded to second order. ◇

7-7 STEADY-STATE AVAILABILITY

In many risk analyses, the system can be presumed to be in steady-state operation so that only the steady-state availability $A(\infty)$ or unavailability $Q(\infty)$ is really needed. In such situations, the time-dependent matrix equation (7-11) need not be solved, but rather only the equation

$$\mathbf{M}_a \mathbf{P}(\infty) = 0, \tag{7-37}$$

which is obtained by setting $d\mathbf{P}(t)/dt = 0$.

The constraint for the set of homogeneous algebraic equations in Eq. (7-37) continues to be Eq. (7-10). The constraint can be incorporated into Eq. (7-37) by formulating a modified set of equations

$$\mathbf{M}_{ss} \mathbf{P}(\infty) = \mathbf{P}(0), \tag{7-38}$$

where $\mathbf{P}(0)$ is the initial state vector of Eq. (7-12) and where $\mathbf{M}_{ss}$ is obtained by replacing every element of the first row of $\mathbf{M}_a$ by unity. [The equation which was omitted can always be used as a check of the correctness of the algebra once the $P_n(\infty)$ are all calculated.]

Equation (7-38) may be solved in the same way as Eq. (7-19); the elements of $\mathbf{P}(\infty)$ are given in terms of matrix cofactors by the equation

$$P_n(\infty) = \frac{(\text{cof } \mathbf{M}_{ss}^T)_{n0}}{|\mathbf{M}_{ss}|}. \tag{7-39}$$

The cofactors may be evaluated as explained in Appendix C. It is usually most convenient to solve for the unavailability using Eqs. (7-29) and (7-39) since fewer unknown $P_n(\infty)$ must be calculated.

Example 7-11 Derive the steady-state availability for a system of two identical units in which only one is required for the system to function. Each unit fails with a MTTF of λ^{-1} during operation and $(\lambda^*)^{-1}$ during standby; there are n repairmen, $n = 1$ or 2, who complete repairs with a mean time of μ^{-1}.

This system is the first one considered in Section 7-4-3, so from Eq. (7-35), the matrix $\mathbf{M}_{ss}$ is found to be

$$\mathbf{M}_{ss} = \begin{bmatrix} 1 & 1 & 1 \\ (\lambda + \lambda^*) & -(\lambda + \mu) & n\mu \\ 0 & \lambda & -n\mu \end{bmatrix},$$

and we need to solve for $P_2(\infty)$ so we can determine $A(\infty) = 1 - P_2(\infty)$. Equation (7-39) gives

$$P_2(\infty) = \frac{\lambda(\lambda + \lambda^*)}{\lambda(\lambda + \lambda^*) + n\mu(\lambda + \lambda^* + \mu)},$$

so therefore

$$A(\infty) = \frac{n\mu(\lambda + \lambda^* + \mu)}{\lambda(\lambda + \lambda^*) + n\mu(\lambda + \lambda^* + \mu)}.$$

By letting $\lambda^* = 0$, the steady-state availabilities in Table 7-1 for one and two repairmen are recovered; similarly, by letting $\lambda^* = \lambda$, the results in Table 7-1 for the case of active-parallel operation are obtained. ◇

7-8 MEAN TIME TO FAILURE

A final quantity sometimes of interest in systems that have subsystems that can be repaired is the mean time to failure (MTTF), which is calculated from the equation

$$\text{MTTF} = \int_0^\infty R(t)\, dt, \tag{7-40}$$

where $R(t)$ is the reliability. The easiest way to calculate the MTTF is to form a modified transition matrix $\mathbf{M}_u$, which consists of *only the up states* of $\mathbf{M}_a$. (An identical result is obtained by using only the up states of $\mathbf{M}_r$.) Thus $\mathbf{M}_u$ is just the submatrix formed from the first $(N_u + 1)$ rows and $(N_u + 1)$ columns of either $\mathbf{M}_a$ or $\mathbf{M}_r$.

The probability that the system is in the nth upstate of the system is $P_n^u(t)$. In analogy with Eq. (7-28), the reliability $R(t)$ is

$$R(t) = \sum_{n=0}^{N_u} P_n^u(t). \tag{7-41}$$

The system of equations for $\mathbf{P}^u(t)$ may be written in the form

$$d\mathbf{P}^u(t)/dt = \mathbf{M}_u \mathbf{P}^u(t), \tag{7-42}$$

where $\mathbf{P}^u(t)$ and its initial condition $\mathbf{P}^u(0)$ satisfy the equations

$$\mathbf{P}^u(t) = \begin{bmatrix} P_0^u(t) \\ P_1^u(t) \\ \vdots \\ P_{N_u}^u(t) \end{bmatrix}, \qquad \mathbf{P}^u(0) = \begin{bmatrix} 1 \\ 0 \\ \vdots \\ 0 \end{bmatrix}. \tag{7-43}$$

The new feature about $\mathbf{P}^u(t)$ is that it also must satisfy the *final* condition

$$\mathbf{P}^u(\infty) = \mathbf{0}, \tag{7-44}$$

since whenever the system enters the absorbing state $(N_u + 1)$ it has failed and there is no mechanism for recovery.

To calculate the MTTF, Eqs. (7-40) and (7-41) show that we need to calculate

$$\text{MTTF} = \sum_{n=0}^{N_u} K_n, \tag{7-45}$$

where the constants

$$K_n = \int_0^\infty P_n^u(t)dt, \qquad n = 0 \text{ to } N_u, \tag{7-46}$$

are the elements of a vector $\mathbf{K}$. The $\mathbf{K}$ follows by integrating Eq. (7-42) over time to obtain

$$\mathbf{M}_u\mathbf{K} = -\mathbf{P}^u(0) \tag{7-47}$$

after using Eq. (7-44). Equation (7-46) is solved in the same way as Eq. (7-19) to obtain the K_n as

$$K_n = -\frac{(\text{cof } \mathbf{M}_u^T)_{n0}}{|\mathbf{M}_u|}, \tag{7-48}$$

where the cofactors are calculated as explained in Appendix C. The MTTF follows from use of Eqs. (7-45) and (7-48).

Example 7-12 Calculate the MTTF for a system of two identical units in which only one is required to function for the system to operate. Each unit fails during operation and during standby; there are n repairmen, $n = 1$ or 2.

This system is the first one we considered in Section 7-4-3, and we also studied it in Example 7-10. Using Eq. (7-35), we see the matrix $\mathbf{M}_u$ is

$$\mathbf{M}_u = \begin{bmatrix} -(\lambda + \lambda^*) & \mu \\ (\lambda + \lambda^*) & -(\lambda + \mu) \end{bmatrix}.$$

The use of Eq. (7-48) gives

$$K_0 = \frac{\lambda + \mu}{\lambda(\lambda + \lambda^*)} \qquad K_1 = \frac{\lambda + \lambda^*}{\lambda(\lambda + \lambda^*)}$$

so therefore Eq. (7-45) gives

$$\text{MTTF} = K_0 + K_1 = \frac{2\lambda + \lambda^* + \mu}{\lambda(\lambda + \lambda^*)}.$$

We observe that the MTTF is independent of the number of repairmen since if two repairmen were working the system would already have failed. The case of active-parallel operation again follows by setting $\lambda^* = \lambda$. ◇

7-9 MODELS FOR COMMON CAUSE FAILURES

There are several ways to treat common cause failures with Markov models [14]. One of these, the Marshall–Olkin or beta factor method, was illustrated in Example 7-6. In this case, a common cause failure can cause extra transitions between any state with more than one unit not in repair and a state in which all units are failed [13].

Example 7-13 Two identical units comprise a system that is to operate with one unit in standby whenever both are up, as described in Example 7-6. Use a Marshall–Olkin model and the rare-event approximation method of Section 7-6 to estimate the system unavailability if a unit can fail at a rate λ_c by common cause failures.

From Example 7-6, the system unavailability is $Q(t) = P_2(t)$ and the transition matrix is

$$\mathbf{M}_a = \begin{bmatrix} -(\lambda + \lambda^* + \lambda_c) & \mu & 0 \\ \lambda + \lambda^* & -(\lambda + \lambda_c + \mu) & n\mu \\ \lambda_c & (\lambda + \lambda_c) & -n\mu \end{bmatrix}.$$

After using Eq. (7-36), we find the system unavailability with the Marshall–Olkin approach to be

$$Q(t) \approx [\mathbf{I} + \mathbf{M}_a t]_{20} = \lambda_c t. \ \Diamond$$

Another method of estimating the effects of common cause failures, the *geometric-mean technique*, was used in the Reactor Safety Study [19] (to be discussed in Chapter 12). In this method, the effect of potential common cause failures on system unavailability is taken to be the geometric mean of the upper and lower bounds on the system unavailability assuming the units of the system are completely coupled and completely independent, respectively. That is, the system unavailability $Q(t)$ is given in terms of the lower bound for the unavailability, $Q(t)$, and the upper bound $Q(t)$, by

$$Q(t) = [Q_l(t)\, Q_u(t)]^{1/2}. \quad (7\text{-}49)$$

Example 7-14 Derive the effect of potential common cause failures on the unavailability using the geometric-mean technique for a system of two

The lower bound for unavailability is $Q_l(t)$ and the upper bound is $Q_u(t)$

identical units in which only one is required to function for the system to operate. Each unit rarely fails during operation and there are no failures during standby; there are n repairmen, $n = 1$ or 2.

We first construct the state transition diagram for the system with completely coupled and completely uncoupled units, as in Fig. E7-14(a) and (b).

For completely coupled failures, the transition matrix is

$$\mathbf{M}_a = \begin{bmatrix} -(\lambda + \lambda^*) & \mu & 0 \\ 0 & -\mu & n\mu \\ (\lambda + \lambda^*) & 0 & -n\mu \end{bmatrix}.$$

This $\mathbf{M}_a$ and the rare-event approximation of Eq. (7-36) give

$$Q(t) = P_2(t) = (\lambda + \lambda^*)t.$$

The unavailability for completely independent failures was derived in Example 7-10 as

$$Q(t) = \lambda(\lambda + \lambda^*)t^2/2.$$

Hence from Eq. (7-49) the system unavailability with potential common cause failures treated with the geometric mean technique can be estimated to be

$$Q(t) = [\lambda(\lambda + \lambda^*)^2 t^3/2]^{1/2}. \ \diamond$$

The results for $Q(t)$ from Examples 7-13 and 7-14 for the same system analyzed by the Marshall–Olkin and geometric-mean methods, respec-

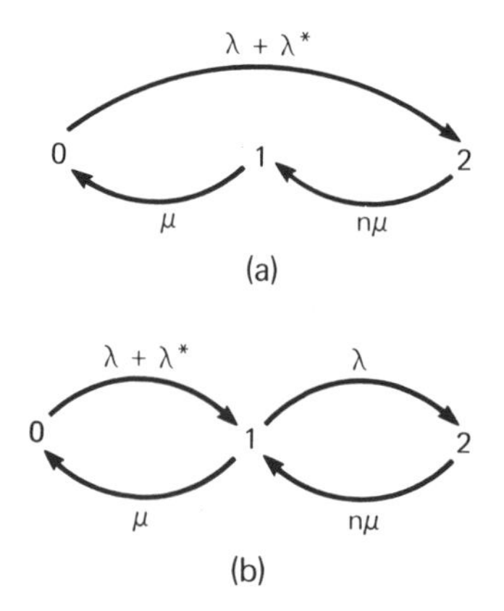

Fig. E7-14 *State transition diagram for availability analysis of two identical, completely independent units (a) for upper bound on unavailability (b) for lower bound on unavailability.*

tively, are quite different in form. It has been shown, however, that the two methods give comparable predictions in most cases so long as λt is not very small [14].

In the analysis of some systems subject to common cause failures, the contribution to the reliability from such failures can be ignored. This is the case, for example, for the M-out-of-N system with N repairmen (so that every failed component is under repair) if the common cause hazard rate λ_c of the Marshall–Olkin approach satisfies the equation [20]

$$\lambda_c \ll \frac{N!}{(M-1)!\,(N-M)!}\frac{\lambda^{N-M+1}}{\mu^{N-M}}, \tag{7-50}$$

provided $\mu \gg \lambda$, as is the usual case for a repairable system.

7-10 OTHER ANALYSIS TECHNIQUES

The Markov approach is quite general for determining the availability of systems with random failures and repairs. However, the number of possible system states can number up to 2^n for a system of n independent units, each of which can be working or failed. Therefore a brute-force analysis on systems with many units is not promising, but techniques are available that reduce the numerical complexity of the method [21].

An extension of the Markov model that is useful for analyzing the impact of aging mechanisms on safety system performance has been developed [22]. The technique treats a system that contains a component that operates (without being renewed) until it attains an age at which sufficient degradation has occurred so that the component is unable to perform its function upon demand in an abnormal environment. This failure is defined as an *undetectable age failure* since the component functions normally except during a condition (environment) that a periodic test does not simulate.

An important step in analyzing the uncertainty in the Markov-model reliability has been accomplished for the case in which each component's failure and repair rate can be considered as a random variable over a range of values that have different probabilities associated with each value in the range [23]. In this case, the reliability of the system itself becomes a random variable. This model permits:

1. Common cause failures, and
2. Inspection and maintenance procedures that depend upon the state of the system and include the possibility of human errors.

This technique could be especially helpful for analyzing newly designed systems for which there is considerable uncertainty about the hazard and repair rates.

Another entirely different approach from the Markov method is sometimes used to analyze systems that can be repaired. *Renewal theory* involves an investigation of the number of renewals in a time period of interest [4, 24–26]; application of this method for complex nuclear power systems has not been particularly promising to date.

EXERCISES

7-1 A system consists of two identical units in active operation, with one repairman. The mean time to failure of each unit is 1000 hr and the mean time to repair is 10 hr. (a) What is the MTTF if there is a repairman? (b) What is the MTTF if there is not a repairman?

7-2 A system with four sequentially operated units (three cold standby units) is operated with no repair and no switch failures. The four units fail randomly with constant instantaneous rates λ_i, $i = 1$ to 4. (a) Define the five states of the system and draw a state transition diagram. (b) Construct the transition matrix $\mathbf{M}_r$. (c) Derive the time-dependent reliability of the system and reduce your answer to a highly symmetric form. (d) Determine the MTTF for the system and reduce it to the simplest form.

7-3 A *single*-unit system can fail by either of two independent failure modes at a constant rate λ_i for mode i, $i = 1$ or 2. The repair of the system after failure occurs with the constant rate μ_i, $i = 1$ or 2, depending upon the mode of system failure. (a) Define the three states for an availability analysis of the system and draw a state transition diagram. (b) Construct the transition matrix $\mathbf{M}_a$. (c) Determine the system availability with time. (d) Determine the steady-state availability. (e) Derive the MTTF.

7-4 A particular nuclear reactor cooling system has three identical loops in normal operation. Although repairs to such loops actually will occur during reactor shutdown, assume for the sake of this exercise that the reactor can be briefly operated while one loop is under repair, but not if two loops have failed. If each coolant loop has a MTTF of λ^{-1}, and if the MTTR of any loop is μ^{-1}, (a) Define the three states of the system for a reliability analysis and draw a state transition diagram. (b) Construct the transition matrix $\mathbf{M}_r$. (c) Determine the reliability of the system with repair. (d) Obtain the MTTF.

7-5 For the system of the preceding exercise, in which there are n repairmen, $n = 1$ or 2, (a) Define the three states of the system for an availability analysis and draw a state transition diagram. (b) Construct the transition matrix $\mathbf{M}_a$. (c) Determine the time-dependent unavailability for rare failures. (d) Determine the steady-state availability.

7-6 You are given a reactor with two heat exchangers, only one of which is to be operated at any one time. The constant hazard rate for an on-line exchanger is λ and the repair rate for each is μ; you should also account for accidents to *either* exchanger via a constant common cause accident rate λ_c present whenever a unit is not undergoing repair. Use the $\mathbf{M}_a$ of Example 7-13 to (a) Determine the steady-state availability. (b) Determine the MTTF.

7-7 A nuclear steam supply system has two turbogenerator units; unit 1 operates and unit 2 is in standby whenever both are up. The units have a MTTF of λ_i^{-1} during active operation, $i = 1$ and 2, and unit 2 has a MTTF during standby of $(\lambda_2^*)^{-1}$. There are two repairmen who can fix a unit during a mean time interval of μ_i^{-1}, $i = 1$ and 2. Use the transition matrix $\mathbf{M}_a$ of Example 7-4 to (a) Determine the steady-state availability. (b) Determine the MTTF.

7-8 The nuclear steam supply system of Exercise 7-7 is operated with only one repairman who will always work on unit 1 whenever both units are failed. (Hence there is only one system failure state during which unit 2 is down and unit 1 is under repair.) (a) Modify the transition matrix $\mathbf{M}_a$ of Example 7-4 to make it appropriate for this system. (b) Determine the steady-state availability. (c) Determine the MTTF.

7-9 For the system having a state transition diagram as in Fig. 7-7, determine the MTTF.

7-10 The system shown in the figure requires that one-out-of-two units of type 1 in active-parallel, and two-out-of-three units of type 2 in active-parallel, be operable at any time. The constant hazard rates for these units are λ_1 and λ_2, respectively. Repairs to either a type 1 or 2 unit are completed at a rate μ by one of three repairmen who do not work together. (a) Define the five possible states of the system for an availability analysis and draw a state transition diagram. (b) Construct the transition matrix $\mathbf{M}_a$. (c) Modify the transition matrix of part b to calculate the system reliability (when there are only two repairmen needed).

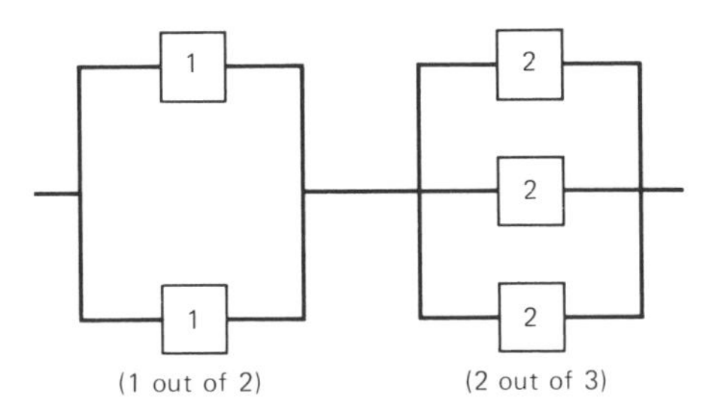

7-11 A system has three identical pumps, only one of which operates at a time. During operation each fails randomly at a rate λ; there are no failures during standby. The repair of any pump is completed with a mean time of μ^{-1} once work is begun; however, there is a mean time delay δ^{-1} before beginning repair of each unit after it fails. (a) Define the seven possible states of the system for a reliability analysis and draw a state transition diagram. (b) Construct the transition matrix $\mathbf{M}_r$.

7-12 A reactor has two identical coolant loops, each with two identical pumps connected in active-parallel. Only one loop is operated at a time, and at least one pump must be functional for a loop to be functional. All pumps fail randomly when in active operation with a MTTF of λ^{-1}; there are no failures during standby, and no failures in switching from one loop to the other. There are two repairmen who never work on the same pump and who never perform repairs on a pump when the loop containing the failed pump is in operation; the mean time to repair a pump is μ^{-1}. Assume there is no switching between loops unless the operating loop fails. (a) Define the eight possible states of the system for an availability analysis and draw a state transition diagram. (b) Construct the transition matrix $\mathbf{M}_a$.

7-13 For the system of the preceding exercise, assume the loop with the most working pumps is always in operation. (a) Define the five possible states of the system for an availability analysis and draw a state transition diagram. (b) Construct the transition matrix $\mathbf{M}_a$.

REFERENCES

1. A. K. S. Jardine, "Maintenance, Replacement, and Reliability." Wiley, New York, 1973.
2. C. O. Smith, "Introduction to Reliability in Design." McGraw-Hill, New York, 1976.
3. A. J. Bourne and A. E. Green, "Reliability Technology." Wiley (Interscience), New York, 1972.
4. M. L. Shooman, "Probabilistic Reliability: An Engineering Approach." McGraw-Hill, New York, 1968.
5. G. H. Sandler, "System Reliability Engineering." Prentice-Hall, Englewood Cliffs, New Jersey, 1963.
6. J. A. Buzacott, Markov approach to finding failure times of repairable systems, *IEEE Trans. Reliability* **R-19,** 128 (1970).
7. R. D. Guild and E. G. Tourigny, Reliability, reliability with repair, and availability of four identical element multiplex systems, *Nucl. Technol.* **41,** 97 (1978).
8. J. B. Fussell and J. S. Arendt, System reliability engineering methodology: A discussion of the state of the art, *Nucl. Safety* **20,** 541 (1979).
9. M. J. Bell, ORIGEN, The ORNL Isotope Generation and Depletion Code. Oak Ridge National Laboratory Rep. ORNL-4628 (1973).

10. E. Kreyszig, "Advanced Engineering Mathematics," 4th ed. Wiley, New York, 1978.
11. C. R. Wylie, "Advanced Engineering Mathematics," 3rd ed. McGraw-Hill, New York, 1966.
12. M. Kodama, H. Nakamichi, and S. Takamatsu, Analysis of 7 models for the 2-dissimilar-unit warm standby, redundant system, *IEEE Trans. Reliability* **R-25,** 273 (1976).
13. A. W. Marshall and I. Olkin, A multivariate exponential distribution, *J. Am. Statist. Assoc.* **62,** 30 (1967).
14. K. N. Fleming and P. H. Raabe, A comparison of three methods for the quantitative analysis of common cause failures, *in* "Probabilistic Analysis of Nuclear Reactor Safety," Vol. 3, p. X.3-1. American Nuclear Society, LaGrange Park, Illinois, 1978.
15. J. Skakala and B. Rohal-Ilkiv, 2-unit redundant systems with replacement and repair, *IEEE Trans. Reliability* **R-26,** 294 (1977).
16. T. Nakagawa, A 2-unit repairable redundant system with switching failures, *IEEE Trans. Reliability* **R-26,** 128 (1977).
17. M. Sasaki and T. Hiramatsu, Reliability of intermittently used systems, *IEEE Trans. Reliability* **R-25,** 208 (1976).
18. R. H. Myers, K. L. Wong, and H. M. Gordy (eds.), "Reliability Engineering for Electronic Systems," Wiley, New York, 1964.
19. Reactor Safety Study—An Assessment of Accident Risks in U.S. Commercial Nuclear Power Plants. U.S. Nuclear Regulatory Commission Rep. WASH-1400, NUREG 75/014 (October 1975).
20. G. E. Apostolakis, The effect of a certain class of potential common mode failures on the reliability of redundant systems, *Nucl. Eng. Design* **36,** 123 (1976).
21. I. A. Papazoglou and E. P. Gyftopoulos, Markov processes for reliability analyses of large systems, *IEEE Trans. Reliability* **R-26,** 232 (1977).
22. E. Oelkers and W. W. Weaver, The impact of aging mechanisms on reactor safety system performance, *Nucl. Sci. Eng.* **68,** 299 (1978).
23. I. A. Papazoglou and E. P. Gyftopoulos, Markovian reliability analysis under uncertainty with an application on the shutdown system of the Clinch River breeder reactor, *Nucl. Sci. Eng.* **73,** 1 (1980).
24. G. E. Apostolakis, Mathematical Methods of Probabilistic Safety Analysis. Univ. of California at Los Angeles Rep. UCLA-ENG-7464 (1974).
25. N. R. Mann, R. E. Schafer, and N. D. Singpurwalla, "Methods for Statistical Analysis of Reliability and Life Data." Wiley, New York, 1974.
26. H. E. Lambert, Fault Trees for Decision Making in Systems Analysis. Lawrence Livermore Laboratory Rep. UCRL-51829 (1975).

CHAPTER 8

Fault Tree Analysis

8-1 INTRODUCTION

In multicomponent systems as complicated as nuclear reactors, it is important to analyze the possible mechanisms for failure and to perform probabilistic analyses for the expected rate of such failures. Rarely can one apply only the techniques of Chapters 6 and 7, however, primarily because the number and linking of the different subsystems is too complex. Fault tree analysis is a technique by which many events that interact to produce other events can be related using simple logical relationships (AND, OR, etc.); these relationships permit a methodical building of a structure that represents the system.

To complete the construction of a fault tree for a complicated system, it is necessary to first understand how the system functions. A system function diagram (or flow diagram) is used for this purpose to depict the pathways by which signals or materials are transmitted between components comprising the system; the reliability block diagrams of Chapter 6 are examples of signal function diagrams. A second diagram, a functional logic diagram, is sometimes needed to depict the logical relationships of the components. This logic diagram is usually different from the function diagram; for example, a component of a function diagram might show a 2-out-of-4 comparison of signals while on a logic diagram the 2-out-of-4 logic would be explicitly displayed.

A third tool frequently used to understand the failure modes of the system is a *Failure Modes and Effects Analysis* (FMEA). A FMEA is one of several variations of *inductive* safety analysis and is frequently used for systems for which the expenses and refinements of a fault tree analysis are not warranted. An introduction to this inductive method is provided in Appendix D.

Only after the functioning of the system is fully understood should an analyst construct a fault tree. Of course, for simpler systems, the function and logic diagrams and an FMEA are unnecessary and fault tree construction can begin immediately.

In fault tree construction, the system failure event that is to be studied is called the *top event*. Successive subordinate (i.e., subsystem) failure events that may contribute to the occurrence of the top event are then identified and linked to the top event by logical connective functions. The subordinate events themselves are then broken down to their logical contributions and, in this manner, a failure event tree structure is created. Progress in the synthesis of the tree is recorded graphically by arranging the events into a tree structure using connecting symbols called *gates*, to be discussed in Section 8-2.

When a contributing failure event can be divided no further, or when it is decided to limit further analysis of a subsystem, the corresponding branch is terminated with a *basic event*. The basic event for a branch is termed a *primary fault* event if the subsystem failed because of a basic mode such as a structural fault, or failure to open or close, or to start or stop. The basic event is a *secondary fault* event if the subsystem is out of tolerance so that it fails because of excessive operational or environmental stress placed on the system element.

In practice, all basic events are taken to be statistically independent unless they are "common cause failures." Such failures are those that arise from a common cause or initiating event, in which case, two or more primary events are no longer independent. (See Sections 5-5, 6-6, 7-9, and 10-2.)

Once the tree structure has been established, subsequent analysis is deductive and takes two forms. The purpose of a *qualitative analysis* is to reduce the tree to a logically equivalent form in terms of the specific combinations of basic events sufficient to cause the undesired top event to occur. Each combination will be a "minimal cut set" of failure modes for the tree. (A minimal cut set is a set of events, which cannot be reduced in number, whose occurrence causes the top event; see Section 6-5.)

One procedure for reducing the tree to a logically equivalent form is by the use of Boolean algebra, as illustrated in Section 6-5. A second procedure is numerical, in which case, the logical structure is used as a model for trial and error testing of the effects of selected combinations of primary event failures.

Quantitative analysis of the fault tree consists of transforming its established logical structure into an equivalent probability form and numerically calculating the probability of occurrence of the top event from the probabilities of occurrence of the basic events. The probability of a basic event is the failure probability of the component or subsystem during the mission time of interest, T. The failure probability is either constant with time, such as a demand failure probability (see Section 5-3) or a steady-state unavailability (see Section 7-7), or the probability changes with time.

This time-dependent failure probability for a nonrepairable subsystem is just the system unreliability, $1 - R(T)$. For any repairable subsystem used in nuclear power applications, repairs would be continued even if the system failed, so the time-dependent failure probability for a terminating event is the unavailability, $1 - A(T)$. The calculation of a time-dependent failure probability for a subsystem frequently can be simplified; for example, if the device fails randomly with constant hazard rate λ, then for sufficiently small values of λT,

$$1 - R(T) = 1 - e^{-\lambda T} \approx \lambda T. \tag{8-1}$$

There are two approaches that can be used to calculate the probability of the top event from the probabilities of the basic events, depending upon whether a qualitative analysis has been performed to determine the minimal cut sets. If these cut sets are available, then the failure probabilities for each basic event that comprises each set are multiplied together to give the probability of that minimal cut set, and then the probabilities of all such sets are combined by using Eqs. (2-15) or (2-25).

Even without knowledge of the minimal cut sets, if the size and complexity of the tree are not too large, then the probability for the top event can be calculated by hand, provided that the failure events are infrequent enough that the rare-events approximation may be used. (This usually occurs if all probabilities are smaller than 0.1.) In this case, the probabilities for events are obtained from Eq. (8-1) and are combined using

$$P(A_1A_2 \cdots A_N) = P(A_1)P(A_2) \cdots P(A_N) \quad \text{(AND)} \tag{8-2}$$

$$P(A_1 + A_2 + \cdots A_N) \approx \sum_{n=1}^{N} P(A_n) \quad \text{(OR)}. \tag{8-3}$$

The resulting probability from each calculation is used as an input to the calculation of the event corresponding to the next gate higher in the tree. In this way, one proceeds in an orderly fashion from the bottom to the top of the tree.

For trees containing several tens of events, even when the rare-events approximation may be used, computer help will be needed. Certainly for all large trees (e.g., 100 or more basic events), or for the exact analysis of trees containing more than a dozen or so basic events, a relatively sophisticated general computer program is required (see Chapter 10).

8-2 FAULT TREE CONSTRUCTION

The first step in fault tree construction is selection of the top failure event that is to be the subject of the analysis. Every following event will be considered in terms of its effect upon that top event.

The next step is to identify contributing events that may directly cause the top event to occur. At least four possibilities exist:

1. No input to the device was received, such as a signal to operate.
2. The device itself has experienced some failure so that it will not operate.
3. There is a human error, such as failure to actuate a switch or to properly install the device.
4. Some external event may have occurred that prevents operation of the device, such as a common cause failure.

If it is decided that any one of the contributing events can cause the top failure, this occurrence corresponds to the logical OR function of the events. This conclusion is then recorded graphically by means of the standard OR gate logic symbol, as illustrated in the following example.

Example 8-1 The top of the tree for an electrical system for which the top event is failure of a circuit breaker to trip is shown in Fig. E8-1. ◇

Once the first level of contributing (branch) events has been established, each branch must be examined to decide whether it is to be further divided. The following question must be answered: Is the input event to be treated as a basic failure event or to be broken down into its component

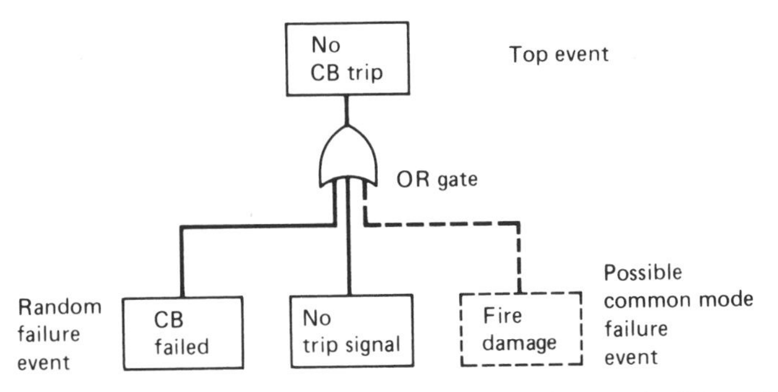

Fig. E8-1 *Top of the tree for failure of a circuit breaker* (CB) *to trip.* [*From Reliability Manual for Liquid Metal Fast Reactor* (*LMFBR*) *Safety Programs. General Electric Company Internal Rep. SRD-74-113* (*1974*).]

failure events? The decision may be influenced by a lack of knowledge regarding modes of failure of the components or a desire to limit the degree of detail to be included.

If it is decided that a given contributing failure event is a primary failure, the corresponding branch of the tree is stopped and this basic event is shown graphically by a circle. This choice carries with it the implication that the terminating event is independent of other subsequent terminating events. There is also an implied requirement that a numerical value for the probability of occurrence of the primary event be available as an input if quantitative analysis of the tree is to be performed.

If a contributing failure event is not identified as basic, it must be examined for its subordinate contributors, and then their logical relationship must be identified.

Example 8-2 In the electrical system of Example 8-1, if the circuit breaker trip signal is conveyed by the opening of one or both of two relay contacts A and B in series, then both relays must fail to prevent the trip signal. The "No Trip Signal" event is then described by the logical AND function and is shown graphically by the standard AND gate symbol; see Fig. E8-2. ◇

This procedure of analyzing every event is continued until all branches have been terminated in independent basic events. In this process, certain terminal events may be viewed as temporary or underdeveloped. These underdeveloped events are those that properly complete the tree structure in its present state but that may require further breakdown to adequately describe the system failure. Such underdeveloped primary events are customarily identified by a diamond symbol rather than the circle symbol.

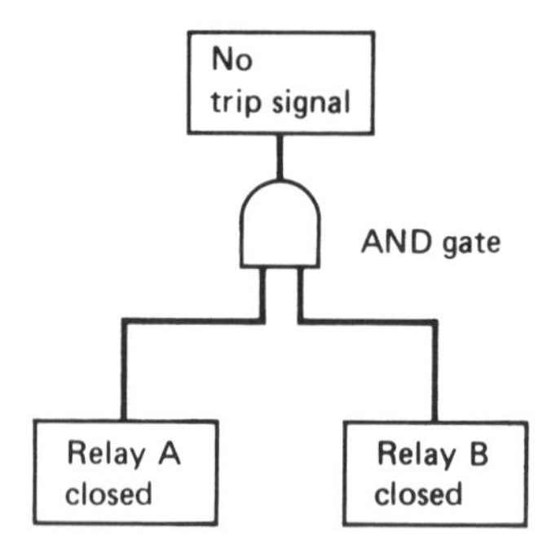

Fig. E8-2 AND *function example for electrical system of Example 8-1* [*From Reliability Manual for Liquid Metal Fast Breeder Reactor (LMFBR) Safety Programs. General Electric Company Internal Rep. SRD-74-113, (1974).*]

Example 8-3 Figure E8-3a illustrates a simple fault tree for failure of the circuit breaker system, schematically shown in Fig. E8-3b, to trip. An identical arrangement of primary failures and gates is shown in Fig. E8-3c for failure of a mechanical holding latch, depicted in Fig. E8-3d, to release [1]. This example illustrates the fact that two systems that may appear entirely different can fail according to the same logic. ◇

Example 8-3 is elementary in scope compared with fault trees for large systems. These trees may grow to a size represented by hundreds of basic events and may also require more complex logical connections to adequately express the true relationships among events. Fault tree symbols commonly encountered in most trees are shown and described in Table 8-1; a more extensive listing of symbols is found elsewhere [2].

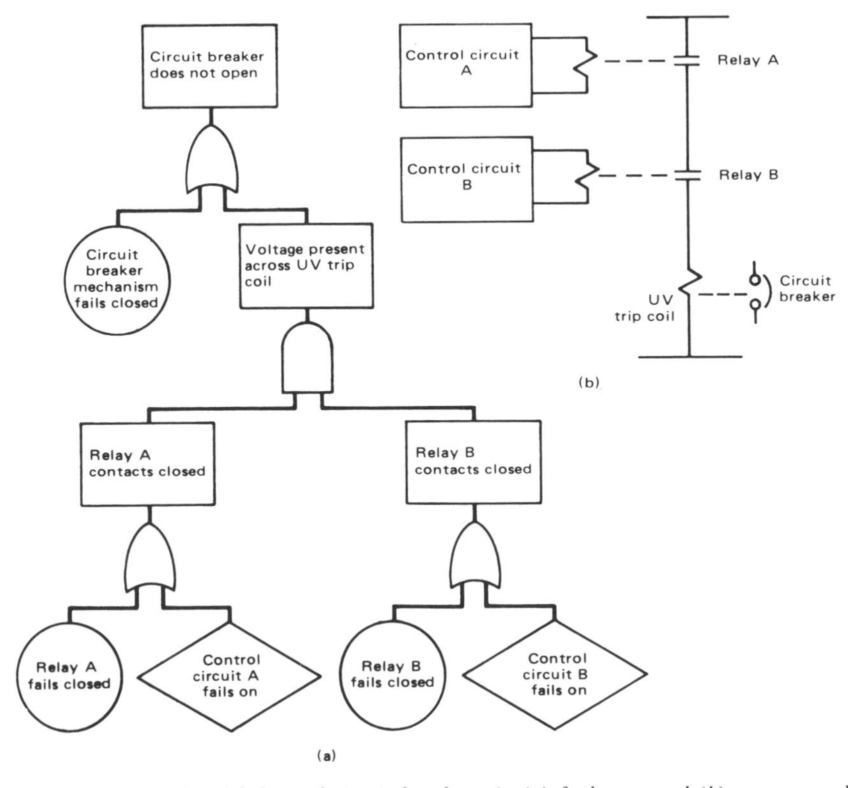

Fig. E8-3 *Example of failure of circuit breaker trip* (*a*) *fault tree and* (*b*) *system, and example of failure of mechanical holding latch* (*c*) *fault tree and* (*d*) *system.* [*From Reliability Manual for Liquid Metal Fast Breeder Reactor* (*LMFBR*) *Safety Programs. General Electric Company Internal Rep. SRD-74-113* (*1974*).]
(*Continued*)

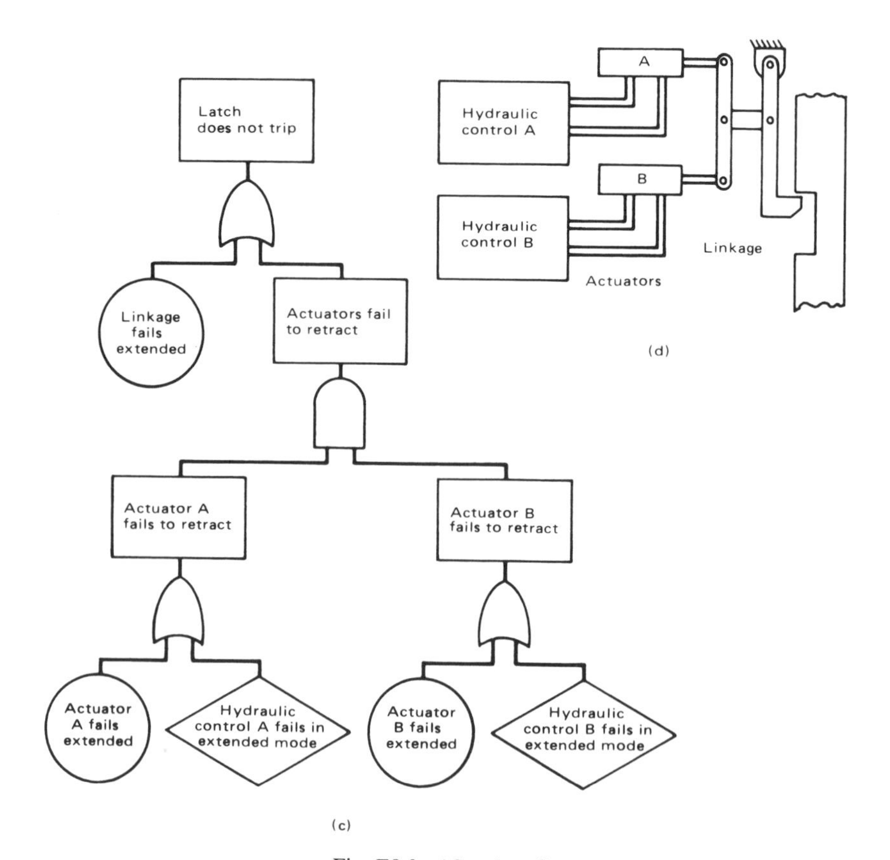

Fig. E8-3 (*Continued*)

A complete safety analysis on an extensive system such as a nuclear power plant normally requires three levels of fault tree development. The top level includes the top undesired event and the key sub-events that are immediate causes of the top event. In the second level, the events are modeled from the point of view of how the system functions. Finally, at the third level, one is faced with the difficult task of verifying that the basic events are statistically independent. This is done by examining common cause events that can simultaneously fail two or more system elements, such as any common environmental or operational stresses, as well as the effects of human errors in the testing, maintenance, and operation of the system.

Actual construction of fault trees is an art as well as a science and comes only through experience. Construction of fault trees is discussed in more detail elsewhere [2], but the following guidelines are helpful [3].

Table 8-1 *Fault Tree Symbols Commonly Used*

Symbol	Name	Description
	Rectangle	Fault event; it is usually the result of the logical combination of other events
	Circle	Independent primary fault event
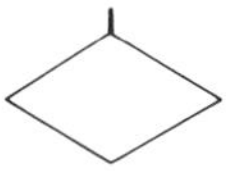	Diamond	Fault event not fully developed as to its causes; it is only an assumed primary fault event
	House	Normally occurring basic event; it is not a fault event
	OR Gate	The union operation of events; i.e., the output event occurs if one or more of the inputs occur
	AND Gate	The intersection operation of events; i.e., the output event occurs if and only if all the inputs occur
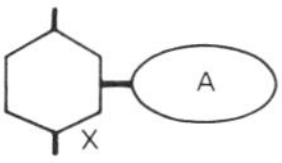 A X	INHIBIT Gate	Output exists when X exists and condition A is present; this gate functions somewhat like an AND gate and is used for a secondary fault event X
 A	Triangle-in	Triangle symbols provide a tool to avoid repeating sections of a fault tree, or to transfer the tree construction from one sheet to the next. The triangle-in appears at the bottom of a tree and represents that branch of the tree (in this case "*A*") shown someplace else. The triangle-out appears at the top of a tree and denotes that the tree "*A*" is a subtree to one shown someplace else.
 A	Triangle-out	

Rule 1. State the fault event as a fault, including the description and timing of a fault condition at some particular time. Include

(a) what the fault state of that system or component is,
(b) when that system or component is in the fault state.

Test the fault event by asking

(c) Is it a fault?

(d) Is the what-and-when portion included in the fault statement?

Rule 2. There are two basic types of fault statements, state-of-system and state-of-component. To continue the tree:

(a) If the fault statement is a state-of-system statement, use Rule 3.

(b) If the fault statement is a state-of-component statement, use Rule 4.

Rule 3. A state-of-system fault may use an AND, OR, or INHIBIT gate or no gate at all. To determine which gate to use, the faults must be the

(a) minimum necessary and sufficient fault events,

(b) immediate fault events.

To continue, state the fault events input into the appropriate gate.

Rule 4. A state-of-component fault always uses an OR gate.

To continue, look for the primary, secondary, and command failure fault events. Then state those fault events.

(a) Primary failure is failure of that component within the design envelope or environment.

(b) Secondary failures are failures of that component due to excessive environments exceeding the design environment.

(c) Command faults are inadvertent operation of the component because of a failure of a control element.

Rule 5. No gate-to-gate relationships, i.e., put an event statement between any two gates.

Rule 6. Expect no miracles; those things that would normally occur as the result of a fault will occur, and only those things. Also, normal system operation may be expected to occur when faults occur.

Rule 7. In an OR gate, the input does not cause output. If any input exists, the output exists. Fault events under the gate may be a restatement of the output events.

Rule 8. An AND gate defines a causal relationship. If the input events coexist, the output is produced.

Rule 9. An INHIBIT gate describes a causal relationship between one fault and another, but the indicated condition must be present. The fault is the direct and sole cause of the output when that specified condition is present. Inhibit conditions may be faults or situations, which is why AND and INHIBIT gates differ.

[From H. E. Lambert, Systems Safety Analysis and Fault Tree Analysis. Lawrence Livermore Laboratory Rep. UCID-16238 (1973).]

8-3 FAULT TREE EVALUATION

Once constructed, a fault tree can be described by a set of Boolean algebraic equations, one for each gate of the tree. For each gate, the input events (such as primary events) are the independent variables, and the output event (such as an intermediate event) is the dependent variable. Utilizing the rules for Boolean algebra given in Section 6-5, it is then possible to solve these equations so that the top and intermediate events are individually expressed in terms of minimal cut sets that involve only basic events.

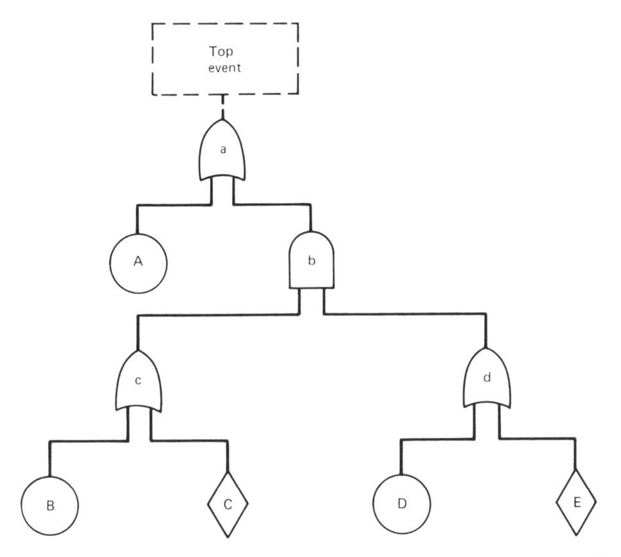

Fig. E8-4 *Coded tree for Fig.* E*8-3.* [*From Reliability Manual for Liquid Metal Fast Breeder Reactors* (*LMFBR*) *Safety Programs. General Electric Company Internal Rep. SRD-74-113* (*1974*).]

Example 8-4 Construct the minimal cut sets for Example 8-3.

If we use small letters to code gates and capital letters to code basic events, as shown in Fig. E8-4, then for the mechanical latch the events imply the following: (A) linkage fails extended, (B) actuator A fails extended, (C) control A fails in extended mode, (D) actuator B fails extended, and (E) control B fails in extended mode. The tree is completely specified by the information given in Table E8-4.

The next step is to obtain Boolean logic statements for the gates in terms of the input events. For the gates in the table,

$$a = A + b \qquad c = B + C$$

$$b = cd \qquad d = D + E.$$

Table E8-4 *Fault Tree Specifications for Fig.* E*8-4*

Gate	Type	Inputs	Gate	Type	Inputs
a	OR	A, b	c	OR	B, C
b	AND	c, d	d	OR	D, E

The top event a is expressed in terms of the primary events by substitution:

$$\begin{aligned} a &= A + b \\ &= A + cd \\ &= A + (B + C)(D + E). \end{aligned}$$

If the rules of Boolean algebra in Section 6-5 are used, the result is

$$a = A + BD + BE + CD + CE.$$

Each of the terms on the right-hand side is a "cut set" and, in this case, also a minimal cut set.

After the minimal cut sets have been obtained by eliminating all the redundancies in the events, the qualitative analysis is complete and the failure modes contributing to the top event have been identified. For this example the modes of failure are: (A) linkage fails, (BD) actuator A and actuator B fail, (BE) actuator A and control B fail, (CD) control A and actuator B fail, (CE) control A and control B fail. ◇

The importance of a minimal cut set to the failure of the top event can be illustrated with two general rules-of-thumb:

1. The importance of each minimal cut set is inversely proportional to the number of basic events in the path. (That is, if each basic event in the fault tree has a probability of 10^{-5} of occurring, then the one-event cut sets will have a probability of 10^{-5} of occurring, the two-event cut sets will have a probability of 10^{-10} of occurring, etc.)
2. Any one-event minimal cut sets are to be avoided if possible by a system redesign, especially if the component is a dynamic component (such as a switch, valve, etc.) rather than a quasi-static component (such as a pipe, a heat exchanger, etc.). This is because failure rates for dynamic components are characteristically an order of magnitude, or more, larger than for quasi-static components.

In summary, the qualitative analysis portion of fault tree analysis is a Boolean algebraic reduction of a fault tree, and its procedure can be summarized as:

1. Code gates and primary events.
2. List gate types and inputs.
3. Write Boolean equation for each gate.
4. Use Boolean algebra to solve for top event in terms of cut sets.
5. Eliminate the cut set redundancies (using Boolean algebra) to obtain the minimal cut sets.

The quantitative analysis of a fault tree is directed at calculation of the probability of occurrence of the top event, given the fault tree and the probability of occurrence of the basic events. As discussed in Section 8-1, there are two possible approaches that can be followed, depending upon the size of the tree and the desired information. If the minimal cut sets are known, then the probability of the top event is the probability of the union of the minimal cut sets, and this probability must be expressed in terms of the probabilities of each basic event. An example of this for the minimal cut sets of Example 8-4 is given in Example 6-9, and a numerical illustration of the results is in Example 6-10.

The second approach is suitable when the rare-events approximation is valid and consists of marching through the tree from bottom to top by means of Eqs. (8-2) and (8-3), as illustrated in the following example.

Example 8-5 Calculate the probability of the top event for Example 8-3 assuming the terminal event probabilities are:

$$P(A) = 0.01$$

$$P(B) = P(C) = P(D) = P(E) = 0.1.$$

Figure E8-5 shows the tree with the rare-event probabilities filled in. As we can see, the approximate probability for the top event is found to be $P(a) \approx 0.05$; the exact result from Example 6-10 is 0.045739. ◇

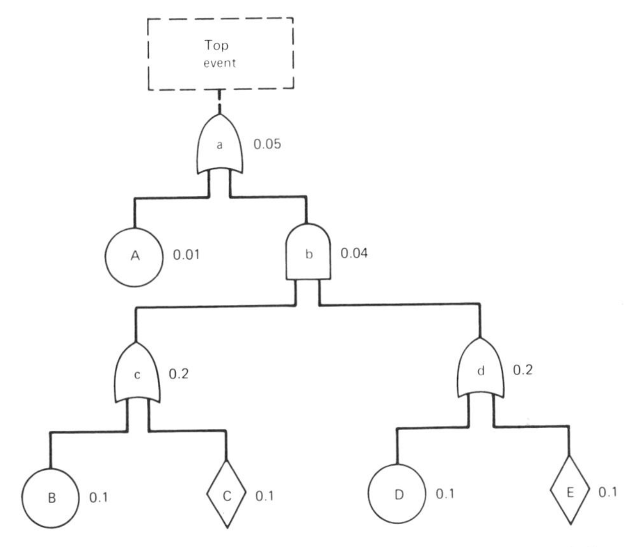

Fig. E8-5 *Fault tree of Fig. E8-3 showing probabilities of failure.*

8-4 EXAMPLES OF SIMPLE FAULT TREES

8-4-1 A SIMPLE WARNING SYSTEM

It is assumed that three sensors are to be used to provide a reliable warning of whether a system is on or off. Each sensor output is connected to an indicator light. When it is working properly, each sensor activates the indicator light as soon as the system starts up, and extinguishes the light at the moment the system is shut down. The indicator lights are considered perfect, but with probability r it is possible for the operator to misinterpret them. Suppose that each sensor may fail in either of two modes: (1) it may fail "on," so that its output lights the indicator lamp regardless of whether the system is on or off, the probability of this failure being q_{on} ; or (2) it may fail "off," so that the indicator lamp is out whether the system is on or off, the probability of this failure being $q_{off} = 1 - p - q_{on}$, where p is the probability of sensor success.

Assume operator A judges the system to be on if he or she perceives one, two, or three indicator lights are on and that it is off if all lights are out. Figure 8-1 shows a fault tree to compute the probability P(on) that the lights indicate on but are actually off, while Fig. 8-2 shows the fault tree to

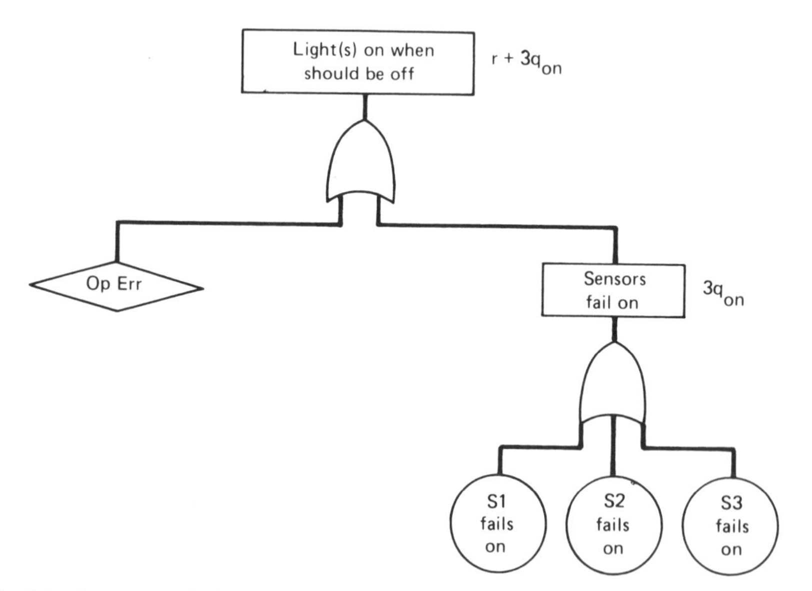

Fig. 8-1 *Operator A fault tree for incorrect information that system is on (1, 2, or 3 lights on).*

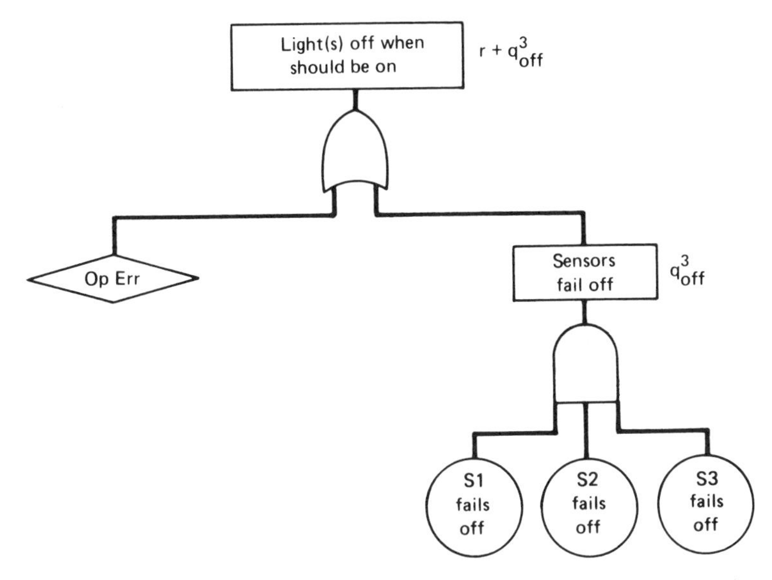

Fig. 8-2 *Operator A fault tree for incorrect information that system is off (all lights out).*

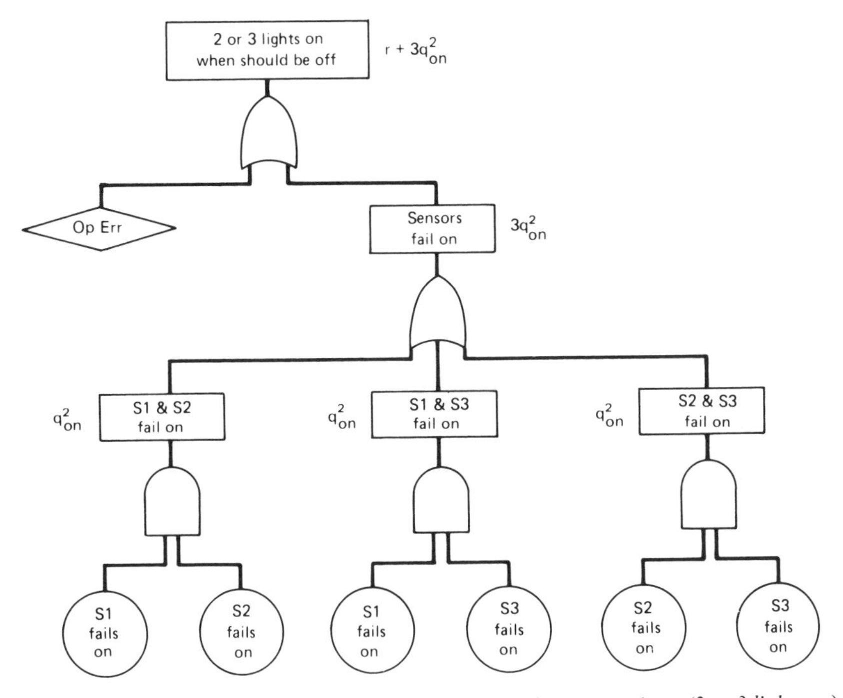

Fig. 8-3 *Operator B fault tree for incorrect information that system is on (2 or 3 lights on).*

compute the probability $P(\text{off})$ that the lights indicate off but are actually on.

Figures 8-3 and 8-4 are the fault trees for a second operator B who judges the reactor to be on if he or she perceives two or three indicator lights are on and the reactor to be off if two or three of the indicator lights are out.

The results of Figs. 8-1 through 8-4 can be summarized by the following probabilities:

$$\left.\begin{aligned} P(\text{on}) &= r + 3\,q_{\text{on}} \\ P(\text{off}) &= r + q_{\text{off}}^3 \end{aligned}\right\} \text{Operator } A \qquad \left.\begin{aligned} P(\text{on}) &= r + 3\,q_{\text{on}}^2 \\ P(\text{off}) &= r + 3\,q_{\text{off}}^2 \end{aligned}\right\} \text{Operator } B \tag{8-4}$$

For example, if $p = 0.88$, $q_{\text{on}} = 0.02$, $q_{\text{off}} = 0.10$, and $r = 0.01$, then the results are

$$\left.\begin{aligned} P(\text{on}) &= 0.07 \\ P(\text{off}) &= 0.011 \end{aligned}\right\} \text{Operator } A \qquad \left.\begin{aligned} P(\text{on}) &= 0.0112 \\ P(\text{off}) &= 0.04 \end{aligned}\right\} \text{Operator } B \tag{8-5}$$

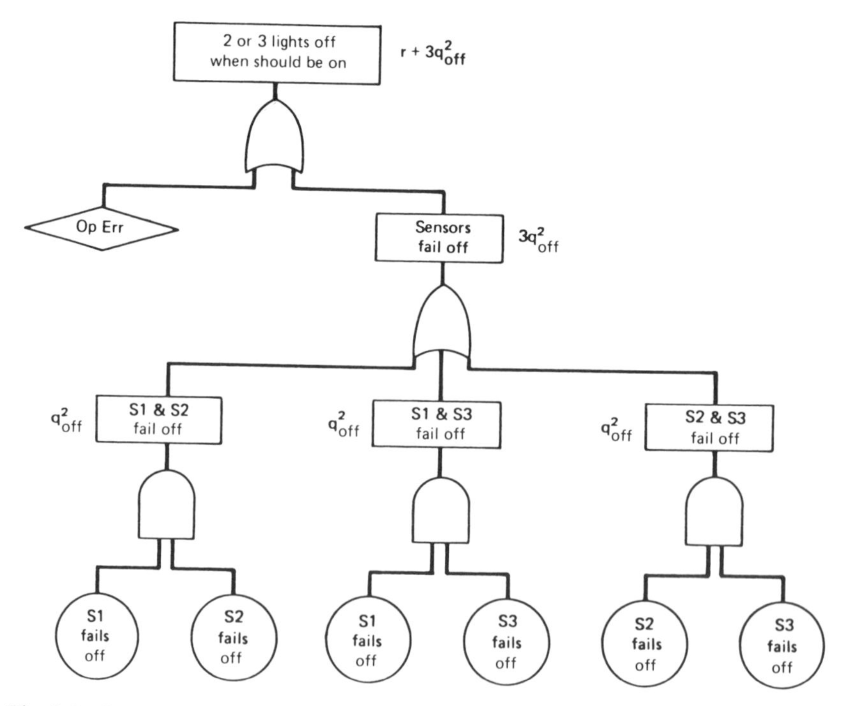

Fig. 8-4 *Operator B fault tree for incorrect information that system is off (2 or 3 lights off).*

In this instance, from a safety standpoint, the system with operator A is best because P(off) is smaller. From a reactor availability standpoint, where false alarms are to be minimized, the system with operator B is best because P(on) is smaller.

8-4-2 Coolant Supply System

The system shown in Fig. 8-5 is a simple coolant supply system consisting of a constant speed pump, heat exchanger, control valve, reservoir, and piping [4]. The function of the system is to supply enough cooling to the prime equipment. The fault tree in Fig. 8-6 is for the top event that consists of loss of minimum (coolant) flow to (the) heat exchanger. Such a loss could arise either by a break in the primary coolant line or loss of flow from the coolant valve, so these events are connected by an OR gate. Rupture of the pipe is a primary fault, and therefore that event is not developed further. The three events that could directly cause loss of flow from the control valve also are connected by an OR gate. These and other fault events are detailed in Fig. 8-6.

8-4-3 Electric Motor Circuit

For the simple electric motor circuit shown in Fig. 8-7, a fault tree for motor failure [5] is shown in Fig. 8-8. One primary failure event is that of failure of the motor itself because, for example, of a wiring failure within the motor or loss of lubrication to the bearings. The diamond symbol used

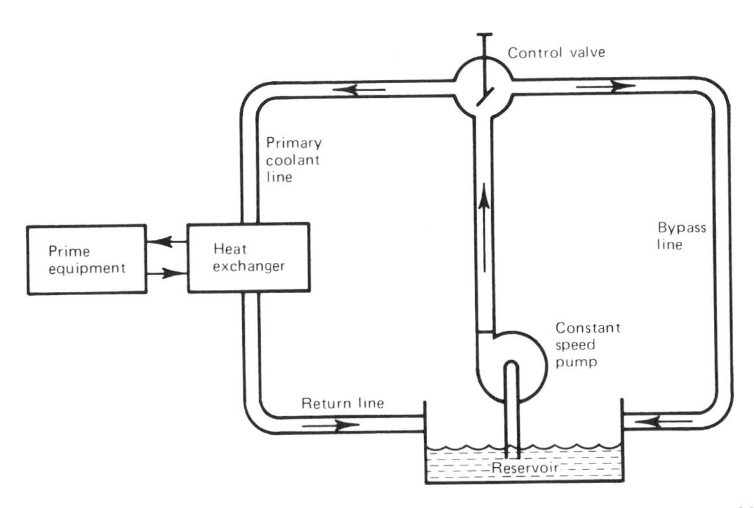

Fig. 8-5 *Coolant supply system.* [*From J. A. Burgess, Machine Design* **42,** *No. 23, 150 (1970).*]

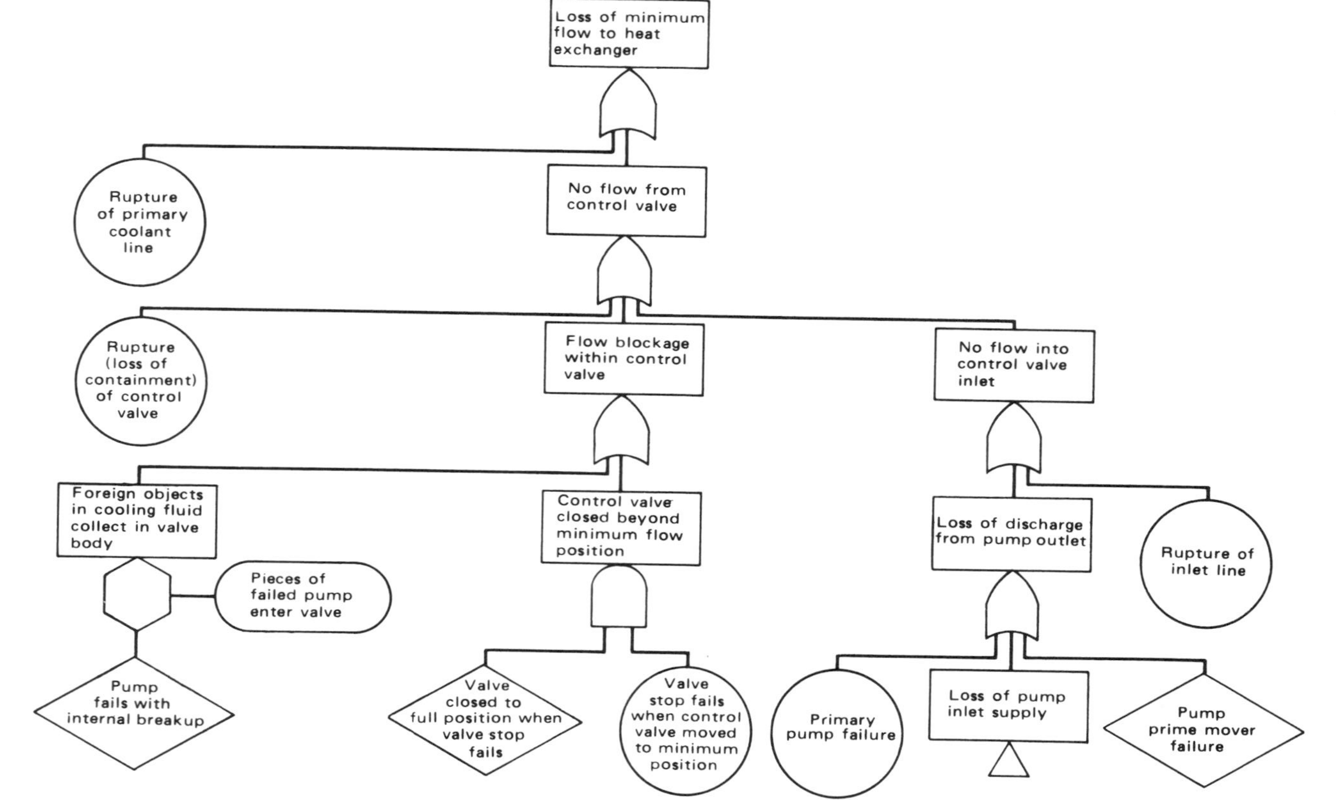

Fig. 8-6 *Fault tree for coolant supply system of Fig. 8-5. [From J. A. Burgess, Machine Design* **42,** *No. 23, 150 (1970).]*

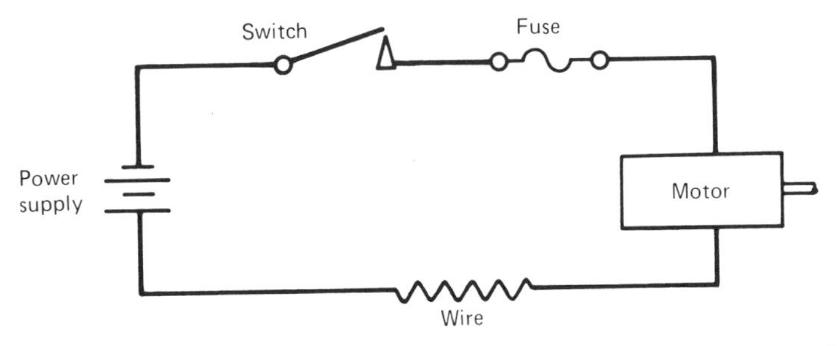

Fig. 8-7 *Electric motor circuit. (From J. B. Fussell,* in *"Generic Techniques in Systems Reliability Assessment," E. J. Henley and J. W. Lynn, eds., pp. 133–162. Noordhoff, Leyden, 1976).*

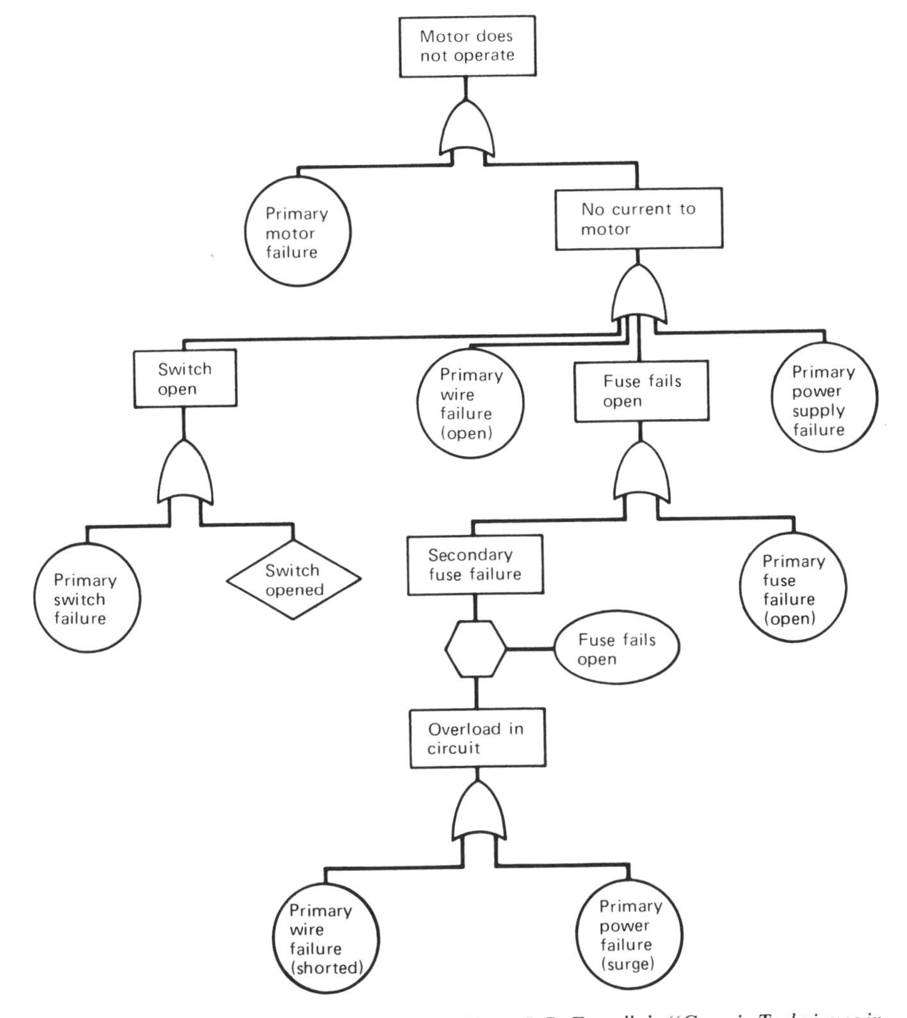

Fig. 8-8 *Fault tree for electric motor circuit. (From J. B. Fussell,* in *"Generic Techniques in Systems Reliability Assessment," E. J. Henley and J. W. Lynn, eds., pp. 133–162. Noordhoff, Leyden, 1976).*

to indicate the event "switch opened" is not developed since insufficient information is available to develop the causes of the human error in allowing the switch to be opened.

The event "fuse fails open" occurs if a primary or secondary fuse failure occurs. Secondary fuse failure can occur if the fuse does not open every time an overload is present in the circuit (because all conditions of an overload do not necessarily result in sufficient overcurrent to open the fuse). The INHIBIT gate is used to treat the secondary failure.

8-4-4 AN ELECTRICAL CIRCUIT

It is desired to analyze the electrical circuit shown in Fig. 8-9 for the overheating of wire *AB* [6]. Figure 8-10 shows the fault tree for the electrical circuit; the tree uses the transfer-in and transfer-out symbols to avoid needless repetition of portions of the fault tree. In this circuit, secondary failure events exist for the motor and the relay contacts, so INHIBIT gates again are needed; for example, the output of INHIBIT gate G_1 in Fig. 8-10 is the event represented by the output of OR gate A_3 and event Y_1.

The reader may verify that the top event B_1 in Fig. 8-10 may be expressed in terms of events X_1 to X_{10}, Y_1, and Y_2 as

$$B_1 = (X_1 + X_2)[X_4 + X_3Y_1 + (X_7 + X_8)(X_9 + X_{10})Y_1] \times [(X_1 + X_2)X_5Y_2 + X_6 + (X_7 + X_8)(X_9 + X_{10})]. \tag{8-6}$$

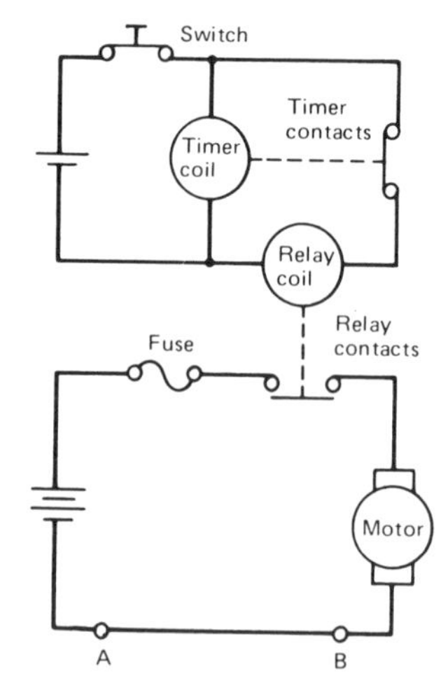

Fig. 8-9 *An electrical circuit. (From W. Hammer, "Handbook of System and Product Safety," Prentice-Hall, Englewood Cliffs, New Jersey, 1972.)*

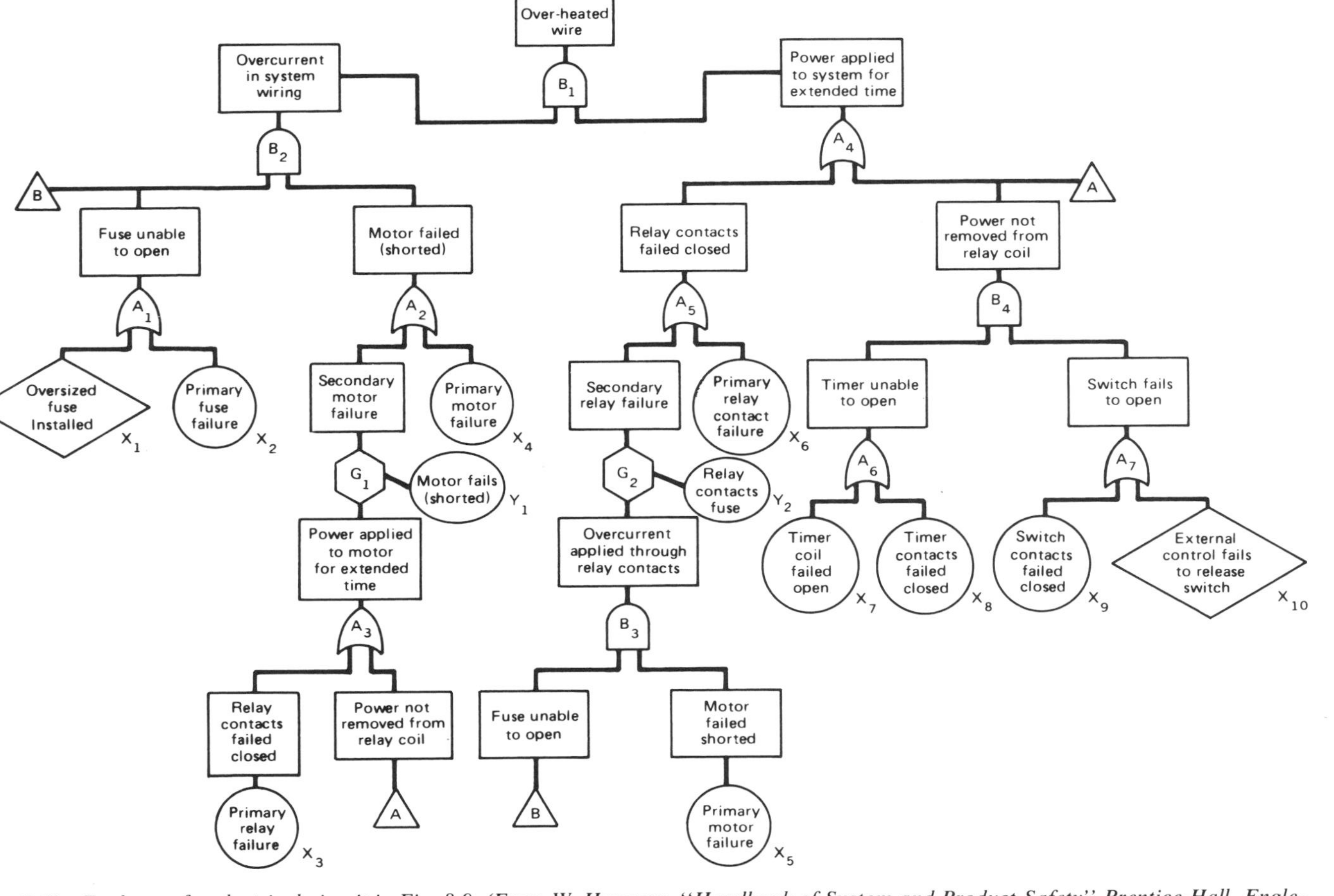

Fig. 8-10 *Fault tree for electrical circuit in Fig. 8-9. (From W. Hammer, ''Handbook of System and Product Safety'' Prentice-Hall, Englewood Cliffs, New Jersey, 1972.)*

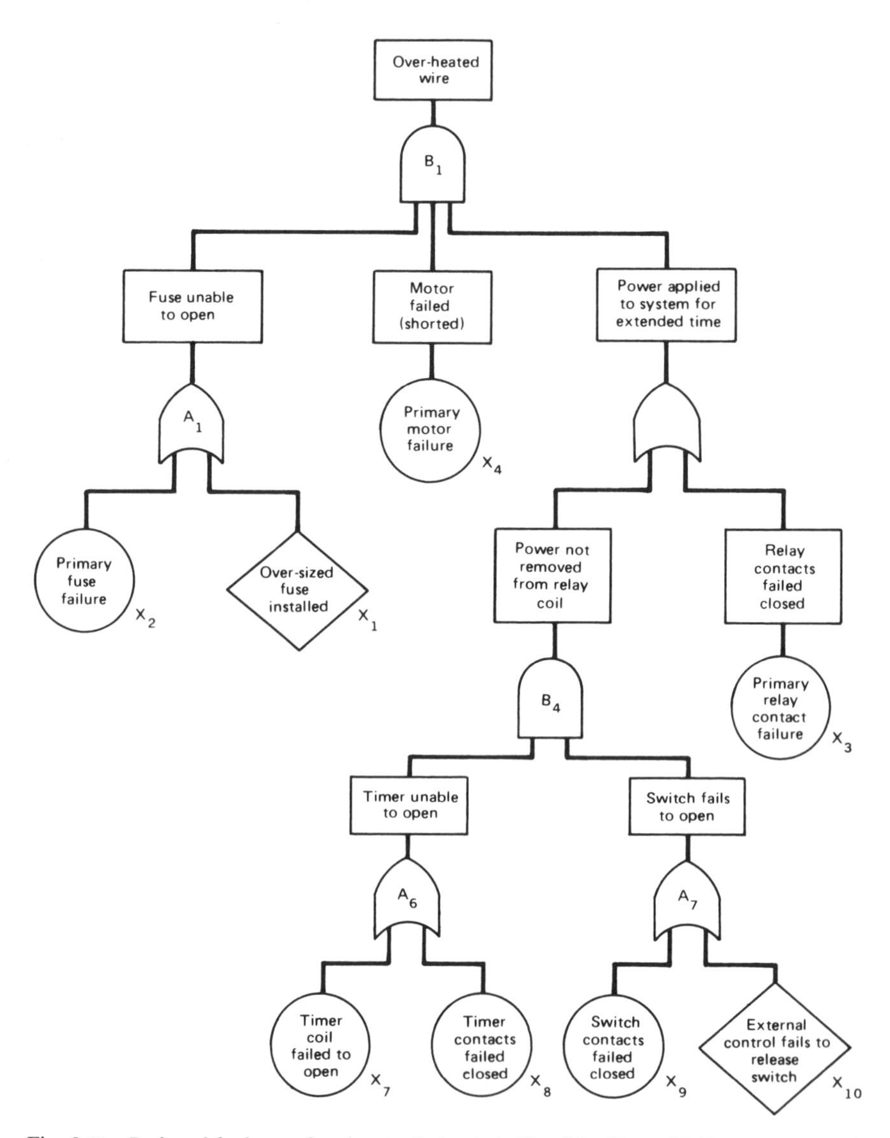

Fig. 8-11 *Reduced fault tree for electrical circuit in Fig. 8-9. (From W. Hammer, "Handbook of System and Product Safety," Prentice-Hall, Englewood Cliffs, New Jersey, 1972.)*

Since some of the events are identical, i.e.,

$$X_4 = X_5 = Y_1$$
$$X_3 = X_6 = Y_2, \tag{8-7}$$

it therefore follows from the idempotent law of Boolean algebra (Table 6-1) that

$$B_1 = (X_1 + X_2)\, X_4\, [X_3 + (X_7 + X_8)(X_9 + X_{10})]. \tag{8-8}$$

This Boolean expression can be used to redraw Fig. 8-10 into the reduced fault tree shown in Fig. 8-11.

8-5 LIGHT WATER REACTOR TRIP SYSTEM FAULT TREES

A light water reactor is designed to trip (or "scram") whenever an out-of-tolerance signal is received by an appropriate combination of sensors, not all of which need be of the same type. The number of sensors and the trip logic for a typical pressurized water reactor are given in Table 8-2.

The fault trees for failure of the reactor to trip become quite complex and depend upon the reactor-operating conditions at the time the trip is required. Here we shall consider only an example fault tree for failure of a sufficient number of pressurized water reactor control rods (more than 2 out of 48) to enter the core when a small loss-of-coolant accident (LOCA) would require a reactor trip or scram [7].

The trip system is defined to consist of control rods and their magnetic jack assemblies, breakers and motor generator sets that provide power to the magnetic jacks, and electronic logic. A simplified schematic diagram of the reactor protection system is shown in Fig. 8-12. This diagram is beneficial when one constructs the fault tree of complicated systems. Figure 8-12 shows only a portion of the signals that feed the two relay logic trains that can trip the reactor: a trip occurs for a 2-out-of-3 coincidence monitoring either of low reactor pressurizer pressure or of high reactor coolant over-temperature, or a scram results from coincidence of low pressurizer pressure and low pressurizer water level, or the Safety Injection Control System (SICS) initiates the trip.

The entire rod control system gets its power from the reactor trip circuit breakers that open upon a reactor scram and remove all power to the magnetic jacks. Without power, all magnetic jacks release the control rods and allow them to fall into the core unless prior mechanical damage restrains them.

The two reactor trip breakers RTA and RTB are each bypassed by a special test breaker of the same type as the trip breakers. The BYA-

Table 8-2 *Trojan Reactor Trip Sensors Used in Each of Two Relay Logic Trains*[a]

Reactor trip	Trip logic
1. Power range high neutron flux	2/4
2. Intermediate range neutron flux	1/2
3. Source range neutron flux	1/2
4. Power range high positive neutron flux rate	2/4
5. Power range high negative neutron flux rate	2/4
6. Over temperature ΔT	2/4
7. Over power ΔT	2/4
8. Pressurizer low pressure	2/4
9. Pressurizer high pressure	2/4
10. Pressurizer high water level	2/3
11. Low reactor coolant flow	2/3 per loop
12. Reactor coolant pump breakers open	2/4
13. Reactor coolant pump bus undervoltage	1/2 per bus
14. Reactor coolant pump bus underfrequency	1/2 per bus
15. Low feedwater flow	1/2 per loop
16. Low–low steam generator water level	2/3 per loop
17. Safety-injection signal	—
18. Turbine–generator trip	
(a) Low emergency trip fluid pressure	2/3
(b) Turbine stop valve closed	4/4
19. Manual	1/2

[a] From Portland General Electric, Atomic Energy Commission Docket 50-344, Vol. 5, Section 7.2 (1973).

bypass breaker is connected across RTA, and BYB-bypass is connected across RTB; both bypass breakers are normally open. The bypass breakers are used to test the operation of the trip breakers and for repair of the trip breakers whenever necessary; repair can be done without reactor shutdown.

The control rod assemblies are dropped by the opening of either reactor trip breaker (52/RTA) or a second breaker (52/RTB). Breaker 52/RTA is controlled by Reactor Protection System Logic Train A and Breaker 52/RTB is controlled by Train B. The two series-connected trip breakers RTA and RTB control power from two parallel-connected motor generator sets which are powered by 480-volt buses.

Figure 8-13 shows the reduced fault tree for failure of the reactor to trip following a small loss of coolant accident (LOCA). This "top-of-the-tree" includes possible failures other than failure of the trip system itself, such as distortion of the core so that the rods cannot drop. The failure to remove power to the trip bus because of failure, e.g., of Train A occurs if

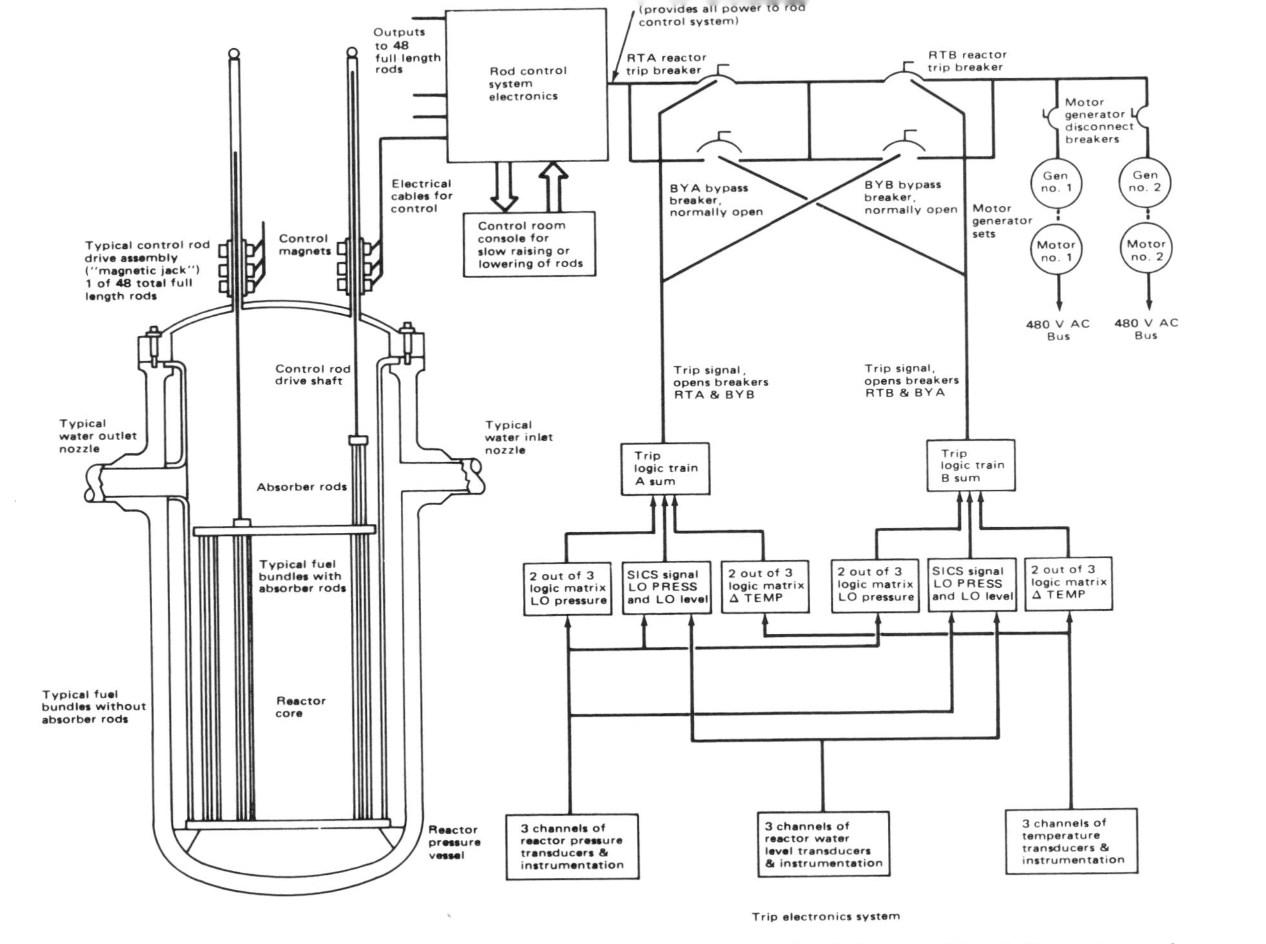

Fig. 8-12 *Reactor protection system simplified schematic diagram.* [*From J. E. Kelly et al., Electric Power Research Institute Rep. EPRI NP-265, Part II, Vol. 3* (1976).]

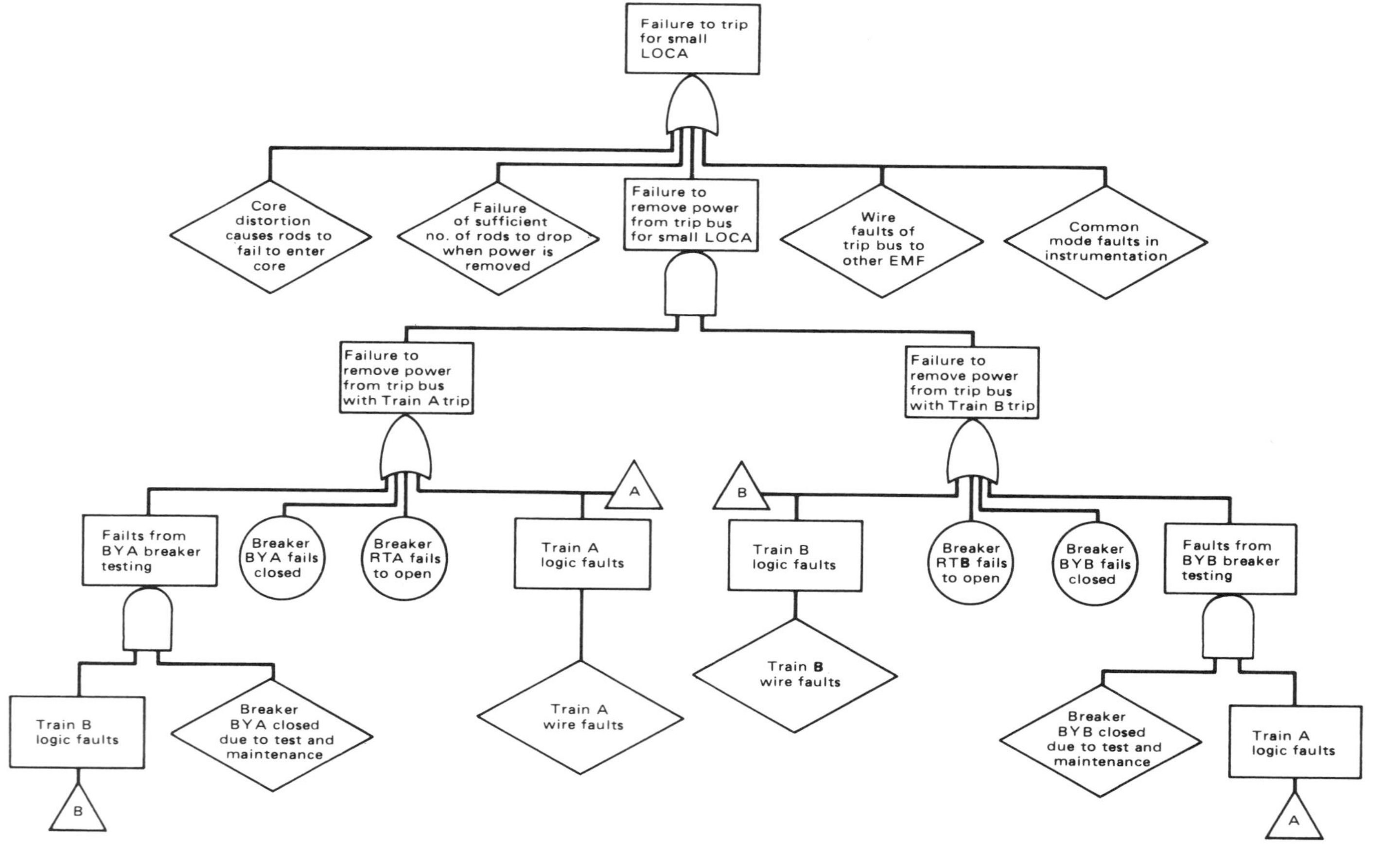

Fig. 8-13 *Reactor protection system reduced fault tree*. [*From J. E. Kelly et al., Electric Power Research Institute Report EPRI NP-265, Part II, Vol. 3, (1976).*]

the breaker RTA fails to open, if the bypass breaker BYA fails closed, or if there are failures in the logic train itself. Also considered are faults arising from the testing of the bypass breaker BYA.

8-6 LIGHT WATER REACTOR SAFETY FAULT TREES

The Reactor Safety Study (RSS) for United States commercial light water reactors [8] was directed at assessing the effectiveness of the Engineered Safety Features (ESFs) of a typical Boiling Water Reactor (BWR) and a Pressurized Water Reactor (PWR). Appendix E provides a brief introduction to light water reactor safety systems and gives some of the labeling used to develop fault trees for failure of the safety features to perform when needed.

The RSS required the construction of many fault trees. As an example, we consider the tree for failure of a PWR containment spray-injection system (CSIS) to operate following a reactor coolant system break. In this case, the top event of the fault tree becomes "Insufficient Fluid Flow Through CSIS Nozzles." To analyze this failure, a flow diagram as shown in Fig. 8-14 is needed.

The first page of the CSIS fault tree is shown in Fig. 8-15. There are two spray subsystems associated with the PWR, each of which is capable of supplying sufficient spray fluid to the containment. Since both of these spray subsystems must fail in order to fail the system, the second-level events, "Insufficient fluid from spray subsystem A header-nozzles" and "Insufficient fluid from spray subsystem B header-nozzles," are related to the top undesired event by an AND gate. The first of these second-level events can be caused by "spray header 8CS-22-153 ruptures," "Containment pressure sufficiently high to reduce spray effectiveness," "Spray subsystem A header-nozzles plugged," or "Insufficient fluid to header 8CS-22-153." In this manner, the tree of logical relationships among fault events is constructed until all components, including control and power components, are identified in terms of basic fault states or fault modes. Appendix F gives the subtrees of the fault tree in Fig. 8-15 to show this development.

The circle-within-a-diamond symbol on some fault trees in Appendix F indicates that a subtree was developed for the event, but the subtree was analyzed separately. That is, the subtree was found to be independent of other parts of the overall fault tree and therefore could be treated as a type of component (so the numerical results of the subtree can thus be applied as input data to the system tree).

Another observation about the fault trees in Appendix F is that, be-

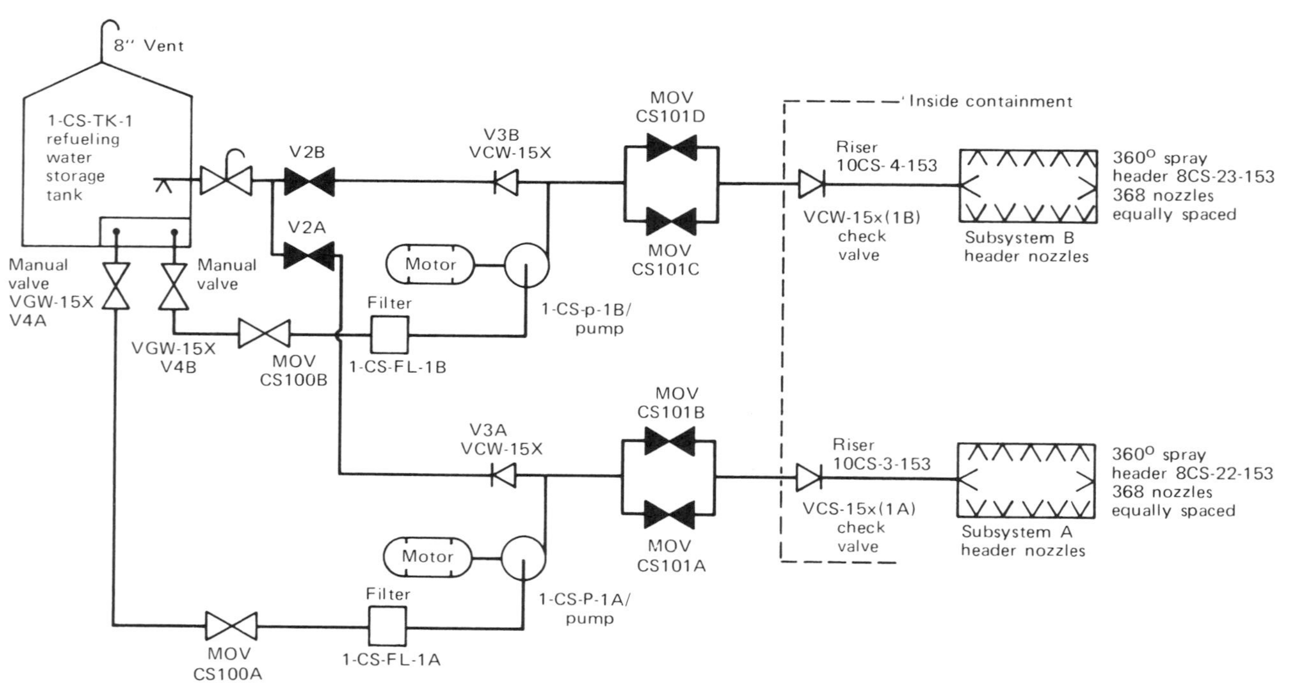

Fig. 8-14 *PWR containment spray injection system flow diagram.* [*From Reactor Safety Study, Appendix II. U.S. Nuclear Regulatory Commission Rep. WASH-1400, NUREG 75/014 (October 1975).*]

cause of the similarity between some subsystems, a single fault tree was drawn in Figs. F-3, F-4, and F-5 to represent similar subsystems. In these cases, multiple input and output transfers were used. For example, Fig. F-3 was developed for the event "Insufficient fluid from pipe between parallel valves and 4-inch test line" for CSIS subsystem A. Since subsystem B is the same as subsystem A, except for component names, two output transfers, C002 and C003, were used for that event.

The fault probability of a passive event (e.g., pipe rupture, wire shorts, normally closed switch contacts opening when not actuated, etc.) is usually considerably less than the fault probability of an active event (e.g., valve not closing, pump not starting, contacts not closing, operator error, etc.). The significance of a passive event becomes even smaller when it must coexist with other active and passive events to produce system failure. In order to make the calculations tractable and to keep the significant events consistent with the resolution of the data available, the fault trees of the RSS were reduced before the final, detailed calculations were performed. The procedures and rules used in reducing the detailed trees in the RSS were [8]:

(a) All single passive events were retained on the reduced tree for quantification. Higher-order passive events (i.e., those that must coexist with other passive or active events) were discarded except in those cases of potential common cause failures or when single events were nonexistent or did not clearly dominate the system failure probability.

(b) All single active and double active events were retained. A number of higher-order active events were retained involving pumps, diesel generators, and other components where they were known to have relatively high failure probabilities when compared with other active events.

(c) Where necessary, several events identified on the detailed tree were coalesced into a single event on the reduced tree. For example, a fluid system may contain pumps, valves, and other components connected in series by segments of pipe; on a detailed tree the pipe segments might appear as individual events, while on the reduced tree the individual pipe segments would be lumped into one event. This combining of events is necessary for the events to be consistent with the level of detail in the available data.

A few systems could not be reduced in their entirety according to the foregoing rules. For example, the BWR scram system contains no single passive faults and no double active faults. The rules for this system were changed so that double passive faults and triple active faults were retained and then the dominant contributions could be obtained.

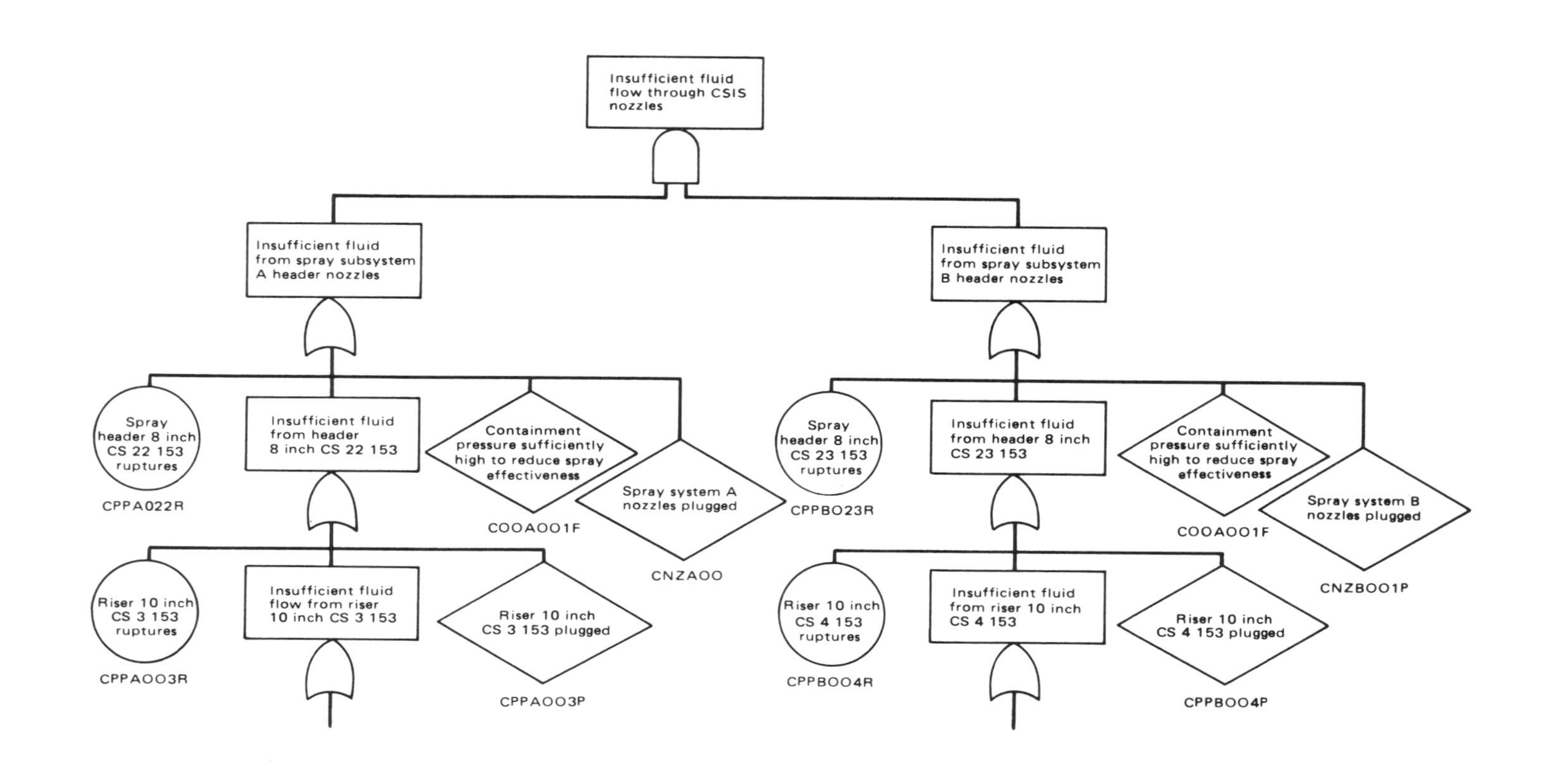

Insufficient fluid flow through CSIS nozzles
Insufficient fluid from spray subsystem A header nozzles
Insufficient fluid from spray subsystem B header nozzles
Spray header 8 inch CS 22 153 ruptures
CPPA022R
Insufficient fluid from header 8 inch CS 22 153
Containment pressure sufficiently high to reduce spray effectiveness
COOAOO1F
Spray system A nozzles plugged
CNZAOO
Riser 10 inch CS 3 153 ruptures
CPPAOO3R
Insufficient fluid flow from riser 10 inch CS 3 153
Riser 10 inch CS 3 153 plugged
CPPAOO3P
Spray header 8 inch CS 23 153 ruptures
CPPBO23R
Insufficient fluid from header 8 inch CS 23 153
Containment pressure sufficiently high to reduce spray effectiveness
COOAOO1F
Spray system B nozzles plugged
CNZBOO1P
Riser 10 inch CS 4 153 ruptures
CPPBOO4R
Insufficient fluid from riser 10 inch CS 4 153
Riser 10 inch CS 4 153 plugged
CPPBOO4P

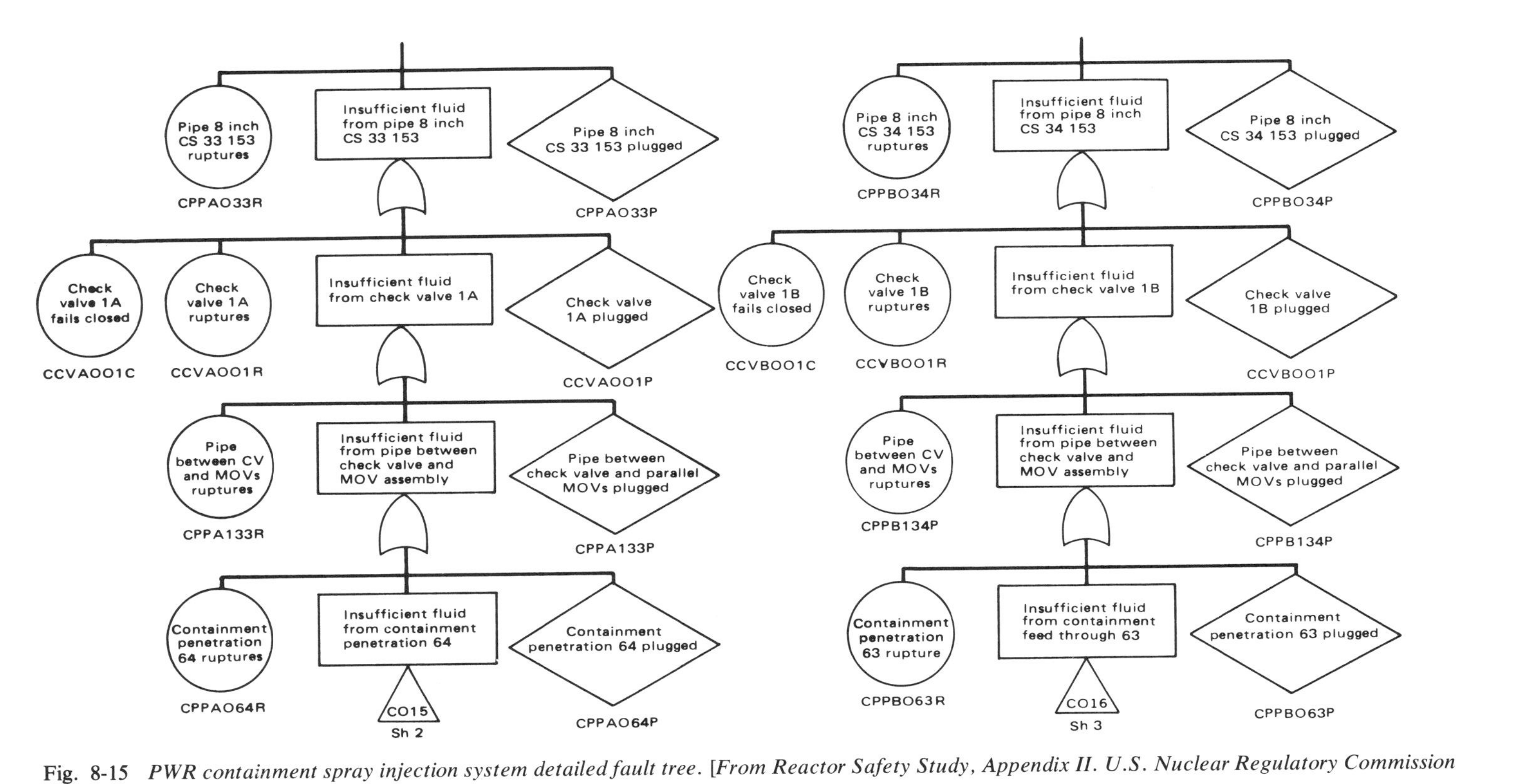

Fig. 8-15 *PWR containment spray injection system detailed fault tree*. [*From Reactor Safety Study, Appendix II. U.S. Nuclear Regulatory Commission Rep. WASH-1400, NUREG 75/014* (*October 1975*).]

8-7 SPENT NUCLEAR FUEL TRUCK TRANSPORT FAULT TREES

Fault trees have been developed for the release of radioactive material from truck casks used to ship spent fuel assemblies [9]. The fault trees were used to help evaluate the risks from shipping spent nuclear fuel to interim storage facilities in the United States (see Section 14-2). The analysis considered the combined effects of the truck accident environment and the packaging condition; the effects of sabotage or theft were not considered.

All significant releases of radioactivity from a water-filled cask would have to breach both the cask and the cladding of the rods in the fuel assemblies. The cask components that could fail during an accident are the wall, lid, closure seal, drain valve, pressure relief valve, and vent valve. Lesser amounts of radioactivity could be released from small leaks of the cask coolant water (because of closure errors). Construction of a fault tree for the release of radioactive material to environment during spent fuel truck transport is left as Exercise 8-7; the interested reader may also wish to consult the detailed trees of ref. 9.

Fault trees for spent fuel transport are needed in an analysis of the risks from such transport. These risks are considered in Chapter 14.

8-8 GEOLOGIC WASTE DISPOSAL FAULT TREES

For the release of geologically stored waste, normally we are concerned only with the movement of the waste into our environment, although it is also possible to treat the case where our human environment expands to interact with the waste (as in mining or exploration operations). Two possible pathways for release are failure of man-made environmental seals and of natural geologic barriers, and fault trees have been constructed for both [10].

In order for a man-made seal to fail, it must be ruptured and not be detected or repaired properly, or the seal must be incorrectly installed or fabricated. Also, for a release to occur following failure of the seal, there must be a failure to prevent active transport mechanisms from causing flow through the failed pathway; some of the possible mechanisms for this transport are ground water, oil, or molten (volcanic) rock. Construction of a fault tree for the release of geologically stored waste from a man-made environmental seal covering a shaft is left as Exercise 8-8; the interested reader may also wish to consult the more extensive tree developed in [10].

Fault trees for the failure of systems to isolate radioactive wastes in interim storage or permanent disposal are useful in the analysis of the risks from such storage. Such risks are considered in Chapter 15.

EXERCISES

8-1 Construct a fault tree for the failure of a two-tube fluorescent fixture to emit light. The events which should be considered are human error H (the switch is in the "off" position, rather than the "on" position), E (external electrical power outage), S (switch failure in the "off" position), and T_i (failure of tube i). Also include primary events for failure of the wiring between the switch and tubes (W_{st}) and between the power supply and the switch (W_{es}).

8-2 Consider the combustion engine shown in the figure. (a) Construct a fault tree for the top event "low cylinder compression" (LCC) in terms of head gasket leaks (HG), spark plug leaks (SP), piston hole (P), intake valve leaks (IV), exhaust valve leaks (EV), and piston ring leaks (PR1 and PR2). [This tree was published by R. R. Fullwood and R. C. Erdmann in *Proc. Fast Reactor Safety Meeting,* p. 1493, U.S. Atomic Energy Commission Rep. CONF-740401-P3 (1974)]. (b) Obtain the reduced Boolean algebraic expression for LCC.

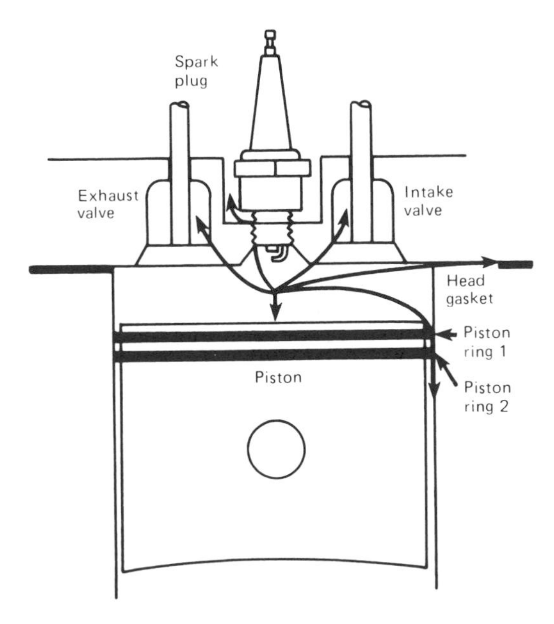

8-3 Construct fault trees for failure of systems a through d to operate and obtain the minimal cut sets.

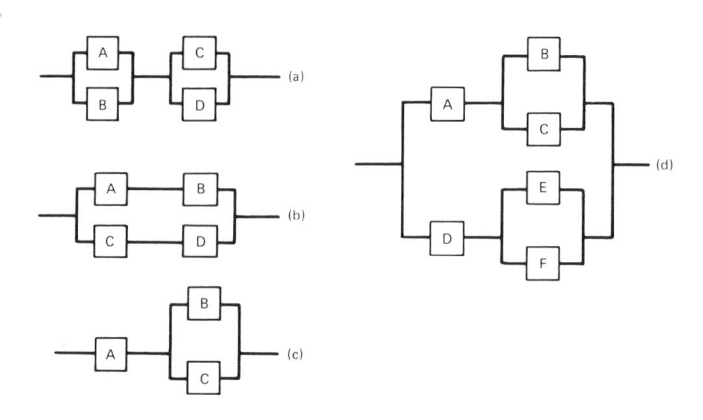

8-4 You are given a system of switches connected as shown in the figure.

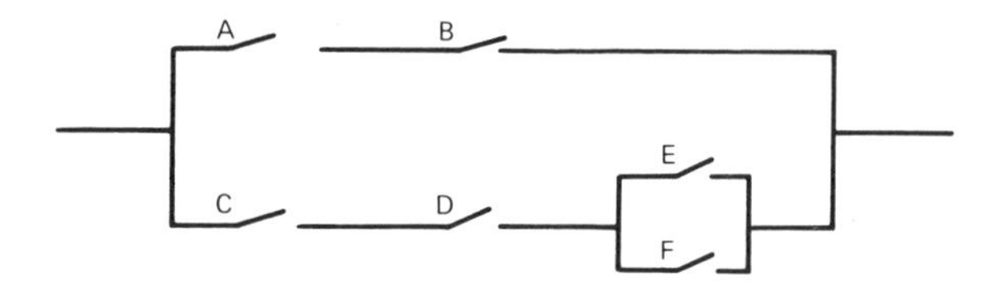

The probability per demand that a switch fails in the closed position is 10^{-4} and in the open position is 10^{-3}. There are no other causes of failure. (a) Construct a fault tree for the top event "T_C, the circuit is closed when it should be open." (b) Identify the minimal cut sets. (c) Evaluate the probability of the top event T_C. (d) Construct a second fault tree for the top event "T_O, the circuit is open when it should be closed." (e) Identify the minimal cut sets. (f) Evaluate the probability of the top event T_O.

8-5 In the flow system shown in the figure, (a) construct a fault tree for failure of flow, (b) determine the minimal cut sets, and (c) calculate the failure probability of the system if each component has a failure probability of 10^{-3}.

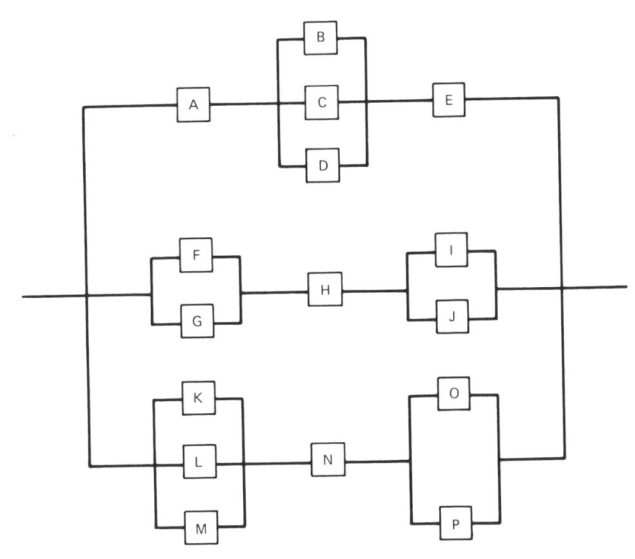

8-6 Construct any simple fault tree for top event T in terms of two types of gates G_n [$n = 1$ to N are AND gates and $n = (N + 1)$ to M are OR gates] and basic fault events E_1, E_2, etc. Use your tree and De Morgan's Theorems in Table 6-1 to show that the corresponding *success* tree for top event $\bar{T}$ can be constructed from the same tree structure, provided the basic events are success events $\bar{E}_1$, $\bar{E}_2$, etc., and the gates $n = 1$ to N are changed to OR gates, and those for $n = (N + 1)$ to M are changed to AND gates.

8-7 Construct a fault tree for "release of radioactive material to environment during spent fuel truck transport," as described in Section 8-7.

8-8 Construct a fault tree for release of geologically stored waste from a man-made environmental seal covering a shaft, as described in Section 8-8.

8-9 A certain reactor has three coolant loops, each with two pumps in parallel. Only one loop is needed to cool the core and a loop can fail if one pump fails or the pipe between the two pumps and core ruptures. Failure of the external and internal system power supplies to the loops also will cause system failure. Construct a fault tree for failure to cool the core.

8-10 For a certain submarine reactor safety circuit, there are six reactor period meters connected to three sensor systems. Each sensor system is wired by connecting two of the period meters in parallel and is designed so that it is activated whenever the reactor power level is increasing with a period of less than 10 sec. A sensor can fail either by failure of the meters or wiring, or by failure of its own power source. An automatic scram of the reactor is activated if two out of three sensor systems are activated and powered. (a) Construct the fault tree to compute the failure of activation of the automatic scram system when the reactor period is less than 10 sec. (b) The hazard rate for the reactor period meter is 10^{-2}/month, for the wiring is 10^{-6}/month, and for the external power source is 10^{-4}/month. Calculate the unavailability of the automatic period scram system for a 10-month period using the rare-event approximation. (c) Repeat part (b) without using the rare-event approximation.

8-11 Consider the domestic hot water system of Example D-1 of Appendix D.
(a) Construct a fault tree for the top event "rupture of water tank when system is in operation and faucet is closed." You should consider the following:

1. Tank failure due to rupture or damage during installation and shipping.
2. Pressure relief valve failures due to its being jammed closed, or to piping to and/or from the valve being blocked, or to the use of an incorrect size of valve or piping or to incorrect installation (human error), or to an improper setting of the relief valve (human error).
3. Excess inlet water supply pressure due to the normal pressure of the water supply exceeding the design pressure limit of tank (human error), or to the gas valve being jammed open, or to the gas valve being damaged during installation and shipping, or to the controller failing open, or to the temperature measuring and comparing device failing, or to the temperature measuring and comparing device and controller being disconnected from tank (human error).

(You may wish to compare your tree to that in [3]).
(b) Evaluate the probability of the top event occurring during the first year of operation if:

1. all human errors have a probability of 1×10^{-3}/demand,
2. all damage during installation and shipping has a probability of 2×10^{-3}/demand,
3. blockage of pipe has a probability of 3×10^{-4}/yr,
4. rupture of tank has a probability of 3×10^{-2}/yr,
5. relief valve fails closed has a probability of 1×10^{-5}/day,
6. controller failure has a probability of 2×10^{-2}/yr,
7. gas valve fails open has a probability of 1×10^{-3}/yr, and

8. temperature measuring and comparing device fails has a probability of 1×10^{-2}/yr.

(c) What failure probability most strongly affects your result from part (b)?

REFERENCES

1. Reliability Manual for Liquid Metal Fast Breeder Reactors (LMFBR) Safety Program. General Electric Company Internal Rep. SRD-74-113 (1974).
2. W. E. Vesely, F. F. Goldberg, N. H. Roberts and D. F. Haasl, Fault Tree Handbook. U.S. Nuclear Regulatory Commission Rep. NUREG-0492 (1981).
3. H. E. Lambert, System Safety Analysis and Fault Tree Analysis. Lawrence Livermore Laboratory Rep. UCID-16238 (1973).
4. J. A. Burgess, Spotting trouble before it happens, *Machine Design* **42,** No. 23, 150 (1970).
5. J. B. Fussell, Fault tree analysis—concepts and techniques, *in* "Generic Techniques in Systems Reliability Assessment," (E. J. Henley and J. W. Lynn, eds.), pp. 133–162. Noordhoff, Leyden, 1976.
6. W. Hammer, "Handbook of System and Product Safety," pp. 242–244. Prentice-Hall, Englewood Cliffs, New Jersey, 1972.
7. J. E. Kelly, R. C. Erdmann, F. L. Leverenz, and E. T. Rumble, ATWS: A Reappraisal. Part II, Evaluation of Societal Risks due to Reactor Protection System Failure, Vol. 3 PWR Risk Analysis. Electric Power Research Institute Rep. EPRI NP-265 (August 1976).
8. Reactor Safety Study, An Assessment of Accident Risks in U.S. Nuclear Power Plants. U.S. Nuclear Regulatory Commission Rep. WASH-1400, NUREG 75/014 (October 1975).
9. H. K. Elder et al., An Assessment of the Risk of Transporting Spent Nuclear Fuel by Truck. Battelle Pacific Northwest Laboratories Rep. PNL-2588 (1978).
10. K. J. Schneider and A. M. Platt (eds.), High-Level Radioactive Waste Management Alternatives. Battelle Pacific Northwest Laboratories Rep. BNWL-1900, Vol. 1 (May 1974).

CHAPTER 9

Event Tree Analysis

9-1 EVENT TREE CONSTRUCTION

Event trees and decision trees are inductive logic methods for identifying the various possible outcomes of a given initiating event, but they differ depending upon whether human control can influence the outcomes (as in decision trees) or whether the outcomes depend only upon the laws of science (as in event trees). The initiating event in a decision tree typically is a particular business or risk acceptance decision, and the various outcomes depend upon subsequent decisions [1, 2]. In risk analysis applications, the initiating event of an event tree is typically a system failure, and the subsequent events are determined by the system characteristics [3].

An event tree begins with a defined accident-initiating event. This event could arise from failure of a system component, or it could be initiated externally to the system. Different event trees must be constructed and evaluated to analyze a set of accidents.

Once an initiating event is defined, all the safety systems that can be utilized after the accident must be defined and identified. These safety systems are then structured in the form of headings for the event tree. This is illustrated in Fig. 9-1 for two safety systems that can be involved after the defined initiating event has occurred.

Once the systems for a given initiating event have been identified, the set of possible failure and success states for each system must be defined and enumerated. Careful effort is required in defining success and failure states for the systems to ensure that potential failure states are not included in the success definitions; much of this analysis is done with the fault tree technique discussed in Chapter 8. If bifurcation (two-state) modeling is employed, for example, then one failed state and one success state are defined for each system, and each gives a branch of the tree; if a greater number of discrete states are defined for each system (such as

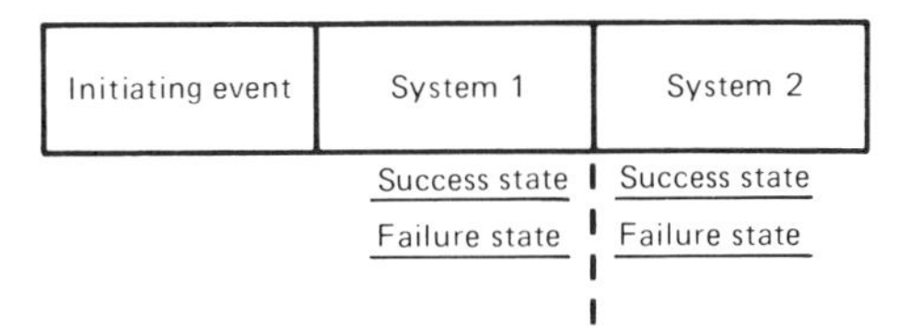

Fig. 9-1 *System state definitions for system 1 and system 2.* [From *Reactor Safety Study. U. S. Nuclear Regulatory Commission Rep. WASH-1400, NUREG 75-014* (*October 1975*).]

would be used when including partial failures), then a branch must be included for each state.

Once the system failure and success states have been properly defined, the states are then combined through the decision-tree branching logic to obtain the various accident sequences that are associated with the given initiating event. As illustrated in Fig. 9-2, the initiating event is depicted by the initial horizontal line and the system states are then connected in a stepwise, branching fashion; system success and failure states have been denoted by S and F, respectively. The format illustrated follows the standard tree structure characteristic of event tree methodology, although sometimes the fault states are located above the success states.

The accident sequences that result from the tree structure are shown in the last column of Fig. 9-2. Each branch of the tree yields one particular accident sequence; for example, IS_1F_2 denotes the accident sequence in which the initiating event (I) occurs, system 1 is called upon and succeeds (S_1), and system 2 is called upon but is in a failed state so that it does not perform its defined function. For larger event trees, this stepwise branching would simply be continued.

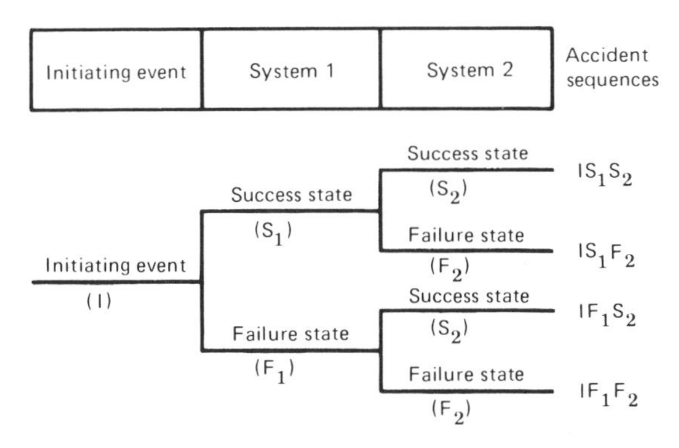

Fig. 9-2 *Illustration of event tree branching.* [From *Reactor Safety Study. U. S. Nuclear Regulatory Commission Rep. WASH-1400, NUREG 75/014* (*October 1975*).]

The system states on a given branch of the event tree are conditional on the previous states having already occurred. For example, in Fig. 9-2, the success and failure of system 1 must be defined under the condition that the initiating event has occurred; likewise, in the upper branch of the tree corresponding to system 1 success, the success and failure of system 2 must be defined under the conditions that the initiating event has occurred and system 1 has succeeded.

It is possible to "prune" an event tree by eliminating all the branches that have a zero conditional probability for at least one event. For example, in Fig. 9-2 if the failure of system 1 caused system 2 to fail, then instead of considering the accident sequences IF_1S_2 and IF_1F_2, we would consider only the sequence IF_1. A summary of the basic steps to be followed in constructing an event tree is in Fig. 9-3 [4].

A major concern in event tree construction involves accounting for the

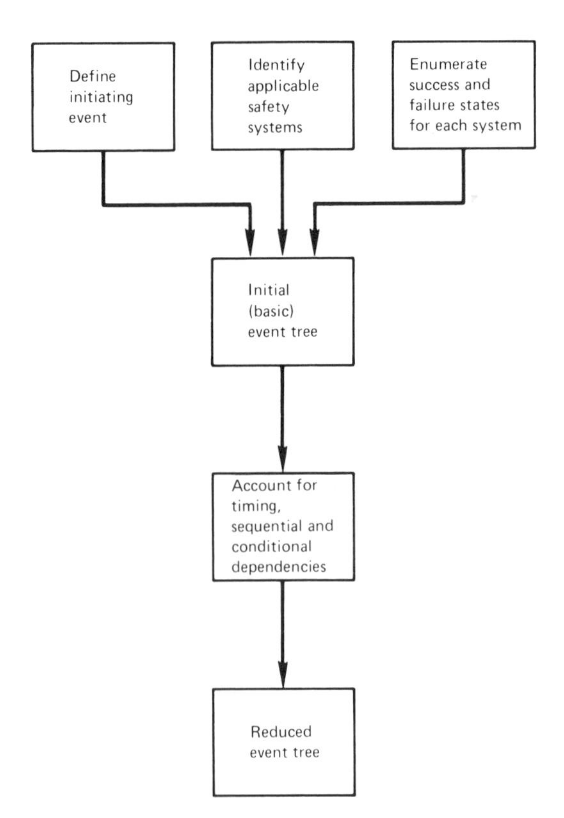

Fig. 9-3 *Steps in construction of an event tree.* [From *R. C. Erdmann (ed.), A Risk Methodology Presentation. Electric Power Research Institute Informal Rep. NP-79-1-LD (January 1979).*]

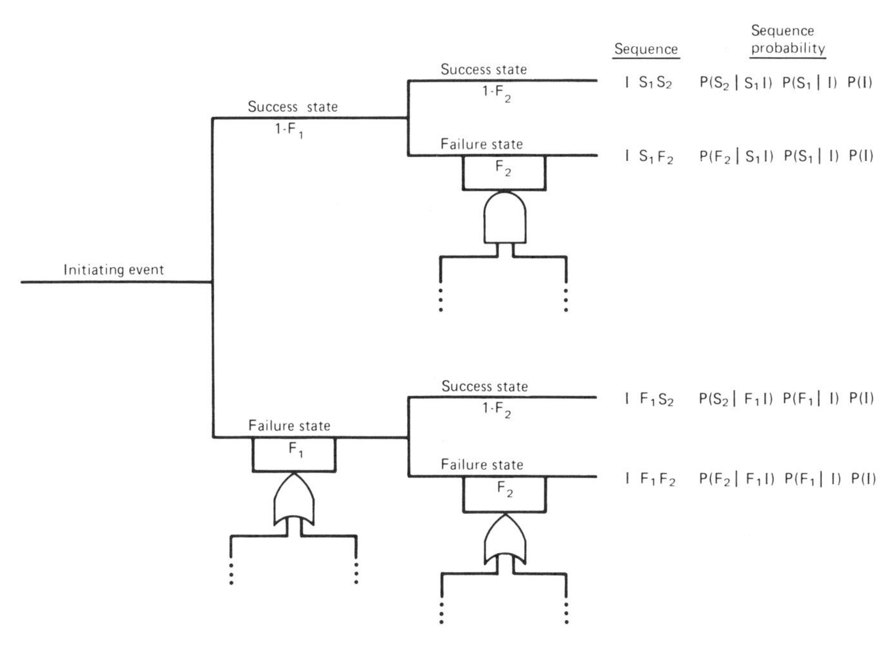

Fig. 9-4 *Schematic of an event tree shown with fault trees used to evaluate probabilities of different events.*

timing of the events. In some instances, the failure logic changes depending upon the time at which the events take place; such a case occurs, for example, in the operation of emergency core cooling systems in nuclear plants. Then *phased mission* analysis techniques are needed to model the system that changes during the accident, even though the safety-system components remain the same [5, 6].

Once the final event tree has been constructed so that the results associated with each accident sequence have been defined, the final task is to compute the probabilities of system failure. Fault tree analyses are used to calculate the *conditional* probabilities needed for each branch of the event tree. Multiplication of the conditional probabilities for each branch in a sequence gives the probability of that sequence, as shown in Fig. 9-4.

9-2 EVENT TREES FOR REACTOR SAFETY ANALYSIS

9-2-1 INTRODUCTION

In the Reactor Safety Study (RSS), the application of event trees was limited to the analysis of potential accidents involving the reactor core [7].

It was found to be convenient to separate the event trees into two types: one type was for assessing how potential accidents were affected by failures in major systems, particularly the engineered safety systems described in Appendix E, and the other type was for accident sequences leading to the release of radioactivity from the containment. The event trees for the failures of engineered safety systems covered all significant loss of coolant accidents (as initiated by pipe ruptures, for example) and reactor transient initiating events (such as a turbine trip or loss of main feedwater pumps).

The application of event trees is illustrated by an example for which the initiating event is a large pipe break in the primary system of a reactor. The first step in developing the event tree for this loss of coolant accident is to determine which systems might affect the subsequent course of events. In Fig. 9-5, these are station electric power, the emergency core cooling system, the radioactivity removal system, and the containment system. From a knowledge of these systems it is possible to order them (across the top in Fig. 9-5) in the time sequence in which they are expected to influence the events in the reactor.

If there is no information about the way in which the initiating event could propagate, as in the *basic* tree of Fig. 9-5, then there are $2^{(n-1)}$ accident sequences for the n functional headings. (For example, $n = 5$ in Fig. 9-5 and there are 16 accident sequences in the basic tree.) Note that only binary outcome events, success or failure, have been utilized in the figure; if triple outcome events were possible (such as success, partial success, and failure) then $3^{(n-1)}$ accident sequences could arise in the trifurcation event tree.

In the RSS, event trees were developed in which each branch point provided only two options—system success or system failure; the upper branches of each figure represented successes and the lower branches represented failures of a system to perform its intended function. No consideration was given to the fact that partial system success might occur within an accident sequence. Thus, a sequence of events was conservatively assumed to lead to total core melt or no core melt, but never to a partial core melt. Similarly, because of the uncertainty in calculating the effects of partial system failure, the RSS treated all questionable cases of system operability as complete system failures.

Fortunately there are constraints among different system functions that tend to reduce the number of possible accident sequences. Such constraints arise whenever the probability of a branch occurring vanishes, in which case, that branch and all following sub-branches connected to it need not be displayed. The constraints are due to functional and hardware relationships; for example, if station electric power fails none of the other

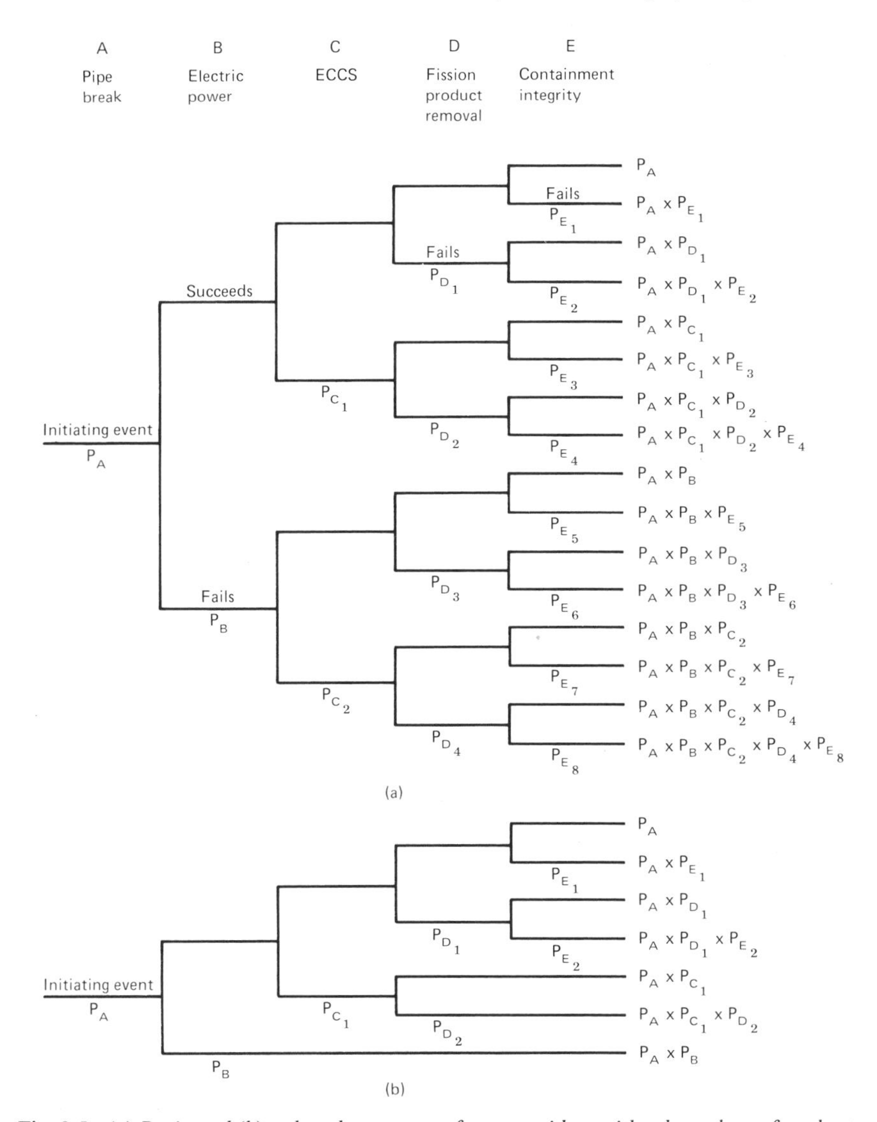

Fig. 9-5 *(a) Basic and (b) reduced event trees for an accident with a large loss of coolant. Note that since the probability of failure* P *is generally less than 0.1, the probability of success (1 − P) is always close to 1. Thus, the probability associated with the upper (success) branches in the tree is assumed to be 1.* [From *Reactor Safety Study, U. S. Nuclear Regulatory Commission Rep. WASH-1400, NUREG 75/014 (October 1975).*]

systems can operate because they depend upon power. Functional relationships were used to prune the basic tree results into a reduced tree, shown in Fig. 9-5(b), with only about half as many accident sequences.

The probability of failure of each system in Fig. 9-5 is indicated by the P values noted. The probability of success is $(1 - P)$ since it is assumed that a system is successful if it does not fail; since the failure probabilities are almost always 0.1 or less, it was the practice in the RSS to make the conservative approximation $(1 - P) \approx 1$, as shown in Fig. 9-5.

The probability of occurrence of events in a sequence is the product of the conditional probabilities of the individual events in that chain. In Fig. 9-5, the probability of occurrence of each system failure is shown to be different in each accident sequence in which it appears (P_{C_1} and P_{C_2}, e.g.); this accounts for differences that may arise due to differing dependencies in each accident sequence. If the successive events in a sequence are independent, then the probability of a sequence is the product of unconditional probabilities of the individual events (so that $P_{C_1} = P_{C_2}$, for example).

Once an event tree had been constructed with sufficiently detailed information, the series of events in each RSS accident sequence were defined well enough so that it was possible to calculate the consequences for that particular series of events. For example, the bottom sequence in Fig. 9-5(b), where no core cooling would be available, would result in melting of the core. The molten core would violate the containment, so the accident could produce a release of radioactivity outside of the containment. The mode of containment failure would affect the overall probability of the sequences, however, as well as the magnitude of the release.

In the RSS, event trees first were utilized to identify the many possible significant accident sequences. Then, through an iterative process involving successive improvements in the definition of failure probabilities, the incorporation of system interactions, and the resolution of physical process descriptions, the trees enabled identification of those accident sequences for which the radioactivity releases were important. The probability for each important radioactivity release sequence was obtained by multiplying the initiating event probability by the system and containment failure probabilities for the sequence.

9-2-2 Event Trees for Loss of Coolant Accidents

This section illustrates the evolutionary process used to construct an event tree for the Loss of Coolant Accident (LOCA). The Engineered Safety Features (ESF) to mitigate such an accident are discussed in Appendix E. The reader may wish to review this material to again become familiar with the different systems and with the notation used (see especially the glossary of terms in Table E-1).

The drawing of the event tree in Fig. 9-6 for a LOCA is started by indicating the system functions RT, ECC, PARR, PAHR, and CI, as well as the initiating pipe break event (PB), as event tree headings [8, 9]. The rationale for the ordering of events is as follows:

(a) RT is listed first because failure to shut down the fission process during a LOCA could result in high core temperatures and thus nullify the effectiveness of ECC even if cooling water were provided.

(b) ECC is listed next because cooling determines whether the core will melt. If it does not, the consequences of pipe break will be very small, but if the core does melt, the potential consequences can be great and are strongly affected by PARR, PAHR, and CI.

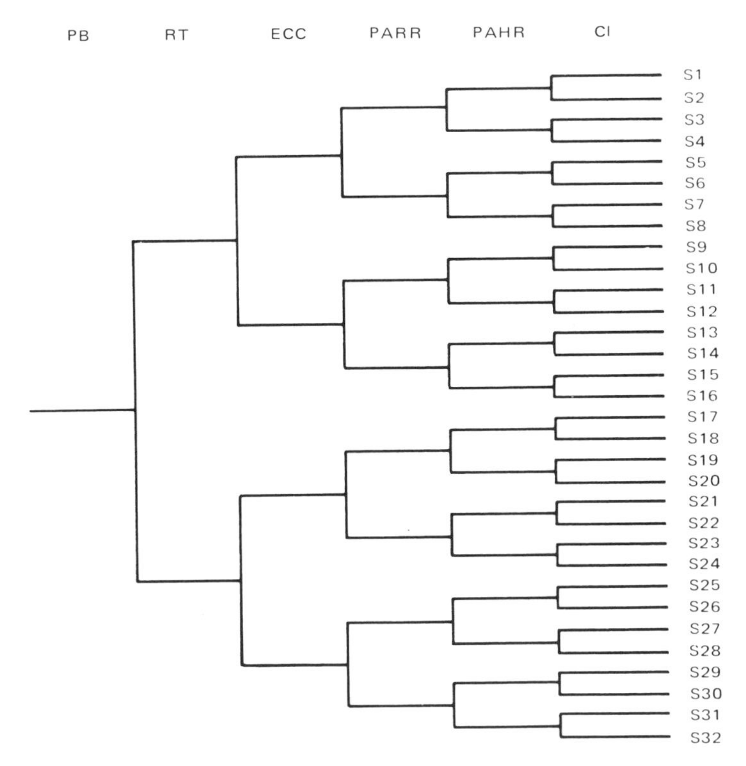

Fig. 9-6 *Preliminary event tree for* LOCA *functions*. [From *Reactor Safety Study, Appendix* I. *U. S. Nuclear Regulatory Commission Rep. WASH-1400, NUREG 75/014* (*October 1975*).]

(c) PARR comes after ECC because its function is to remove any radioactivity released from the fuel into the containment.

(d) PAHR is put just before CI because the containment has failure modes that depend on the performance of PAHR (as well as on ECC).

The preliminary event tree in Fig. 9-6 does not account for the following:

(a) The time-dependent performance requirements for the physical systems needed to perform the various ESF functions.

(b) ESF functional interrelationships such that failure of one function eliminates the need for another.

(c) ESF functional failures producing physical processes that cause other functions to fail.

(d) The effect of accident characteristics such as pipe-break size and location on the event tree and on the operability requirements for the systems providing ECC.

A consideration of the ECC shows that the performance requirements change greatly with time. The initial few minutes after pipe break are most demanding because ECC must provide high flow rates at relatively high pressures to refill the core in the presence of blowdown of the reactor coolant system and at high levels of core decay heat. Once the core has been reflooded with cold water, the flow and pressure requirements are much reduced. The configuration of the physical systems required for successful performance also changes significantly with time.

It is therefore convenient to separate ECC into two distinct time phases: an emergency cooling injection mode (designated ECI) to cover the initial period, and a long-term recirculation mode (ECR) for the rest of the time, as in Fig. 9-7. ECR is located between PAHR and CI on the tree because failure of PAHR causes failure of ECR, and failure of ECR causes failure of CI.

Because the RT system in Fig. 9-6 has been shown to be unnecessary in some instances, it has been eliminated from the tree of Fig. 9-7, so that the total number of sequences is still 32 after replacement of ECC by ECI and ECR. (This step is taken here merely to simplify the illustrations and later discussions.)

One type of functional interrelationship is such that a failure in one function eliminates the need for another. For instance, if ECI fails, the core will melt and whether ECR functions becomes unimportant since it can no longer prevent the melt. The effect of this on the event tree is to eliminate the ECR choices in sequences S17 to S32 of Fig. 9-7 and to produce sequences S17 to S24 of Fig. 9-8.

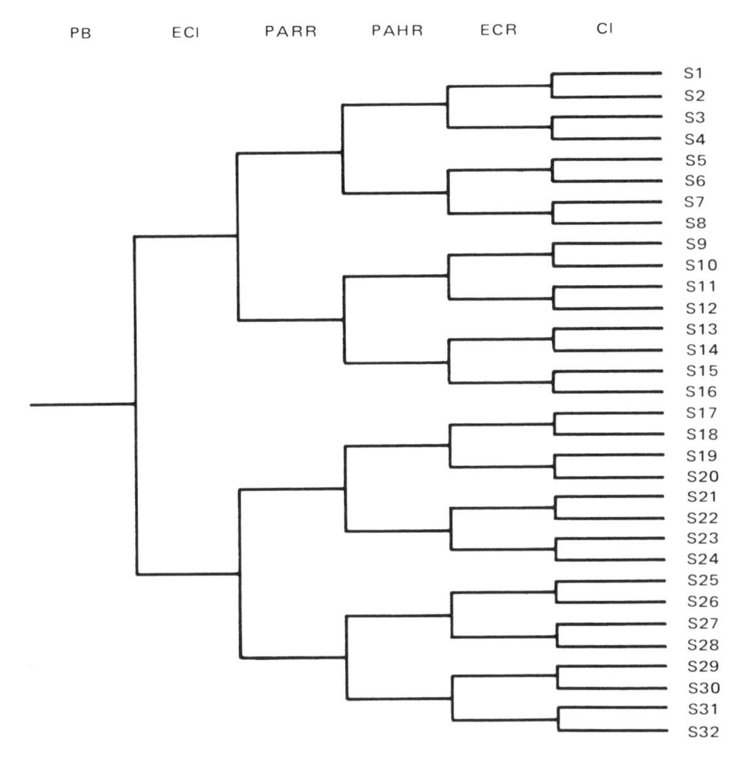

Fig. 9-7 *Preliminary event tree with* RT *removed and* ECC *replaced by* ECI *and* ECR. [From *Reactor Safety Study, Appendix* I, *U. S. Nuclear Regulatory Commission Rep. WASH-1400, NUREG 75/014* (*October 1975*).]

Sequences S1 to S16 of Fig. 9-8 assume availability of ECI and therefore no core melt in the initial period. However, if ECR were to fail, the core would melt and CI would fail; therefore, no CI alternatives should be shown where ECR has failed. If PAHR were to fail, then CI and ECR would both ultimately fail; the first because of overpressure and the second because the ECC water would get hot enough to cause pump cavitation. Thus, with PAHR failed, no ECR or CI alternatives should be shown, and sequences S1 to S16 of Fig. 9-8 reduce to sequences S1 to S8 of Fig. 9-9.

A second type of functional interrelationship is such that the failure of a function causes other functional failures due to physical processes taking place. In general, these physical processes introduce a time delay between failures that can be important in assessing consequences. In Fig. 9-8, sequences S17 to S24 involve ECI failure. Without ECI, the core will melt

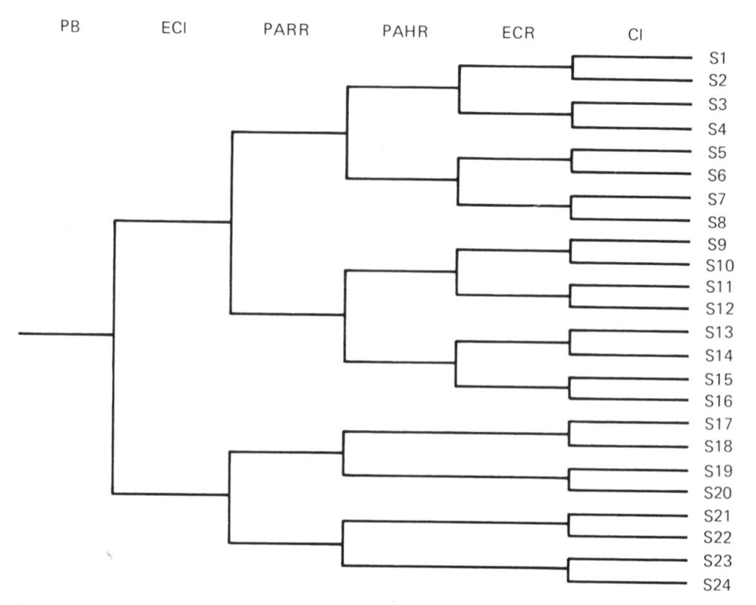

Fig. 9-8 *Reduced* LOCA *event tree showing* ECI–ECR *functional interrelationship.* [From *Reactor Safety Study, Appendix* I. *U. S. Nuclear Regulatory Commission Rep. WASH-1400, NUREG 75/014* (*October 1975*).]

and CI will surely fail; therefore, no CI choices should be shown in these sequences so S17 to S24 of Fig. 9-8 reduce to sequences S9 to S12 of Fig. 9-9.

A third type of interrelationship between the operability of one system and that of other systems is called an operability–operability interrelation-

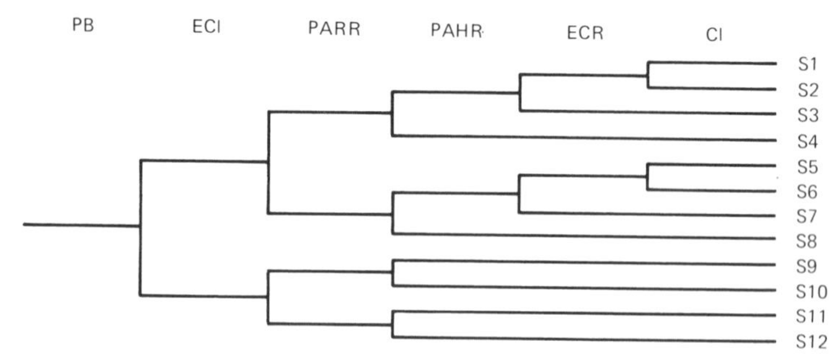

Fig. 9-9 *Reduced* LOCA *event tree showing more functional interrelationships.* [From *Reactor Safety Study, Appendix* I. *U. S. Nuclear Regulatory Commission Rep. WASH-1400, NUREG 75/014* (*October 1975*).]

ship [8]. An example of such an interrelationship involves Electric Power (EP). The ESFs that perform ECI, PARR, PAHR, and ECR functions all depend on the availability of EP. If EP were to fail, all the ESF systems associated with the above-mentioned functions would fail. Unavailability of EP clearly represents a common cause failure of great importance. (The off-site electric power systems of nuclear power plants are therefore backed up with diesel-generator power systems.) Because of its importance, EP availability should appear explicitly on the LOCA event tree; this is shown in Fig. 9-10, which is the same as Fig. 9-9 except that EP failure has been added as sequence S13.

Figure 9-10 is a functional event tree directly applicable to a large LOCA in a PWR since the RT does not, in general, affect the course of a LOCA. (This is because the ECC water is borated and therefore prevents the fission process from restarting even if RT fails.) However, if the pipe break is small, the blowdown is slow and the fission process can generate significant heat after normal heat removal has stopped. In such a small LOCA event, RT is therefore needed since it prevents the core from overheating during plant blowdown. Since RT occurs automatically with loss of EP in both types of reactors, it is proper to place the RT branch after EP, as shown in Fig. 9-11; the added sequences are S13 to S16, in which the success and failure paths for ECI and ECR are omitted because RT failure is assumed to result in core melt. Thus Fig. 9-11 is the final event tree for a small LOCA in a PWR.

In the BWR, ECC water is not borated and its addition would not prevent the fission process from restarting if RT failed. RT is therefore needed after EP in the BWR LOCA trees for all break sizes, so Fig. 9-11 is the appropriate functional event tree rather than Fig. 9-10.

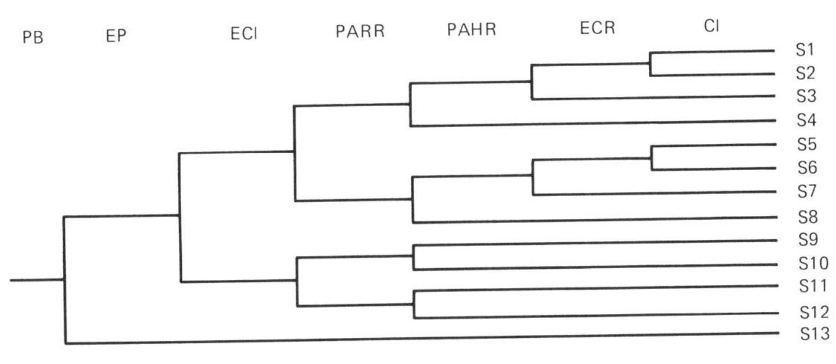

Fig. 9-10 *Final event tree showing functional interrelationships for a large* LOCA *in a* PWR. [From *Reactor Safety Study, Appendix* I. *U. S. Nuclear Regulatory Commission Rep. WASH-1400, NUREG 75/014* (*October 1975*).]

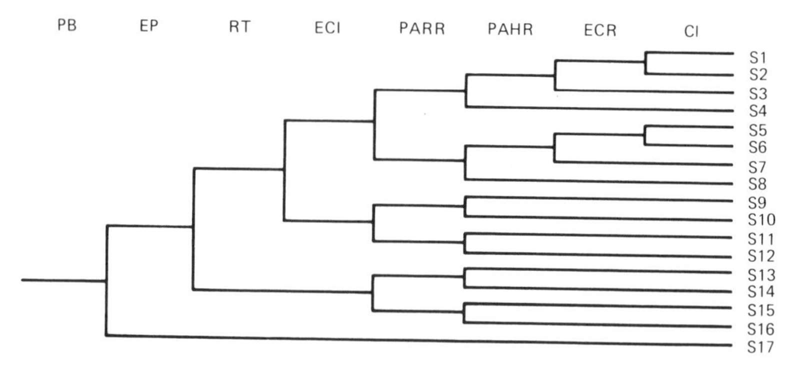

Fig. 9-11 *Final event tree showing functional interrelationships for a small* LOCA *in a* PWR *or any* LOCA *in a* BWR. [From *Reactor Safety Study, Appendix I. U. S. Nuclear Regulatory Commission Rep. WASH-1400, NUREG 75/014* (*October 1975*).]

9-2-3 Containment Event Trees

A significant number of headings would be needed on the LOCA event tree to describe all the various containment failure modes. It is therefore convenient to construct a separate containment event tree for the sequences on the LOCA event tree that cause core melt. The LOCA tree defines the probabilities and amounts of radioactivity in the containment atmosphere, whereas the containment tree defines, for each LOCA tree sequence, the probabilities and amounts of radioactivity released into the environment.

Containment failure can occur in basically three ways: (1) excessive leakage due to lack of adequate isolation of the containment atmosphere from the external environment, (2) gross rupture due to physical processes resulting from core meltdown, and (3) overpressure rupture which precedes the core meltdown processes. In the latter case, because core cooling systems are rendered inoperable by the rapid release of the containment atmosphere to the environment, meltdown occurs and there is excessive leakage through the ruptured containment.

Containment failure due to excessive leakage is harder to define. Clearly there is a spectrum of containment leakage rates with their associated probabilities. To define how containment leakage affects accident sequences on the event trees, three factors must be considered: (1) the effect of containment leakage on the operability of various ESFs, (2) the competition for post-accident radioactivity between removal systems and the leakage to the environment, and (3) the relationship of leakage to physical processes such as pressure buildup that occur during an accident.

In developing a containment event tree, the first step is to define headings representing possible events that can significantly affect the modes of containment failure or the resulting releases. For example, containment rupture (CR) may arise from a steam explosion in the reactor vessel (VSE), which could occur by interaction of finely dispersed molten fuel with water. This would result in a CR–VSE. Alternatively, a steam explosion within containment (CSE), which would occur if molten core material contacted water, could release sufficient energy to result in a CR–CSE. Other possible causes of containment rupture are containment leakage (CL), overpressure of containment (OP), burning of hydrogen in containment (B), core melt-through of containment (MT), and vapor suppression (VS).

An example of the containment event tree for CR–VSE in a PWR is shown in Fig. 9-12. To construct the tree, the events were put into logical order and unnecessary sequences were eliminated. CR–VSE is placed first on the tree because, if such a steam explosion occurs, it will occur before any of the other possible gross containment failure modes. CR–VSE has no branches on the path for its occurrence since it precludes the other events.

CL precedes CR–B and CR–OP because leakage at a sufficient rate precludes both hydrogen burning and pressure buildup sufficient to violate containment. The CL occurrence path therefore has no branches for CR–B or CR–OP.

CR–B is included in the containment event tree because in some accident sequences, the availability of coolant to generate steam imposes a limit on the steam partial pressure that can be developed in the containment atmosphere. In such circumstances, the presence of additional energy to superheat the containment atmosphere can raise the containment pressure into the rupture range; thus this incremental pressure from hydrogen burning leads to containment failure. CR–B precedes CR–OP because if hydrogen burning occurs, it probably will be before sufficient

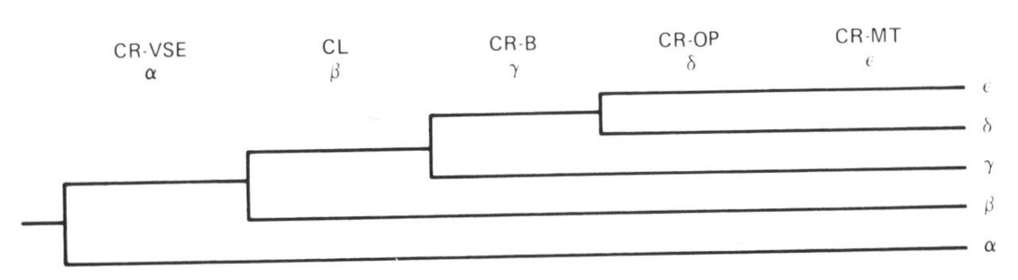

Fig. 9-12 PWR *containment event tree for* CR–VSE. [From *Reactor Safety Study, Appendix I. U. S. Nuclear Regulatory Commission Rep. WASH-1400, NUREG 75/014 (October 1975).*]

Table 9-1 *PWR Containment Failure Modes*[a]

Symbol	Meaning
α	Containment rupture due to a reactor vessel steam explosion
β	Containment failure resulting from inadequate isolation of containment openings and penetrations
γ	Containment failure due to hydrogen burning
δ	Containment failure due to overpressure
ϵ	Containment vessel melt-through

[a] *From* Reactor Safety Study. U.S. Nuclear Regulatory Commission Rep. WASH-1400, NUREG 75/014 (October 1975).

partial pressure of steam can build up to cause containment failure. (With the containment heat removal system (CHRS) operating successfully, CR–OP cannot occur.) CR–MT is placed last because melt-through generally will occur last.

In Fig. 9-12, each heading has been assigned a Greek letter. The sequences on the tree are assigned the same letters to indicate their modes of containment failure. These letter designations, summarized in Table 9-1, are also used to indicate the possible failure modes in accident sequences described in Chapter 12.

A similar example of the containment event tree for CR–VSE in a BWR is shown in Fig. 9-13; note that the CL has been broken down into the case of a large (>15-cm diameter) equivalent opening and a moderate (2.5 to 15-cm) opening. Each heading in the figure has been assigned a Greek letter, as defined in Table 9-2, to designate containment failure modes.

Table 9-2 *BWR Containment Failure Modes*[a]

Symbol	Meaning
α	Containment failure due to steam explosion in vessel
β	Containment failure due to steam explosion in containment
γ	Containment failure due to overpressure–release through reactor building
γ'	Containment failure due to overpressure–release direct to atmosphere
δ	Containment isolation failure in drywell
ϵ	Containment isolation failure in wetwell
ξ	Containment leakage greater than 2400 volume percent per day
η	Reactor building isolation failure
θ	Standby gas treatment system failure

[a] *From* Reactor Safety Study. U.S. Nuclear Regulatory Commission Rep. WASH-1400, NUREG 75/014 (October 1975).

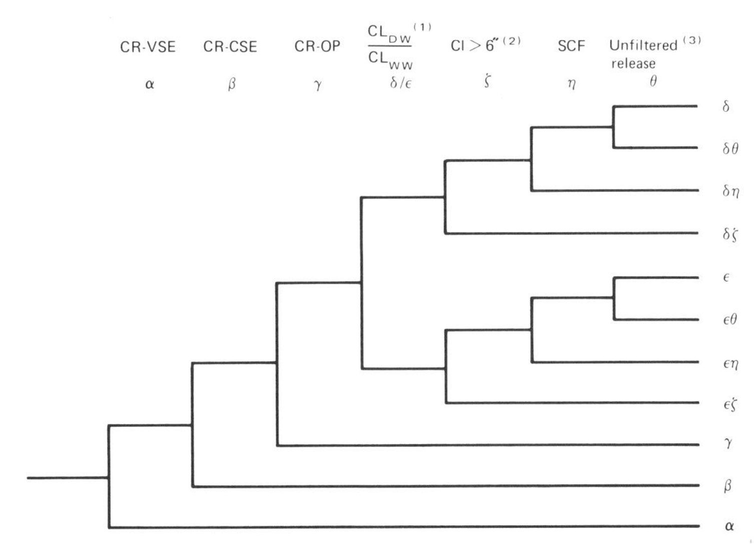

Fig. 9-13 BWR *containment event tree for* CR–VSE. *Note that* (*1*) *upward path indicates containment leakage from the dry well* (δ), *downward path indicates leakage from the wet well* (ε); (*2*) *leakage greater than a 15-cm equivalent diameter hole in the containment;* (*3*) *elevated release bypasses filters.* [From *Reactor Safety Study, Appendix* I. *U. S. Nuclear Regulatory Commission Rep. WASH-1400, NUREG 75/014* (*October 1975*).]

9-2-4 Combination of Accident Event Trees and Containment Event Trees

Figure 9-14 illustrates how event trees are used to combine probabilities of events in estimating the probability of a sequence. The event tree for a large LOCA in a PWR is shown in the figure, except that the CI sub-branches have been omitted because they are treated with the containment event trees. For simplicity, only three sequences are illustrated:

(a) For sequence S1, all functions after the pipe break are successful and the core does not melt; therefore, the only function of interest on the containment tree is CL.

(b) Sequence S3 results in core melt because PAHR fails; therefore the complete PWR containment event tree of Fig. 9-12 is needed.

(c) Sequence S7 results in core melt due to ECI failure, even though PAHR functions successfully. For the PWR, the PAHR success implies negligible probability of containment failure through overpressure with or without hydrogen burning (path δ and γ); thus there are no δ and γ paths included from the containment event tree of Fig. 9-12 for this sequence.

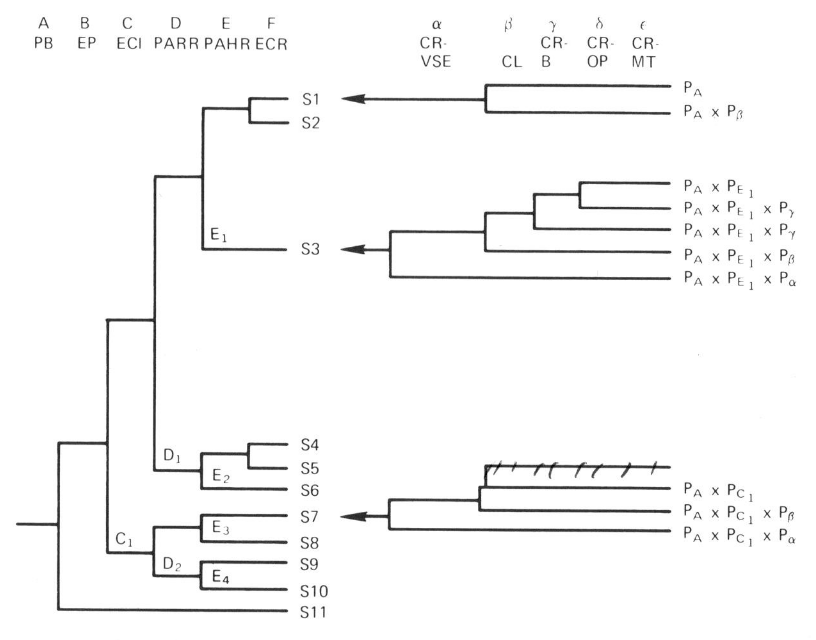

Fig. 9-14 *Linking of accident and containment event trees for a large* LOCA *in* PWR. [From *Reactor Safety Study, Appendix* I. *U. S. Nuclear Regulatory Commission Rep. WASH-1400, NUREG 75/014* (*October 1975*).]

In the RSS, sequences with equivalent consequences were ultimately grouped together, and their probabilities were combined to determine the probability that this consequence level would be reached in an accident. As an example, Fig. 9-15 shows the probabilities as a function of magnitude of consequences for a few sequences in Fig. 9-14. The use of results such as those in Fig. 9-15 will be discussed in more detail in Chapter 12.

9-2-5 Summary of Reactor Safety Event Tree Construction

The initial requirement in construction of event trees for the RSS was the definition of the functions to be performed after an initiating (failure) event and of the interrelationships between the various functions. Next, the systems provided to perform the functions were identified, and the interrelationships among the functions to be performed and the operability states of the systems were analyzed. Finally, the interrelationships among the operability states of the various systems were defined. At each step, dependencies were considered and illogical or meaningless sequence combinations were eliminated. Figures 9-10 and 9-11 are the final RSS

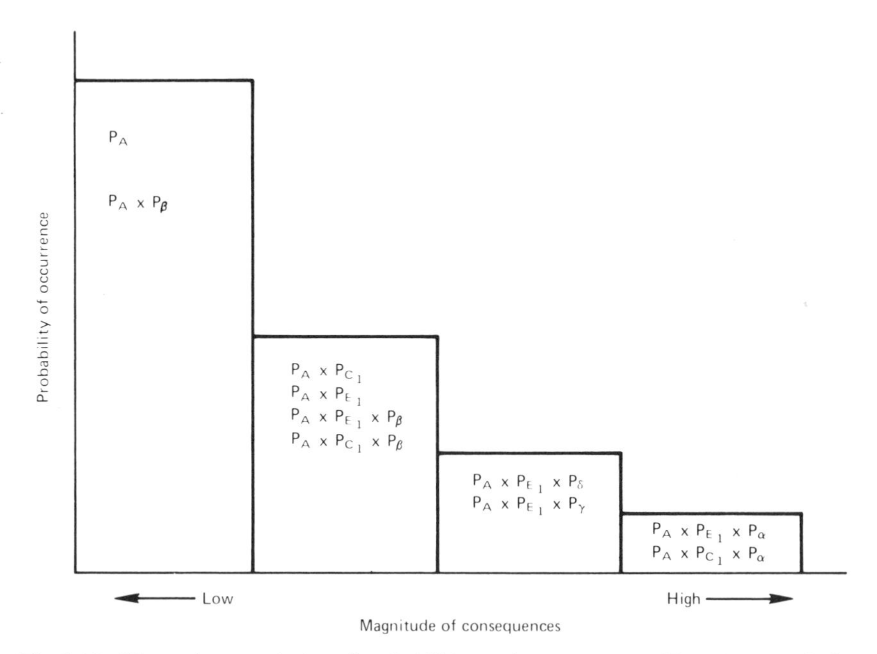

Fig. 9-15 *Illustrative association of probabilities and consequences.* [From *Reactor Safety Study, Appendix* I. *U. S. Nuclear Regulatory Commission Rep. WASH-1400, NUREG 75/014 (October 1975).*]

event trees showing functional interrelationships for safety systems operation following a large LOCA in a PWR and BWR, respectively; these trees were combined with containment event trees, as illustrated in Fig. 9-14, to calculate release sequence probabilities.

The accident sequences that dominate the risks in a Babcock and Wilcox PWR and a newer Westinghouse PWR (more modern LWR power plants than those considered in the RSS) have been analyzed in a follow-up safety study [10]. For both plants, many of the accident sequences were common to the RSS plant, but many were unique to each of the two newer plants, and some were unique to the RSS plant. This suggests that it is necessary to analyze each specific reactor for the functioning of the safety systems following an accident, and such analyses were begun in 1979 for the operating reactors in the United States.

9-3 EVENT TREES FOR SAFEGUARDS ANALYSIS

Event tree techniques are one form of system effectiveness analysis used to study nuclear safeguards, which is the protection of a facility from overt attack or covert diversion of nuclear material. An attack could be either for the purpose of sabotage or the acquisition of nuclear material.

An example is the analysis of an overland external attack by force on the plutonium packaging and shipping (PPS) cell in a hypothetical reprocessing plant [11]. The schematic layout of the plant is shown in Fig. 9-16.

The event tree consists of combinations of the various subsystems comprising the physical security system; the following symbols denote these subsystems:

PB = physical barrier system, including doors, walls, fences, gates, etc.

Gl = guard force at main gate.

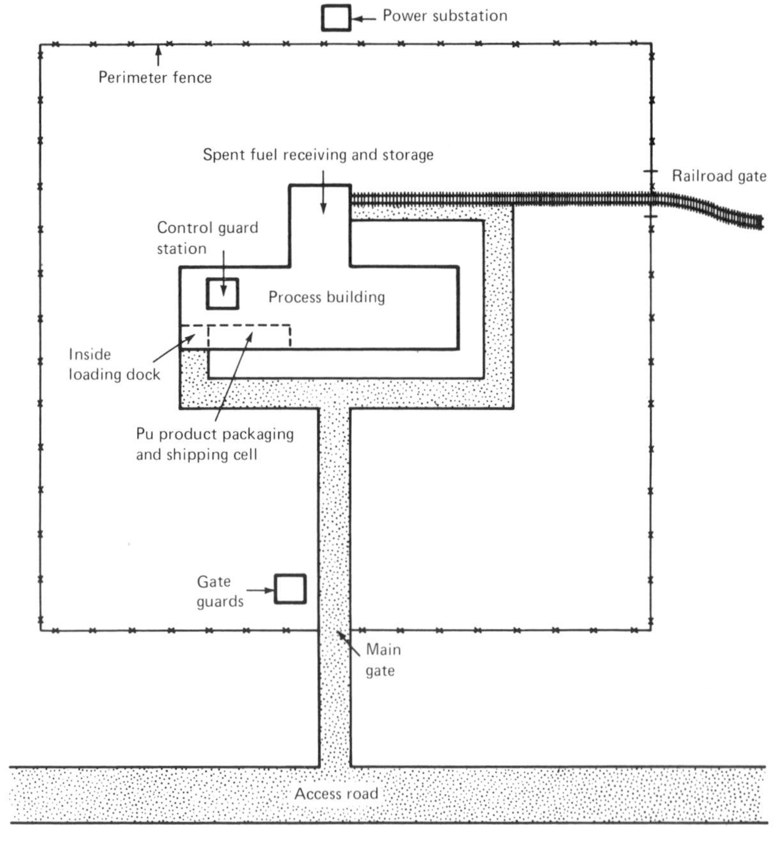

Fig. 9-16 *Hypothesized reprocessing facility showing plutonium product packaging and shipping cell.* [From *H. Kendrick* et al., Nucl. Mater. Management *V, 226 (1976)*.]

G2 = guard force at control station, including guards that may be on patrol.
C1 = communications system at the main gate utilized to communicate to the response force and to the control station; this system includes phones, transmitters, and alarms.
C2 = communications system at the control station utilized to communicate to the response force and to the main gate.
A1 = perimeter alarm system, including microwave intrusion alarm and pressure/movement sensors on perimeter fence.
A2 = intrusion detection devices, door alarms, closed circuit TV, etc. at PPS cell.

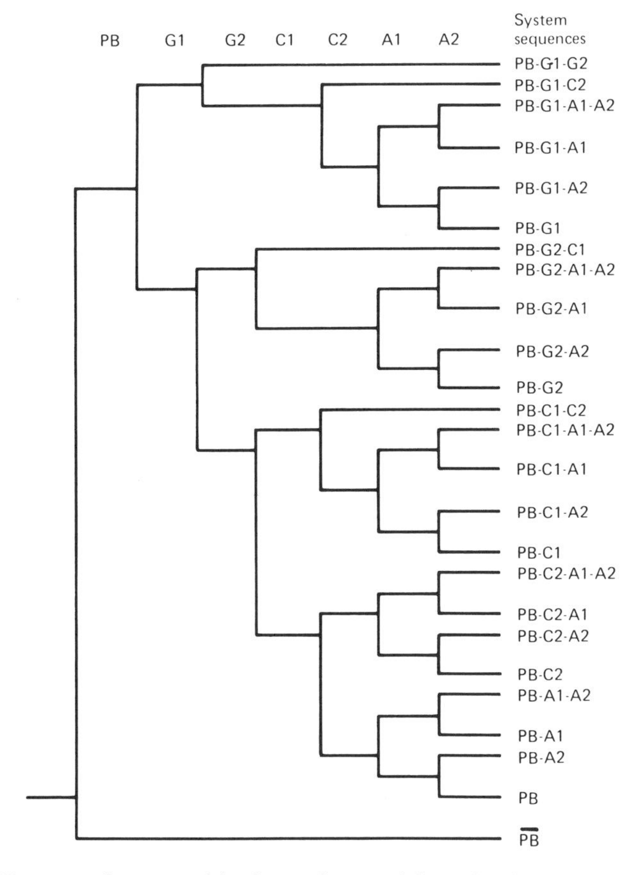

Fig. 9-17 *Event tree for removal by force of material from the plutonium packaging and shipping cell.* [From *H. Kendrick* et al., Nucl. Mater. Management *V, 226 (1976)*.]

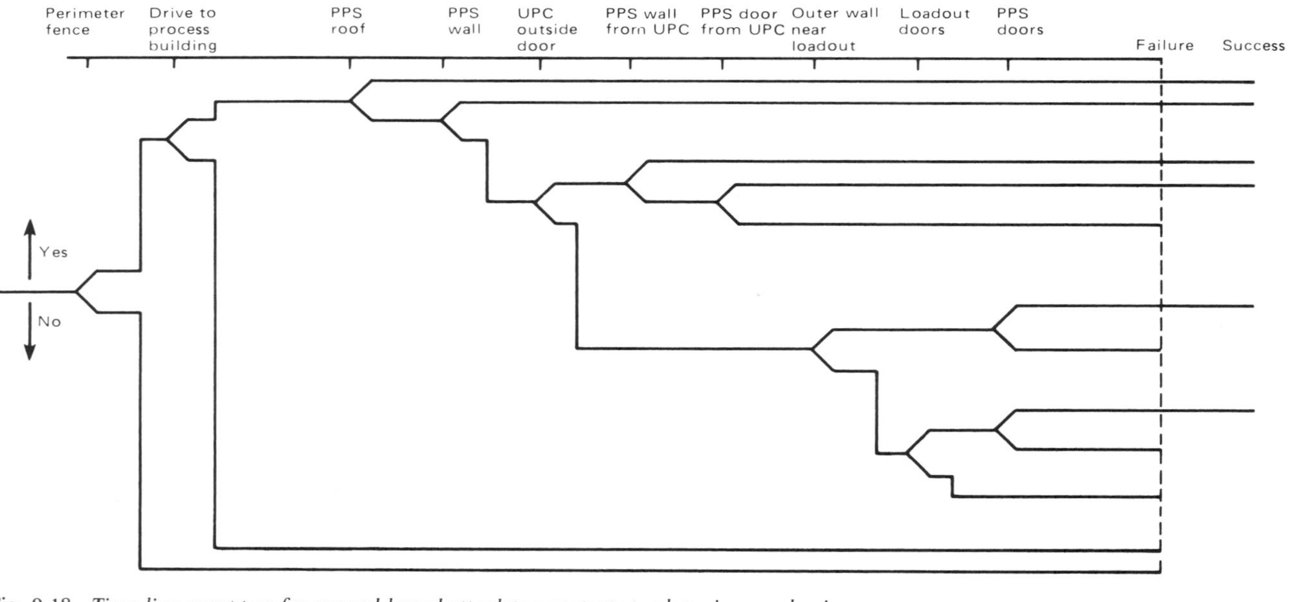

Fig. 9-18 *Time-line event tree for ground-based attack to penetrate to plutonium packaging and shipping cell.* [From *H. Kendrick* et al., Nucl. Mater. Management *V, 226 (1976)*.]

The *adversary action* event tree is shown in Fig. 9-17; an upper branch indicates successful adversary action against that particular component of the physical security system, while a lower branch indicates unsuccessful or no adversary action against that component. Thus, the system sequences that denote a successful attack are those branches whose last junctions give upper branches.

For this analysis it was assumed that escape would be by the same route as penetration [11]. (This is obviously just a convenient assumption and not an essential one.) Therefore, once the barriers had previously been defeated for penetration, there was essentially an unhindered escape.

A second type of event tree for the plutonium packaging and shipping (PPS) cell also can be constructed, namely a "time-line" event tree [11]. Such a tree is shown in Fig. 9-18 for a ground-based adversary attack to penetrate to the PPS cell of Fig. 9-16; the horizontal axis is time and the spacing between events depicts the transit time required.

EXERCISES

9-1 Fission gas G routinely generated in a fuel rod is normally prevented from an abnormal release to the offsite environment by the successful operation of the following barriers that are sequentially encountered by the gas: fuel matrix F, fuel rod cladding C, water coolant W, pressure vessel P, and containment building B. (a) Construct the reduced event tree for the abnormal release of fission gas to the environment. (b) Calculate the probability of an abnormal release to the environment if the probabilities of successful operation of the successive barriers can be considered independent and equal to:

$$P_F = 0.1 \qquad P_P = 0.98$$
$$P_C = 0.99 \qquad P_B = 0.2$$
$$P_W = 0.15$$

9-2 An exhaust ventilation system consists of two active air filters and an alarm system, which is to sound after a filter failure, so that a valve can divert the airflow to a standby filter. The failure probabilities may be taken to be 0.07/yr for an active filter, 0.001/demand for the alarm, 0.002/demand for the diversion valve, and 0.003/demand for the backup filter. There are two independent offsite power supplies and an emergency backup power system, so electrical failures are not a problem. (a) Construct the reduced event tree to describe the events leading to a release of radioactivity following failure of a filter and no diversion of the airflow;

clearly label your output combinations as to whether radioactivity is released. (b) Calculate the frequency of each path. (c) Use the results of part (b) to calculate the frequency of release of radioactivity.

9-3 The physical security system at a nuclear fuel fabrication facility is assumed to consist of the following subsystems:

(a) PB, the perimeter physical barriers (including walls and fences, etc.) excluding gates;
(b) GB, the gate physical barriers;
(c) GG, the (stationary) gate guard force, which engages terrorists only at the gate barriers;
(d) PBA, the physical barrier alarm system, which signals the communications center;
(e) GA, the gate guard alarm system, for notifying the communications center;
(f) C, the communications center used to communicate to the mobile guards inside the plant;
(g) MG, the mobile guards, which respond only to instructions from the communications center.

For a terrorist attack T, which is assumed to occur with equal probability on either PB or GB, assume the following probabilities are for the *terrorist* failure when interacting with the different physical security subsystems:

$$\text{PB} = 0.9;\ \text{PBA} = 0.999;\ \text{GG} = 0.8;\ \text{GB} = 0.1;$$

$$\text{MG} = 0.7;\ \text{GA} = 0.95;\ \text{C} = 0.99.$$

Complete the following:

(a) Construct the reduced event tree beginning with T to describe the system combinations for the plant subject to the terrorist attack. Clearly label your output combinations as to success S or failure F *of the attack* assuming that a successful attack is one in which no part of the security system stops the terrorists. Assume that the terrorists can encounter the MG only if C operates, and that if GG does not stop the attack then the guard alarm system is automatically actuated.
(b) Label each branch of the tree in part (a) with the appropriate probability and calculate the probability of each path.
(c) Use the results of part (b) to calculate the probability of a successful terrorist attack on the plant, once it has been initiated.

REFERENCES

1. H. Raiffa, "Introductory Lectures on Making Choices Under Uncertainty." Addison-Wesley, Reading, Massachusetts, 1968.
2. M. Tribus, "Rational Descriptions, Decisions and Designs." Pergamon, Oxford, 1969.
3. J. B. Fussell and J. S. Arendt, System reliability engineering methodology: A discussion of the state of the art, *Nucl. Safety* **20,** 541 (1979).
4. R. C. Erdmann (ed.), A Risk Methodology Presentation. Electric Power Research Institute Informal Rep. NP-79-1-LD (January 1979).
5. J. D. Esary and H. Ziehms, Reliability analysis of phased missions, *in* "Reliability and Fault Tree Analysis" (R. E. Barlow, J. B. Fussell, and N. D. Singpurwalla, eds.), p. 213. Society for Industrial and Applied Mathematics, Philadelphia, Pennsylvania, 1975.
6. G. R. Burdick, J. B. Fussell, D. M. Rasmuson, and J. R. Wilson, Phased mission analysis: A review of new developments and an application, *IEEE Trans. Reliability* **R-26,** 43 (1977).
7. Reactor Safety Study, An Assessment of Accident Risks in U.S. Commercial Nuclear Power Plants. U.S. Nuclear Regulatory Commission Rep. WASH-1400, NUREG 75/014, (October 1975).
8. Reactor Safety Study, An Assessment of Accident Risks in U.S. Commercial Nuclear Power Plants, Appendix I. U.S. Nuclear Regulatory Commission Rep. WASH-1400, NUREG 75/014 (October 1975).
9. S. Levine and W. E. Vesely, "Important event-tree and fault-tree considerations in the reactor safety study, *IEEE Trans. Reliability* **R-25,** 132 (1976).
10. S. V. Asselin, D. D. Carlson, J. W. Hickman, and M. A. Fedele, System event tree analysis for determining accident sequences that dominate risk in LWR power plants, "Probabilistic Analysis of Nuclear Reactor Safety," Vol. 3. p. XII.2-1. American Nuclear Society, La Grange Park, Illinois, 1978.
11. H. Kendrick, E. Lofgren, D. Rundquist, and R. Fullwood, An approach to the evaluation of safeguards systems effectiveness, *Nucl. Mater. Management* **V,** 226 (1976).

CHAPTER 10

Computer Programs for Fault Tree Analysis

The purpose of this chapter is to provide a brief overview of some of the more widely used computer programs for fault tree analysis. Table 10-1 lists the names and purposes of some of the programs available [1]; only a few will be discussed.

10-1 QUALITATIVE AND QUANTITATIVE EVALUATIONS

10-1-1 PREP-KITT AND FRANTIC

The standard technique for evaluating fault trees, up until 1975 or so, was the PREP–KITT technique [2] used in the Reactor Safety Study [3]. This computer code evaluation of fault trees is restricted to the use of AND, OR, and INHIBIT gates, and hence is limited by its inability to correctly handle common cause failures and such things as test and maintenance in which plant procedures prohibit simultaneous testing of redundant systems or portions of systems.

The PREP–KITT code package actually consists of two parts: PREP (comprised of TREBIL, MINKIT, and COMBO) and KITT. The output of TREBIL gives a listing of all the gates in the fault tree, while MINKIT gives a listing of all the minimal cut sets; the COMBO routine is capable of providing the minimal cut sets in combinations of single-component failures, double-component failures, etc. The KITT code, named for Kinetic Tree Theory, uses the minimal cut sets from PREP to determine the probability of the top event.

A quick means of *estimating* the relative importance of the minimal cut sets and the events is provided in the output of PREP–KITT. The minimal cut sets are ranked according to the value for each cut set of the probability ratio

$$(Q_k)\Big/\left(\sum_{p=1}^{n} Q_p\right),$$

Table 10-1 *A Partial List of Computer Programs for Fault Tree Analysis*[a]

Purpose	Program name	Comments
Qualitative evaluations (for minimal cut sets)	PREP[b]	Uses combinatorial testing in a Boolean equation
	MOCUS[c]	Uses a top-down Boolean substitution method
	SETS[d]	Uses Boolean manipulation to determine prime implicants
	TREEL–MICSUP[e]	A bottom-up algorithm using Boolean substitution
	ALLCUTS[f]	Almost identical to MOCUS
	WAM–CUT[g]	Uses a Boolean manipulation and minimization algorithm
	PL–MOD[h]	Uses modular decomposition and calculates top event quantitative characteristics; especially good for common cause failures
Quantitative evaluations (for unavailabilities)	KITT 1 & 2[b]	Calculates time-dependent availability, reliability, and expected number of failures from component failure rates and repair times; also calculates component importance
	WAM–BAM[i]	Calculates the top event existence probabilities using Boolean manipulation
	FRANTIC[j]	Calculates effect on average unavailability and time-dependent instantaneous unavailability due to periodic system testing; testing characteristics include the test interval, test duration time, repair time, and human-caused failure potential associated with the testing
	IMPORTANCE[k]	Calculates various measures of component importance and other quantitative information
	GO[l]	Uses a Boolean manipulation algorithm; also calculates point estimates of top event characteristics
Common cause failure identification aid	COMCAN[m]	Determines possible common cause failure candidates using "brute force" searching algorithm
	COMCAN II[n]	Uses a reduced fault tree approach to determine minimal cut sets having failure potential due to common causes
	BACFIRE[o]	Similar to COMCAN
	SETS[d]	Similar to COMCAN and BACFIRE

(Continued)

Table 10-1 (*Continued*)

Purpose	Program name	Comments
Uncertainty analysis	SAMPLE[p]	Monte Carlo simulation of failure rate and repair time distributions of basic events in order to determine uncertainty of top event reliability characteristics
	MOCARS[q]	Monte Carlo code for determining distribution and simulation limits of top event reliability characteristics
	SPASM[r]	Similar to SAMPLE
Fault tree construction	CAT[s]	Constructs fault trees for simple systems using decision tables

[a] Modified from J. B. Fussell and J. S. Arendt, *Nucl. Safety* **20**, 541 (1979).

[b] W. E. Vesely and R. E. Narum, PREP and KITT: Computer Codes for the Automatic Evaluation of a Fault Tree, USAEC Rep. IN-1349. Idaho Nuclear Corporation, (August 1970).

[c] J. B. Fussell, E. B. Henry, and N. H. Marshall, MOCUS—A Computer Program to Obtain Minimal Sets from Fault Trees, USAEC Rep. ANCR-1156. Aerojet Nuclear Company, (August 1974).

[d] R. E. Worrell and D. W. Stack, A SETS User's Manual for the Fault Tree Analyst, Rep. SAND-77-2051. Sandia Laboratories (November 1978).

[e] P. K. Pande, M. E. Spector, and P. Chatterjee, Computerized Fault Tree Analysis: TREEL and MICSUP, Rep. ORC-75-3 (AD-A010 146). Operations Research Center, Univ. of California, Berkeley, California, (April 1975).

[f] W. J. Van Slyke and D. E. Griffing, ALLCUTS–A Fast Comprehensive Fault Tree Analysis Code, ERDA Rep. ARH-ST-112. Atlantic Richfield Hanford Company, (July 1975).

[g] R. C. Erdmann, F. L. Leverenz, and H. Kirch, WAMCUT, A Computer Code for Fault Tree Evaluation, Electric Power Research Institute Rep. EPRI-NP-803 (June 1978).

[h] J. Olmos and L. Wolf, A Modular Approach to Fault Tree and Reliability Analysis, Rep. MITNE-209. Department of Nuclear Engineering, Massachusetts Institute of Technology (August 1977).

[i] F. L. Leverenz and H. Kirch, User's Guide for the WAM–BAM Computer Code, Electric Power Research Institute Rep. EPRI-217-2-5 (January 1976).

[j] W. E. Vesely and F. F. Goldberg, FRANTIC—A Computer Code for Time-Dependent Unavailability Analysis, NRC Rep. NUREG-0193, (October 1977).

[k] H. E. Lambert and F. M. Gilman, The IMPORTANCE Computer Code, ERDA Rep. UCRL-79269. Lawrence Livermore Laboratory, Univ. of California, (March 1977).

[l] W. Y. Gateley and R. L. Williams, GO Methodology—System Reliability Assessment and Computer Code Manual, Electric Power Research Institute Rep. EPRI-NP-766 (May 1978).

[m] G. R. Burdick, N. H. Marshall, and J. R. Wilson, COMCAN—A Computer Program for Common Cause Failure Analysis, ERDA Rep. ANCR-1314, Aerojet Nuclear Company, (May 1976).

[n] D. M. Rasmuson *et al.*, COMCAN II: A Computer Program for Common Cause Failure Analysis, USDOE Rep. TREE-1289. Idaho National Engineering Laboratory, (September 1978).

[o] C. L. Cate and J. B. Fussell, BACFIRE—A Computer Program for Common Cause Failure Analysis, Rep. NERS-77-02. Nuclear Engineering Department, University of Tennessee, Knoxville, Tennessee (February 1977).

[p] Reactor Safety Study—An Assessment of Accident Risks in U.S. Commercial Nuclear Power Plants, U.S. Nuclear Regulatory Commission Rep. WASH-1400, NUREG-75/014, (October 1975).

[q] S. D. Matthews, MOCARS: A Monte Carlo Simulation Code for Determining the Distribution and Simulation Limits, ERDA Rep. TREE-1138. EG&G Idaho, (July 1977).

[r] R. C. Erdmann *et al.*, Probabilistic Safety Analysis III, Electric Power Research Institute Rep. EPRI NP-749 (April 1978).

[s] G. E. Apostolakis, S. L. Salem, and J. S. Wu, CAT: A Computer Code for the Automated Construction of Fault Trees, Electric Power Research Institute Rep. EPRI NP-705 (March 1978).

where

Q_k = probability of minimal cut set k,

$\sum_{p=1}^{n} Q_p$ = probability of all minimal cut sets, i.e., the system failure probability.

Also provided is a ranked quantitative listing of all important events, where this ranking is calculated by summing the probability ratios of all minimal cut sets that contain the event.

The FRANTIC computer code [4] is a newer numerical program with which, like PREP–KITT, the time-dependent instantaneous unavailability of a system can be calculated. Even though a system has a low unavailability when averaged over a year, at particular times the system may have a high instantaneous unavailability; the FRANTIC code can incorporate the effects of periodic testing, which tends to decrease the unavailability except in those instances when human errors during testing lead to system failures.

10-1-2 The WAM Codes

The WAM system of fault-tree codes allows for additional modeling capabilities over PREP–KITT since it incorporates the NOT operation with events and with AND and OR gates [5]. The inclusion of NOT-type gates makes possible the explicit modeling of dependent events, such as common cause failures (see Section 10-2). The allowable gate types are:

AND is A AND B (up to 8 inputs),
OR is A OR B (up to 8 inputs),
NOT is NOT A (only 1 input),
NOR is NOT (A OR B),
NAND is NOT (A AND B),
ANOT is A AND (NOT B),
ONOT is A OR (NOT B),
COM is COMBINATION of N of M inputs causes failure ($M \leq 8$).

The introduction of these new events and gates means that new fault tree symbols need to be introduced to graphically illustrate them. The NOT can be indicated by a small circle placed on top of the event or gate, as shown in Table 10-2.

The preprocessor WAM provides numeric input generated from the logic tree in which components and gates are input with alphanumeric names. The set of WAM codes is interrelated as shown in Fig. 10-1 and provides the following capabilities [5]:

Table 10-2 *Fault Tree Symbols for the "Not" Events and Gates*

Symbol	Description
A	NOT *A*
A	NOT *A*
	NOR gate
	NAND gate
A B	ANOT, "*A*" AND (NOT "*B*")
A B	ONOT, "*A*" OR (NOT "*B*")

1. Qualitative assessment of the system in terms of failures (minimal cut sets) that cause the system to fail and that cause any event within the system to occur (WAM–CUT code).

2. Point estimates of the system (top event) reliability (or unreliability), together with the reliability of any event within the system (WAM–BAM code).

3. Point estimates of the system after changes have been made to the probability of occurrence of one or more basic events (WAM–TAP code).

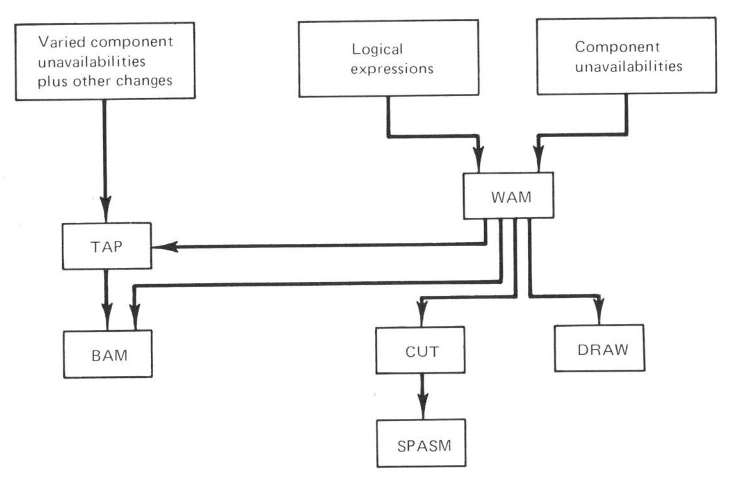

Fig. 10-1 *Interrelationships of* WAM *codes. [Reprinted with permission from R. C. Erdmann, J. E. Kelly, H. R. Kirch, F. L. Leverenz, and E. T. Rumble, A method for quantifying logic models for safety analysis,* in *"Nuclear Systems Reliability Engineering and Risk Assessment"* (*J. B. Fussell and G. R. Burdick, eds.*), *p. 732. Soc. Industrial Applied Math.,* (*Philadelphia, Pennsylvania, 1977.*) *Copyright by SIAM 1977. All rights reserved.*]

4. Qualitative assessment of the system by modeling the basic system components as random variables with a mean and standard deviation (WAM–CUT code) for use by a Monte Carlo code that allows determination of the entire distribution of the system reliability as a function of component distributions (SPASM code).
5. A drawing of the fault tree (WAM–DRAW).

The Boolean Arithmetic Model (BAM) code [5–8] uses the Boolean algebraic techniques of Section 6-5 on the input information from WAM to find the resultant logic expressions for the top and intermediate events; this technique is accomplished in BAM by forming all possible combinations of basic events and then forming a truth table that describes each event and gate as a function of these combinations. Finally, the associated point unavailabilities are calculated.

The computational techniques used in BAM have some advantages over the FRANTIC analysis since, instead of first requiring the calculation of minimal cut sets, BAM solves the entire system while dropping events that are below a user-specified level of significance. FRANTIC, however, provides time-dependent reliability characteristics of the system, whereas BAM gives single point-in-time estimates; that is, BAM requires repeated calculations to treat a general time-dependent problem.

Hence, in a sense, there are features about the two codes that tend to complement one another.

WAM–TAP allows probabilities to be changed for specified components or groups of components that are identified by common alphanumeric characters in the component name [5]. For example, if the event code used letters HE as the first two letters of every Human Error component, WAM–TAP could search for these and change the failure rate for each component starting with HE by a multiplicative factor or to a specific value, then reevaluate the tree via the BAM code. Such a capability allows sensitivity studies or common cause studies to be easily accomplished by changing either the individual component probabilities or the probabilities of all components with one or more similar characteristics (e.g., same manufacturer, same type, same location).

WAM–CUT [5, 9, 10] is a computer program used to determine minimal cut sets. The program can operate in the traditional way of finding all the cut sets containing up to a specified number of basic events (i.e., single-, double-, triple-event failures, etc.), or only those cut sets estimated to have a probability greater than a specified minimum can be retained. In the latter case, the minimum probability is gate-dependent so that valuable information for gates other than the top gate is not lost and, conversely, more information than needed is not stored. WAM–CUT is designed to utilize specific features of the CDC-7600 computer; conversion to any other system could sacrifice much of the versatility of the program [5].

10-1-3 GO

The GO method can be used to compute the probability that a system exists in each of a few states (such as premature operation, normal operation, partially failed, failed, etc.). The system being studied is modeled in the form of a "GO chart" from which the input to the GO computer program is formulated [11–15]. Thus the GO chart is the equivalent of the tree in fault tree analysis.

The development of the GO chart consists of selecting functional operators (also called "building blocks") to represent each component and logical junction, and connecting them with arrows to represent the flow of information. Input events (power, water, pressure, neutron flux, etc.) and human interfaces are entered, where appropriate, with time distributions corresponding to the sequence desired. The construction of the GO chart is vastly different from the construction of fault trees or event trees; an introduction to GO chart construction appears in Appendix

G since the GO method can be considered a competitor of fault tree analysis.

The principal advantages of GO are that the GO chart itself is very similar to the system logic diagram, so that construction of the chart may not be too difficult. Although the method's proponents claim GO has advantages over fault tree and event tree analysis, so far GO has been infrequently utilized for nuclear power applications [16–18].

10-2 FAULT TREE ANALYSIS WITH COMMON-CAUSE FAILURES

Common cause failures (CCFs) add complications to fault tree analysis. The commonality may involve a failure in one system causing a failure in another system or an adverse environment causing simultaneous failures in more than one system; CCFs are frequently due to errors in the design of a system. Overlooked CCFs can lead to errors of orders of magnitude in the calculated failure probability, often overshadowing uncertainties in component failure data. Common cause failures have already been considered in Sections 5-5, 6-6, and 7-9.

Several computer programs have been developed for CCF analysis using minimal cut sets as input; COMCAN [19, 20] and BACFIRE [21] are two such programs (which interface with the PREP–KITT system) for determining the CCFs by a brute force technique. The COMCAN-II code [22] works better for analyzing complex systems since it uses a step-by-step analysis procedure [23]; this is done by dissecting the fault tree at its bottom and determining minimal cut sets for individual branches of the tree (termed *intermediate* minimal cut sets). Then all the intermediate candidates for CCFs can be assigned to a dummy event for the particular branch of the tree, and the process is continued to a higher level in the tree. At the completion of the analysis, the dummy events are expanded and the common cause candidates for the top event are constructed.

Another approach for representing CCF events is to use an irreducible building block [24, 25]. This technique relies upon the use of the Boolean NOT function and the irreducible building block shown in Fig. 10-2. The building block for dependent event B involves the conditional event $B|A$, event B given event A, and conditional event $B|\bar{A}$, event B given "not A." The building block causes the dependency condition to occur between the disjoint events A AND $B|A$ and $\bar{A}$ AND $B|\bar{A}$.

The use of the building block may be illustrated by considering event A to be the possible common cause initiating event and event B to be the

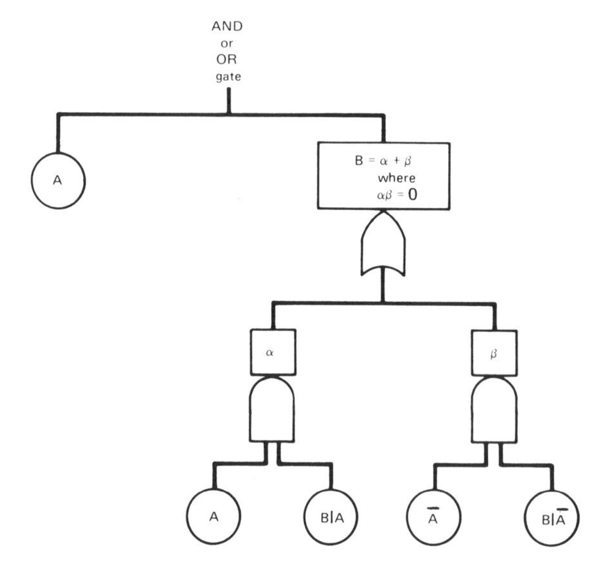

Fig. 10-2 *Irreducible bulding block for the probability of event B dependent on the occurrence of initiating event A.*

dependent event. In the case that the probability of A and B is required, the AND gate is used at the top in Fig. 10-2. Then the probability of the top event is

$$\begin{aligned} P(AB) &= P(A\{A[B|A] + \bar{A}[B|\bar{A}]\}) \\ &= P(A[B|A]) \\ &= P(A)P(B|A). \end{aligned} \tag{10-1}$$

On the other hand, if the probability A or B is required, the OR gate is used at the top in Fig. 10-2 and the probability of the top event is

$$\begin{aligned} P(A + B) &= P(A + \{A[B|A] + \bar{A}[B|\bar{A}]\}) \\ &= P(A) + P(\bar{A})P(B|\bar{A}). \end{aligned} \tag{10-2}$$

To illustrate the importance of properly treating the dependent event B, we assume the following probabilities are valid:

$$\begin{aligned} P(A) &= 0.2 \\ P(B|A) &= 0.3 \\ P(B|\bar{A}) &= 0.8. \end{aligned} \tag{10-3}$$

Using Eqs. (10-1) through (10-3) gives

$$\begin{aligned} P(AB) &= 0.06 \\ P(A + B) &= 0.84. \end{aligned} \tag{10-4}$$

In order to compare these numerical values to those for the case in which event B is *assumed* independent of A, it is first necessary to use the decomposition rule (2-19) to obtain

$$\begin{aligned} P(B) &= P(A)P(B|A) + P(\bar{A})P(B|\bar{A}) \\ &= 0.7. \end{aligned} \tag{10-5}$$

It then follows that

$$\begin{aligned} P(AB) &= P(A)P(B) = 0.14 \\ P(A + B) &= P(A) + P(B) - P(A)P(B) = 0.76. \end{aligned} \tag{10-6}$$

Comparison of Eqs. (10-4) and (10-6) shows the importance of treating the event B as a dependent event in this case.

Generalizations of the irreducible building block are available to treat situations in which the probability of an event is dependent upon the occurrence of multiple events [26].

10-3 ANALYSIS OF DATA UNCERTAINTIES IN A FAULT TREE

There is an uncertainty band surrounding every point estimate of a demand failure or of any of the parameters that specify a time-dependent failure probability of a component. This variation of data exists because of differing conditions that each component may see, for example, due to varying maintenance or different operational demands. Furthermore, differences between similar components in different plants give rise to variations in the failure data, as discussed in Chapter 5.

Since the system unavailability is a function of random variables, it is itself a random variable. To determine the distribution for the system unavailability in terms of the distributions for the components, the SAMPLE-A program was developed during the Reactor Safety Study (RSS) [3]. The program uses Monte Carlo random sampling to obtain the mean, standard deviation, confidence limits, and the probability distribution function for the fault tree top event unavailability.

The input information required includes each component failure probability (mean or median) and dispersion (90% error spread or range factor) estimates, and the number of Monte Carlo trials to be run. Either a normal, lognormal, or log-uniform distribution can be specified for the component failure probabilities; the lognormal distribution was used for the RSS since the raw input data were sparse and the assessed ranges were large, with widths of one or two orders of magnitude [3]. The random sampling process in the RSS was repeated 1200 times for each system in order to obtain a distribution for the top-event unavailability.

An example of the use of SAMPLE-A can be obtained by considering the fault tree in Fig. 10-3. The $X(I), I = 1, 2$, and 4, in the figure represents component failures and $X(6)$ the test and maintenance contribution to system unavailability; the faults $X(3)$ and $X(5)$ are taken to be nonindependent operator errors, and their intersection represents a common cause contribution to system failure. The failure data for the sample problem are given in Table 10-3.

The Boolean expression for the example fault tree is

$$T = X(1) + [X(2) + X(3)][X(4) + X(5)] + X(6) + X_{CC}, \tag{10-7}$$

where an additional common cause contribution $X_{CC} = X(3)X(5)$ has been included in the top event. Using the rare-event approximation, the probability of the top event may be written as

$$P(T) = P_1 + (P_2 + P_3)(P_4 + P_5) + P_6 + P_{CC}. \tag{10-8}$$

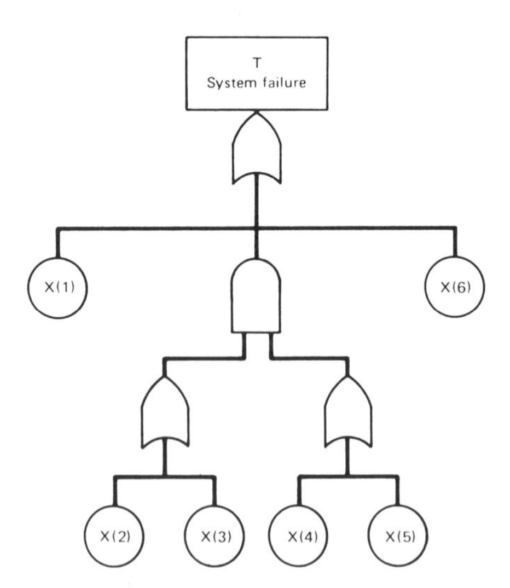

Fig. 10-3 *Example fault tree to illustrate the use of* SAMPLE-A. [*Modified from Reactor Safety Study, Appendix II, WASH-1400 (October 1975).*]

Table 10-3 *Data used for Example Fault Tree to Illustrate the Use of* SAMPLE-A.[a]

Fault	Median probability	Error factor
Component failure: $X(1)$	1.0×10^{-3}	3
Component failure: $X(2)$	3.0×10^{-2}	3
Operator error: $X(3)$	1.0×10^{-2}	3
Component failure: $X(4)$	3.0×10^{-2}	3
Operator error: $X(5)$	1.0×10^{-2}	3
Test and Maintenance: $X(6)$	3.0×10^{-3}	3

[a] From Reactor Safety Study, Appendix II, WASH-1400 (October 1975).

In the RSS, the common cause contribution P_{CC} was estimated by the geometric-mean technique (see Section 7-9) to be the square root of the product of the upper and lower bounds of the probabilities for independent and for completely coupled operation. For the case of this common cause, the upper bound is P_3 or P_5 (both of which happen to equal 1.0×10^{-2}), and the lower bound is the probability X_3 AND X_5, so

$$P_{CC} = [P_3(P_3P_5)]^{1/2}. \tag{10-9}$$

The SAMPLE-A program generated a frequency function in the form of a histogram, as shown in Fig. 10-4. The point value of 6.6×10^{-3} in the figure is that computed by simply substituting the median point values in Table 10-3 into Eq. (10-8); the median value and the 90% probability interval are associated with the distribution.

Another computer program for calculating the uncertainty of the top event is the Systematic Probabilistic Analysis by Sampling Methods (SPASM) program, which is part of the WAM series (see Section 10-1). The SPASM code has evolved from the Monte Carlo code SAMPLE and has been designed to [10]:

1. include the capability of describing different types of failure distributions for components besides the lognormal distribution,
2. provide plots of the resultant distributions, and
3. operate by using the output of WAM–CUT, which minimizes the amount of user interface required with the WAM–CUT and WAM–BAM programs.

SPASM evaluates the probability distribution for the top event of a fault tree by using as input data the first and second moment probability polynomials, which are part of the output of WAM–CUT. Details on how these polynomials are calculated and used are available [10].

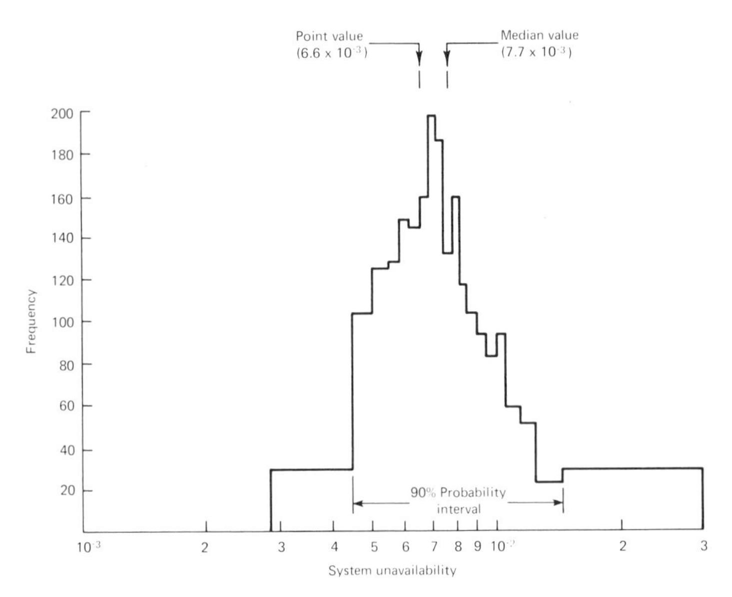

Fig. 10-4 *Frequency distribution for the example* SAMPLE-A *calculation.* [*From Reactor Safety Study, Appendix II, WASH-1400,* (*October 1975*).]

10-4 AUTOMATED FAULT TREE CONSTRUCTION

In an attempt to simplify the construction of fault trees of complex systems and to improve the completeness of safety analyses, automated system logic models have been developed. The synthetic tree method [27, 28] is more useful for electrical systems since it uses transfer functions as models for component failures. Trees for chemical process systems have been developed with the digraph method [29–31], which utilizes the ideas of signal flow graphs (see Section 6-4).

A general, computer-implemented approach has been developed for modeling nuclear and other complex systems involving mechanical, electrical, and hydraulic components and human interactions [32–35]. The methodology for the component models employs an extension of truth tables, known as "decision tables." A decision table is used to describe each possible output state as a complete set of combinations of states of inputs and internal operational or failed states.

The Computer Automated Tree (CAT) code has been developed to utilize the decision table method [35]. As examples of the way in which the CAT code can be usefully employed in analyzing systems, the residual heat removal (RHR) system [32] and a containment spray recirculation system [33] of a nuclear reactor have been studied. To date, however, the CAT code has not been widely implemented because it has not yet shown promise for use in constructing very complicated nuclear power plant fault trees.

REFERENCES

1. J. B. Fussell and J. S. Arendt, System reliability engineering methodology: A discussion of the state of the art, *Nucl. Safety* **20,** 541 (1979).
2. W. E. Vesely and R. E. Narum, PREP and KITT: Computer Codes for the Automatic Evaluation of Fault Trees. Idaho Nuclear Corp., Idaho Falls, Idaho, Rep. IN 1349 (1970).
3. Reactor Safety Study—An Assessment of Accident Risks in U. S. Commercial Nuclear Power Plants. U. S. Nuclear Regulatory Commission Rep. WASH-1400, NUREG 75/014 (October 1975).
4. W. E. Vesely and F. F. Goldberg, FRANTIC–A Computer Code for Time-Dependent Unavailability Analysis. U. S. Nuclear Regulatory Commission Rep. NUREG-0193 (October 1977).
5. R. C. Erdmann, J. E. Kelly, H. R. Kirch, F. L. Leverenz, and E. T. Rumble, A method for quantifying logic models for safety analysis, *in* "Nuclear Systems Reliability Engineering and Risk Assessment" (J. B. Fussell and G. R. Burdick, eds.), p. 732. Soc. Industrial Applied Math., Philadelphia, Pennsylvania, 1977.
6. E. T. Rumble, F. L. Leverenz, and R. C. Erdmann, Generalized Fault Tree Analysis for Reactor Safety. Electric Power Research Institute Rep. EPRI 217-2-2 (June 1975).
7. F. L. Leverenz, and H. Kirch, User's Guide for the WAM–BAM Computer Code. Electric Power Research Institute Rep. 217-2-5 (January 1976).
8. R. C. Erdmann *et al.,* Probabilistic Safety Analysis. Electric Power Research Institute Rep. EPRI NP-424 (April 1977).
9. R. C. Erdmann *et al.,* WAMCUT, A Computer Code for Fault Tree Evaluation. Electric Power Research Institute Rep. EPRI NP-803 (June 1978).
10. R. C. Erdmann *et al.,* Probabilistic Safety Analysis III. Electric Power Research Institute Rep. EPRI NP-749 (April 1978).
11. W. Gateley, D. Stoddard, and R. L. Williams, GO, A Computer Program for the Reliability Analysis of Complex Systems. Kaman Science Corporation Rep. KN-67-704(R) (April 1968).
12. R. L. Williams and W. Y. Gateley, GO Methodology—An Overview. Electric Power Research Institute Rep. EPRI NP-765 (May 1978).
13. W. Y. Gateley and R. L. Williams, GO Methodology—System Reliability Assessment and Computer Code Manual. Electric Power Research Institute Rep. EPRI NP-766 (May 1978).
14. Kaman Sciences Corporation, GO Methodology—Fault Sequence Identification and Computer Code Manual. Electric Power Research Institute Rep. EPRI NP-767 (May 1978).
15. R. L. Williams and W. Y. Gateley, Use of the GO Methodology to Directly Generate Minimal Cut Sets, *in* Nuclear Systems Reliability Engineering and Risk Assessment" (J. B. Fussell and G. R. Burdick, eds.), p. 825. Soc. Industrial Appl. Math., Philadelphia, Pennsylvania, 1977.
16. D. E. Wood and N. J. Becar, A Comparison of Results from the GO Methodology and Fault Tree Analysis. Kaman Sciences Rep. K77-38U(R) (1977).

17. D. E. Wood and N. J. Becar, Risk Analysis of a Spent Fuel Receiving and Storage Facility Using the GO Methodology. Kaman Sciences Rep. KSC-1006-1 (1979).
18. D. E. Wood, Probabilistic calculations of radioactive material release distributions using the GO methodology, *Trans. Am. Nucl. Soc.* **30,** 354 (1978).
19. G. R. Burdick, N. H. Marshall, and J. R. Wilson, COMCAN—A Computer Program for Common Cause Failure Analysis. Aerojet Nuclear Company Rep. ANCR-1314 (May 1976).
20. R. B. Worrell and G. R. Burdick, Quantitative Analysis in Reliability and Safety Studies, *IEEE Trans. Reliability* **R-25,** 164 (1976).
21. C. L. Cate and J. B. Fussell, BACFIRE—A Computer Code for Common Cause Failure Analysis, Rep. NERS-77-02. Nuclear Engineering Department, University of Tennessee, Knoxville, Tennessee (February 1977).
22. D. M. Rasmuson *et al.*, COMCAN-II: A Computer Program for Common Cause Failure Analysis. Idaho National Engineering Laboratory Rep. TREE-1289 (September 1978).
23. D. P. Wagner, C. L. Cate, and J. B. Fussell, Common cause failure analysis methodology for complex systems, *in* "Nuclear Systems Reliability Engineering and Risk Assessment" (J. B. Fussell and G. R. Burdick, eds.), p. 289. Soc. Industrial Applied Math., Philadelphia, Pennsylvania, 1977.
24. F. L. Leverenz, E. T. Rumble, and R. C. Erdmann, A dependent-event model for fault trees, *Trans. Am. Nucl. Soc.* **21,** 212 (1975).
25. E. T. Rumble and J. Olmos, Fault tree analysis incorporating dependent events, *in* "Probabilistic Analysis of Nuclear Reactor Safety," Vol. 3, p. X.4-1. American Nuclear Society, LaGrange Park, Illinois, 1978.
26. E. T. Rumble, F. L. Leverenz, and R. C. Erdmann, Generalized Fault Tree Analysis for Reactor Safety. Electric Power Research Institute Rep. EPRI 217-2-2 (June 1975).
27. J. B. Fussell, A formal methodology for fault tree construction, *Nucl. Sci. Eng.* **52,** 421 (1973).
28. J. B. Fussell, Computer aided fault tree construction for electrical systems, *in* "Reliability and Fault Tree Analysis" (R. E. Barlow, J. B. Fussell, and N. D. Singpurwalla, eds.). Soc. Industrial Applied Math., Philadelphia, Pennsylvania, 1975.
29. G. J. Powers and F. C. Topkins, Fault tree synthesis for chemical processes, *Am. Inst. Chem. Eng. J.* **20,** 376 (1974).
30. S. A. Lapp and G. J. Powers, The Synthesis of Fault Trees, *in* "Nuclear Systems Reliability Engineering and Risk Assessment" (J. B. Fussell and G. R. Burdick, eds.), p. 778. Soc. Industrial Applied Math, Philadelphia, Pennsylvania, 1977.
31. S. A. Lapp and G. J. Powers, Computer-aided synthesis of fault-trees, *IEEE Trans. Reliability* **R-26,** 2 (1977).
32. S. L. Salem, G. E. Apostolakis, and D. Okrent, A new methodology for the computer-aided construction of fault trees, *Ann. Nucl. Energy* **4,** 417 (1977).
33. J. S. Wu, S. L. Salem, and G. E. Apostolakis, The use of decision tables in the systematic construction of fault trees, *in* "Nuclear Systems Reliability Engineering and Risk Assessment" (J. B. Fussell and G. R. Burdick, eds.), p. 800. Soc. Industrial Applied Math., Philadelphia, Pennsylvania, 1977.
34. S. L. Salem, J. S. Wu, and G. Apostolakis, Decision table development and application to the construction of fault trees, *Nucl. Technol.* **42,** 51 (1978).
35. G. E. Apostolakis, S. L. Salem, and J. S. Wu, CAT: A Computer Code for the Automated Construction of Fault Trees. Electric Power Research Institute Rep. EPRI NP-705 (March 1978).

PART II

Nuclear Power Risks

CHAPTER 11

Risk Concepts

11-1 DEFINITION OF RISK

The methods of reliability engineering, including fault tree and event tree analysis, are used to determine the expected frequency of an undesired event (in events per unit time). Such an event might lead to a variety of consequences. In the Reactor Safety Study [1], for example, the types of consequences considered were: (a) early fatalities, (b) early illness, (c) latent cancer fatalities, (d) thyroid nodule formation, and (e) property damage. In nonnuclear accident studies involving events that have fairly immediate consequences (and not a delayed response such as radiation-induced illness), the consequences typically considered from accidents are [2]: (a) fatalities, (b) man-hours or man-days lost,* and (c) dollars lost. Other more intangible consequences are the deterioration of air and water quality.

One way to estimate the risk for an undesired event is to evaluate the *expected frequency* in events per unit time and the *expected damage,* which is the magnitude of a consequence. Then a customary definition of risk $\mathcal{R}$, in terms of both the expected frequency of occurrence $\mathcal{F}$ and the expected damage $\mathcal{D}$, is

$$\mathcal{R} = \mathcal{F}\mathcal{D}. \tag{11-1}$$

It should be noted that there are other possible definitions of risk; for example, if $\mathcal{R}_k = \mathcal{F}\mathcal{D}^k$, then for $k > 1$ this risk function would tend to amplify the importance of events with large damages.

Risk differs from *hazard,* which is a condition with the potential of causing an undesired consequence. There is *danger* from exposure to a hazard.

A more general interpretation of risk than in Eq. (11-1) involves frequency *and* damage in a nonproduct form. This broader definition is

* Man-days includes the time lost by both men and women, i.e., person-days.

useful because it reminds us that the outcome of a major catastrophic event could be a spectrum for damage, with each magnitude having its own corresponding frequency of occurrence. This situation can be illustrated in the form of cumulative distributions or *risk curves*, illustrated in Fig. 11-1, in which the ordinate is the frequency that a damage (i.e., fatalities) of magnitude X or greater will be produced. Since a point estimate from Eq. (11-1) provides less information than a curve, it is preferable whenever possible to obtain a curve for assessing risks to society from major catastrophic events.

The example risk curves in Fig. 11-1 show frequencies obtained from

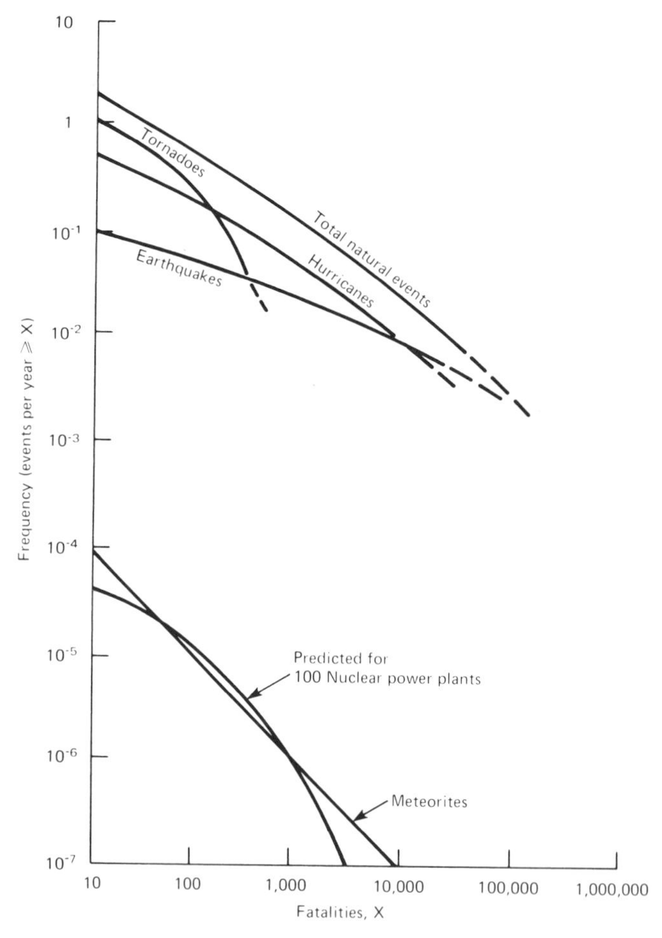

Fig. 11-1 *Frequency of natural events involving fatalities. [From Reactor Safety Study, WASH-1400 (October 1975).]*

statistical data of different natural events involving fatalities in the United States. Each of the natural events curves has an associated uncertainty band (not shown) that increases as the number of fatalities increases; the typical 95% confidence factor is roughly 1.3 (i.e., there is approximately 30% error in the corresponding frequency estimates) [1]. The justification of the predicted results for the 100 nuclear power plants will be discussed in Chapter 12.

It may be pointed out that if there is more than one type of consequence, the risk curves of Fig. 11-1 could be replaced by a "risk surface" [3]. For example, for damage X from one type of consequence and Y from a second type, a three-dimensional figure is obtained for frequency $\mathcal{F}(\geq X, \geq Y)$ versus X and Y, as shown in the three-dimensional Fig. 11-2.

Still another approach for interpreting risk may be developed by using a mathematical notation that somewhat parallels that in Chapter 2. We first define *risk density*, $\mathcal{R}_i(x_j, t)$, which is the frequency (in events per unit time) that an event E_i will occur at time t producing an ultimate damage of consequence type j between x_j and $x_j + dx_j$, per unit dx_j. This risk is considered a density function because it has units of (consequence)$^{-1}$.

As an example, $\mathcal{R}_i(x_j, t)$ could be the frequency of deaths (for $j = 1$) or disability man-hours (for $j = 2$) from an earthquake event E_i with Richter magnitude 6.5 (for $i = 1$) or 8.3 (for $i = 2$). Another example might be the frequency of deaths ($j = 1$) from a nuclear reactor loss of coolant accident E_i arising from a break in a pipe of ≤ 5 cm diameter (for $i = 1$) or ≥ 5 cm diameter (for $i = 2$). In the latter case, if there are changes in the reactor, possibly arising from changes with age in probabilities of a pipe break or from changes in the hazard rates $\lambda(t)$ of the various reactor safety sys-

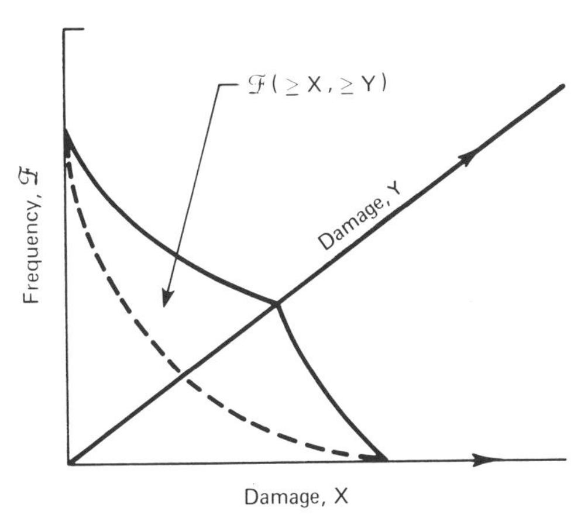

Fig. 11-2 *Risk surface.* [*Adapted from D. Okrent,* Nucl. Safety ***20***, *148* (*1979*).]

tems, then $\mathcal{R}_i(x_j, t)$ will change with time; on the other hand, for purely random system failures with constant values of λ, and with no change with time in the probability of an initiating event, then $\mathcal{R}_i(x_j, t)$ will be independent of time. Note also that the use of the word "ultimate" in the definition of risk density implies that the time delay between the occurrence of an event and the appearance of the consequence is deemed not important; if such time delays were important, the definition would have to be generalized.

In terms of $\mathcal{R}_i(x_j, t)$, the risk $\mathcal{R}_i(\geq X_j, t)$ may be defined as

$$\mathcal{R}_i(\geq X_j, t) = \int_{X_j}^{\infty} \mathcal{R}_i(x_j, t)\, dx_j. \tag{11-2}$$

That is, *risk* $\mathcal{R}_i(\geq X_j, t)$ is the frequency that event E_i will occur at time t having an ultimate damage (of consequence type j) of magnitude equal to or greater than X_j. For example, $\mathcal{R}_i(\geq X, t)$ could be the frequency of deaths greater than magnitude X from an earthquake event E_i.

Often it is desired to estimate the risk arising from all possible initiating events that might occur during operation of a system at a given time t. This *composite risk* $\mathcal{R}(\geq X_j, t)$ is obtained by summing Eq. (11-2) to obtain

$$\mathcal{R}(\geq X_j, t) = \sum_i \mathcal{R}_i(\geq X_j, t). \tag{11-3}$$

An example of $\mathcal{R}(\geq X, t)$ for the predicted fatalities arising from all accidents while operating 100 nuclear reactors is shown in Fig. 11-1.

The *time-integrated composite risk* $\mathcal{R}(\geq X_j, T)$ corresponding to Eq. (11-3) is

$$\mathcal{R}_i(\geq X_j, T) = \int_0^T \mathcal{R}(\geq X_j, t)\, dt. \tag{11-4}$$

For the example of the 100 nuclear reactors, this risk could be the predicted number of deaths arising from all accidents while operating each reactor over its lifetime of (say) $T = 40$ years.

When the operation of a system leads to risks from more than one consequence type, still another risk can be defined. *Aggregated risk* from event E_i, $\mathcal{R}_i^{(k)}(t)$, as measured in terms of consequence type k, is the risk arising from event E_i due to all the possible consequences. Aggregated risk may be written as

$$\mathcal{R}_i^{(k)}(t) = \sum_j a_j^{(k)} \int_0^{\infty} \mathcal{R}_i(x_j, t)\, dx_j, \tag{11-5}$$

where the $a_j^{(k)}$ are aggregation factors. Such an aggregation could involve a combination of all types of consequences (deaths, person-days of injury, property damage, etc.) as measured in dollars, for example.

In summary, the risk density $\mathcal{R}_i(x_j, t)$ is a basic building block for reporting results of a risk analysis although, as we shall see in the following chapters, usually the composite risk $\mathcal{R}(\geq X_j, t)$ from all accident initiating events is what is plotted for the different consequences of type j, $j = 1, 2$, etc.

11-2 PROBABILISTIC RISK ASSESSMENT PROCEDURE

There are two major tasks in a probabilistic risk assessment (PRA), such as performed for the Reactor Safety Study (RSS) [1]: to define the type of accidents that can occur and their frequency of occurrence, and to quantify the consequences of each event. To define an accident, we select an initiating event and then examine the subsequent events that can occur during propagation of the initiating event; a sequence of events leading to a particular accident may be termed an *accident sequence*. For the RSS, the frequency of a particular accident sequence was broken down into three general factors:

$$\begin{aligned} \mathcal{F}[\text{release}] = {} & \mathcal{F}[\text{initiating event}] \\ & \times P[\text{system failure given the initiating event}] \\ & \times P[\text{containment failure given the initiating} \\ & \qquad \text{event and system failure}], \end{aligned} \tag{11-6}$$

where the symbols $\mathcal{F}$ and P denote the frequency and probability of the event described within the brackets, respectively. Note that the second and third factors on the right-hand side of the equation are conditional probabilities. The technique used in the RSS to define a particular accident sequence or a sequence for failure of the reactor containment was the event tree method (as discussed in Chapter 9), while the engineered safety feature system failure probabilities and containment failure probabilities were obtained from fault tree analysis (discussed in Chapter 8).

In engineering systems less sophisticated than a nuclear reactor, an equation similar to Eq. (11-6) is appropriate but will not include any containment failure probability if there is no passively acting containment system serving as a backup to active engineered safety feature systems.

The second major task in a PRA is the quantification of the consequences of an accident. In many engineering systems in which the consequences may be directly calculated, this is a one-step process. In the case of the RSS, where dispersal of radioactivity was the hazard, the analysis was a two-step process: first, the radioactive releases from the core were identified for each key accident sequence; second, the consequences such as deaths, illness, and property damage were inferred from the amounts of radioactivity released.

For the RSS, the radioactivity releases from the reactor core for each key accident sequence were expressed as the fraction of reactor core radioactive inventory released to the external atmosphere as a function of time. The quantified releases were calculated by the CORRAL computer program, which will be described in Chapter 12. With this characterization of the radioactivity release and the results for the frequency of an accident, it is possible to obtain a histogram, such as shown in Fig. 11-3, for each type of accident that could occur. The histogram shows the frequency and magnitude of the various accidental radioactivity releases. By combining the histograms from all the accidents envisioned, one can obtain a composite histogram for all important accident contributors; the accidents then can be grouped into *release categories* according to the magnitude of radioactivity released in each accident.

To infer the consequences in the RSS, the release histograms for the key accident sequences were used as preliminary input data to a model for calculating consequences outside the reactor. To evaluate the consequences of the radioactive releases from each category, it was necessary to consider how the radioactivity would be dispersed in the environment, the number of people and amount of property exposed, and the effects of

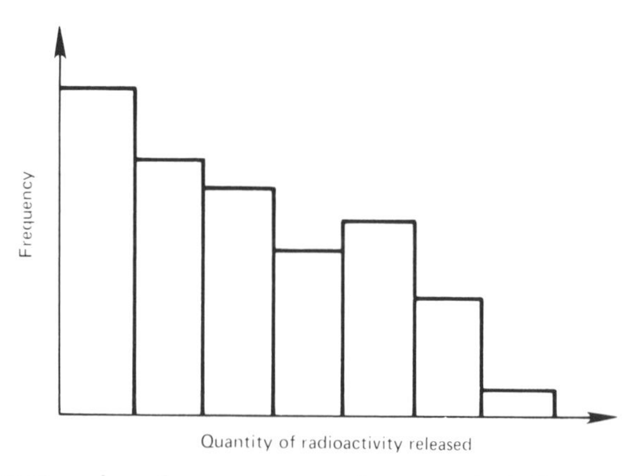

Fig. 11-3 *Illustrative release frequency versus release magnitude histogram.* [*From Reactor Safety Study, WASH-1400 (October 1975).*]

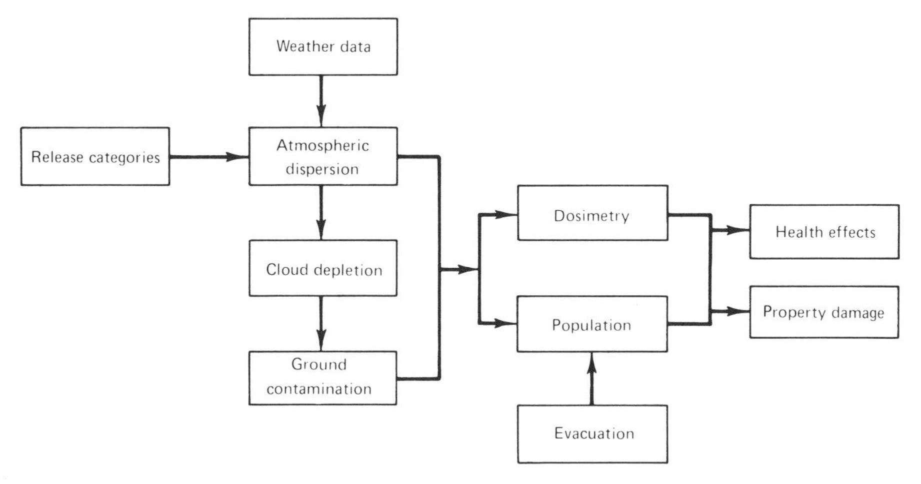

Fig. 11-4 *Schematic outline of the consequence model. [From Reactor Safety Study, Appendix VI, WASH-1400 (October 1975).]*

radiation exposure on people and contamination of property [4]. These major elements of the consequence predictions are indicated in Fig. 11-4.

The frequency for property damage can be inferred from the equation

$$\begin{aligned}\mathscr{F}[\text{property damage at specified location}] &= \mathscr{F}[\text{release}] \\ &\quad \times P[\text{radioactivity is transported to specified location}] \\ &\quad \times P[\text{radioactivity is deposited at specified location}]. \qquad (11\text{-}7)\end{aligned}$$

For the effects on human beings, a similar equation applies except it is necessary to account for the possibility of human evacuation.

When we desire to determine the consequences from a number of possible releases, an iterative scheme, such as the one in Fig. 11-5, can be followed. The essence of such a calculation is that each of the first four boxes—radioactive source, containment leakage, atmospheric dispersion, and population distributions—may contain numerous options with an associated probability for each option. On each pass through the model, we select one option from each of the first four boxes, and evaluate the consequence of that particular combination of events in the fifth box. In order to obtain the complete spectrum of all possible combinations of events, many passes through the iteration sequence are required.

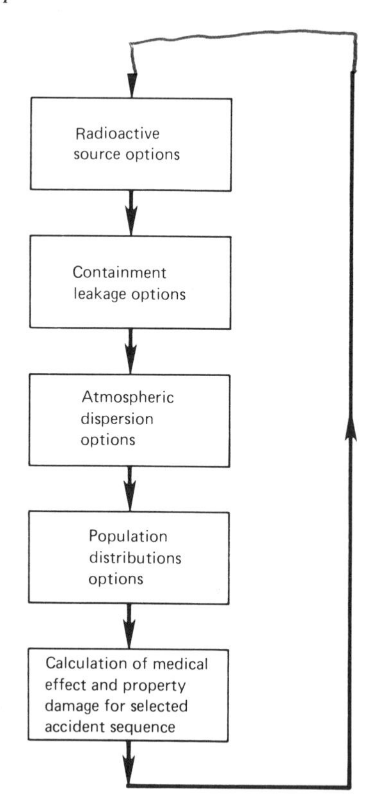

Fig. 11-5 *Schematic view of calculational model. [From Reactor Safety Study,* WASH-*1400, draft (August 1974).]*

EXERCISES

11-1 Use Fig. 11-1 to fit the risk $\mathcal{R}(\geq X, t)$ of death X from all natural events to an approximate analytical expression valid for $10 \leq X \leq 10^3$ fatalities.

11-2 Differentiate the integral in Eq. (11-2) to obtain the risk density $\mathcal{R}_i(x_j, t)$ in terms of the risk $\mathcal{R}_i(\geq X_j, t)$.

REFERENCES

1. Reactor Safety Study—An Assessment of Accident Risks in U.S. Commercial Nuclear Power Plants, U. S. Nuclear Regulatory Commission Rep. WASH-1400, NUREG 75/014 (October 1975).

2. H. Inhaber, Risk of Energy Production. Atomic Energy Control Board (Ottawa) Rep. AECB-1119/REV-3 (September 1980).
3. D. Okrent, Risk-benefit evaluation for large technological systems, *Nucl. Safety* **20**, 148 (1979).
4. Reactor Safety Study—An Assessment of Accident Risks in U.S. Commercial Nuclear Power Plants, Appendix VI Calculation of Reactor Accident Consequences, U.S. Nuclear Regulatory Commission Rep. WASH-1400, NUREG 75/014 (October 1975).

CHAPTER 12

Risks for Light Water Reactors

12-1 INTRODUCTION

The Reactor Safety Study (RSS) for United States commercial light water reactors [1] was a major advancement in the systematic study of risks for technological devices because it emphasized the importance of accidents other than the "maximum credible accident." The RSS is also called WASH-1400 or the "Rasmussen Report," named after the man who directed the study. Although it was completed in 1975, the RSS still is the benchmark study for risks from nuclear power applications and has been the source of data and techniques for several follow-up studies.

In this chapter, we consider many details and results of the RSS, as well as some of its limitations and the resulting criticisms of the study. Portions of the RSS have been discussed in Chapters 8 through 10 covering event trees and fault trees and their numerical evaluation. In addition, the methodology of risk assessment for nuclear reactors has been discussed in Chapter 11. Data for describing failures have been dealt with in Chapter 5 and appear in Appendix B.

The pressurized water reactor (PWR) plant selected for the study was the 778 MW_e Surrey Power Station, Unit I; the boiling water reactor (BWR) was the 1065 MW_e Peach Bottom Atomic Power Station, Unit II. These plants were the largest of each type that were about to start operation in 1972 when the study was begun. The fault tree and event tree analyses to determine the probabilities of accidents and the amounts of radioactivity released were done for these two reactors.

In order to make the results of the study applicable to other PWR and BWR power stations (i.e., to perform a "generic" risk analysis rather than a "site-specific" analysis), the consequences of the accidents were analyzed by using weather and population data for a set of generic sites obtained by averaging such data over the 68 sites of the first 100 nuclear power plants scheduled for operation in the United States.

12-2 RADIOACTIVE INVENTORY

Nuclear power plant accidents differ from those in nonnuclear power plants because they can potentially release significant amounts of radioactivity to the environment. By far the largest amount of radioactivity resides in the reactor core, although a smaller, but still large, amount of radioactivity is located in the spent fuel storage pool. In both of these locations, the fuel is subjected to heating due to absorption of energy from the radioactive decay of fission products. For example, immediately following the shutdown of a reactor that has operated for a month or more, the decay heat amounts to about 7% of the operating power level, although within an hour it has dropped to about 1%, and after a month it is about 0.1%.

Typical magnitudes of the radioactive inventories in the various plant locations are shown in Table 12-1. The core radioactive inventory in curies (Ci)* depends upon the reactor power level, and slightly on the length of reactor operation. For most of the radionuclides, equilibrium has occurred after several months of sustained operation. The reactor core radioactive inventory is based on 550 days of sustained operation and represents the expected equilibrium radioactivity in an operating reactor.

The inventory in Table 12-1 for the Spent Fuel Storage Pool (SFSP) is based on a plant that has a common pool serving two 1000 MW_e reactors. The average number of spent fuel assemblies stored in the SFSP is based on assumed normal unloading and shipment schedules. The radioactive inventory in the shipping cask is based on a full load of fuel in the largest shipping cask licensed, and the shortest decay period (150 days) allowed for fuel shipped in the container. The refueling category represents the radioactivity in a single fuel assembly at three days after reactor shutdown. This time is typical of the earliest time after shutdown that transfer of fuel from the reactor core to the SFSP begins.

The inventory of radioactivity is mostly in the fuel itself, although the gaseous isotopes can migrate to the gap region between the fuel and cladding of a rod or to the rod's gas plenum region. This combined (gap) inventory is generally only about 0.01 to 0.02 of the total inventory.

The RSS treated accidents involving all the radioactive sources listed in Table 12-1, but the main focus was on potential accidents involving fuel in the reactor core and the SFSP. The radioactivity in the waste gas storage tanks and liquid waste storage tanks is very small compared to the other sources and could not result in public consequences nearly as serious as accidents involving melting of the fuel in the reactor core or in the SFSP.

* The SI equivalent, the becquerel (Bq), is related by 1 Ci = 3.7×10^{10} Bq.

Table 12-1 *Typical Radioactive Inventory for a 1000* MW_e *Nuclear Power Reactor*[a]

Location	Total inventory (curies)			Fraction of core inventory		
	Fuel	Gap	Total	Fuel	Gap	Total
Core[b]	8.0×10^9	1.4×10^8	8.1×10^9	9.8×10^{-1}	1.8×10^{-2}	1
Spent fuel storage pool (max)[c]	1.3×10^9	1.3×10^7	1.3×10^9	1.6×10^{-1}	1.6×10^{-3}	1.6×10^{-1}
Spent fuel storage pool (av)[d]	3.6×10^8	3.8×10^6	3.6×10^8	4.5×10^{-2}	4.8×10^{-4}	4.5×10^{-2}
Shipping cask[e]	2.2×10^7	3.1×10^5	2.2×10^7	2.7×10^{-3}	3.8×10^{-5}	2.7×10^{-3}
Refueling[f]	2.2×10^7	2×10^5	2.2×10^7	2.7×10^{-3}	2.5×10^{-5}	2.7×10^{-3}
Waste gas storage tank	—	—	9.3×10^4	—	—	1.2×10^{-5}
Liquid waste storage tank	—	—	9.5×10^1	—	—	1.2×10^{-8}

[a] From Reactor Safety Study, WASH-1400 (October 1975).
[b] Core inventory based on activity ½ hour after shutdown.
[c] Inventory of ⅔ core loading; ⅓ core with three day decay and ⅓ core with 150-day decay.
[d] Inventory of ½ core loading; ⅛ core with 150 day decay and ⅓ core with 60-day decay.
[e] Inventory based on 7 PWR or 17 BWR fuel assemblies with 150-day decay.
[f] Inventory for one fuel assembly with 3-day decay.

12-3 REACTOR ACCIDENTS

12-3-1 Loss of Coolant Accidents

The occurrence of a rupture in the reactor coolant system (RCS) could lead to a loss of coolant accident (LOCA) that allows the fuel in the core to overheat (due to decay heat) unless emergency cooling water is supplied. Alternatively, overheating of fuel could result from transient events that cause the reactor power to increase beyond the heat removal capacity of the reactor cooling system or that cause the heat removal capacity of the reactor cooling system to drop below the core heat generation rate.

The specific LOCA initiating events analyzed in the RSS were:

(a) Large pipe breaks (from approximately 15 cm to 0.9 m equivalent diameter).*

* The actual maximum pipe diameter varies somewhat from plant to plant but is taken to be 0.9 m equivalent diameter. The large pipe break is normally considered to be double-ended, which means that coolant is expelled through both ends of the severed pipe.

(b) Small to intermediate pipe breaks (from about 5 cm to 15 cm quivalent diameter).
(c) Small pipe breaks (from 1 cm to 5 cm equivalent diameter).
(d) Reactor pressure vessel ruptures.
(e) Steam generator ruptures.
(f) Ruptures between systems that interface with the RCS.

The Engineered Safety Feature (ESF) systems described in Appendix E are designed to mitigate the consequences of such accidents. Assuming that all ESF systems operate as designed, a brief description of a LOCA is [1]:

1. A break in the RCS would occur and the high-pressure, high-temperature RCS water would be rapidly discharged into the containment.
2. The emergency core cooling system (ECCS) would operate to keep the core adequately cooled.
3. Any radioactivity released from the core could be largely contained in the low leakage containment building.
4. Natural deposition processes and radioactivity removal systems would remove the bulk of the released radioactivity from the containment atmosphere.
5. Heat removal systems would reduce the containment pressure, thereby reducing leakage of radioactivity to the environment.

If the ESFs operate as designed, the reactor core would be adequately cooled following a LOCA and only minor consequences would result. However, the potential consequences would be much greater if ESF failures were to result in overheating of the reactor core; such situations were of major concern in the RSS risk analysis.

At two key stages in the course of a potential core meltdown there would be conditions that would have the potential to result in a steam explosion that could rupture the reactor vessel and/or the containment. These conditions might occur if molten fuel fell from the core region into water at the bottom of the reactor vessel or if the fuel melted through the bottom of the reactor vessel and fell into water in the bottom of the containment.

If a steam explosion were to occur within the reactor vessel, it was considered possible that both large and small containments could be penetrated by a large missile [1]. Such occurrences might release substantial amounts of radioactivity to the environment. However, these modes of containment failure were predicted to have low probabilities of occurrence.

12-3-2 Transients

Transient events were assumed to include all those situations (except for a LOCA, which was treated separately) that could lead to power-coolant imbalances [1]. When viewed in this way, transients cover the reactor in its shutdown condition as well as in its various operating conditions. The shutdown case is important because many transient conditions result in shutdown of the reactor, in which case decay heat removal systems are needed.

Transients possibly could occur as a consequence of an operator error or the malfunction or failure of equipment. Many transients would be handled by the reactor control system, which would return the reactor to its normal operating condition. Others could be beyond the capability of the reactor control system and require reactor shutdown by the reactor protection system (RPS) in order to avoid damage to the reactor fuel. Regardless of the way in which transients might cause core melting, the consequences would be essentially the same as for a LOCA.

The RSS divided potential transients into anticipated and unanticipated events. Anticipated transients, such as loss of off-site power, were found to occur about ten times per year per reactor. The two anticipated transients found to be important were those involving loss of off-site power and loss of heat removal systems. Unanticipated transients, such as rod ejection, were found to be relatively unimportant because of their relatively low probabilities.

12-3-3 Spent Fuel Storage Pool Accidents

The decay heat levels in freshly unloaded fuel assemblies that may be stored in the SFSP may be sufficiently high to cause fuel melting if the water is completely drained from the pool. Because the maximum amount of fuel stored in the pool immediately after refueling is smaller than that in the core and because it has had time (72 hours minimum) for radioactive decay, it is at most about one-sixth of the heat source of a reactor core. Therefore melt-through of the bottom structure of the pool would occur at a much slower rate; indeed, on the average, fuel in the pool will have undergone about 125 days of decay, and it is questionable that such fuel would melt. However, to assure that the risk would not be underestimated, it was assumed in the RSS that this fuel would melt.

The most probable ways in which SFSP accidents could occur have been determined to be the loss of the pool cooling system or the perforation of the bottom of the pool [1]. The latter could occur, for example, if a shipping cask were dropped in the pool or on the top edge of the pool; this type of accident is very unlikely. The loss of cooling capability, which was determined to be somewhat more probable, requires that a number of audible alarms be inoperative or ignored and that the visual observations be so lax as to permit the lowering of the pool water level to continue uncorrected for about two weeks—the approximate time required to boil off the SFSP water if cooling capability is lost.

12-4 METHODS FOR CONSEQUENCE ANALYSIS

12-4-1 Determination of Fission Product Releases

As discussed in Chapter 11, the first step in the analysis of the consequences of a nuclear power accident is to determine the magnitude of the radioactivity release. This is done for each possible accident sequence by estimating the release of fission products to the reactor containment, their distribution within the containment, and their leakage from the containment. In the RSS, seven classes of fission product species were considered, including the noble gases, the halogens (elemental and organic halide forms), the alkali metals, the tellurium group, the alkaline earths, the noble metals group, and the refractory oxide group [1]. In order to develop the proper release rate terms, the release values for each of these species were expressed as fractions of the total core inventory and then multiplied by the fraction of the core that participates in the release process.

Four basic modes for release of the fission products were considered for both the PWR and BWR systems. These modes consist of (1) the gap release component, (2) the core melt release component, (3) the vaporization release component, and (4) the core steam explosion release component. The first three components occur sequentially in the order listed for accidents involving reactor core meltdown, while the timing of the fourth depends upon predictions of steam explosion events.

The CORRAL computer program was developed for the RSS and consists of a containment-process model that enables one to estimate the behavior of radioactivity in the reactor containment. It is a "multicompartment model" that represents the following processes: natural transport and deposition, removal of radioactivity by aqueous sprays, recirculation filter systems, once-through filter systems, water pool scrubbing, and leakage or exhaust from containment to the outside atmosphere; all these processes may occur simultaneously. The time-dependent analyses done by the CORRAL program require a computer solution of a set of coupled differential equations.

Because of the containment differences between the PWR and the BWR, two versions were developed for the RSS (CORRAL–PWR and CORRAL–BWR). This is because a BWR containment system is not a set of openly connected compartments like a PWR, where the whole containment system usually can be considered as "well mixed." Instead, the compartments of a BWR are usually closed to one another and flows between them occur in complex ways during postulated accidents. Four compartments were used to model the PWR, and as many as six compartments were used in some BWR accident sequences.

A CORRAL calculation begins by specifying input data such as the number of compartments and their arrangement, volumes, surface areas, heights, etc., for the reactor plant. The fission product release component fractions, the time for releases, and the organic iodide formation ratio must be supplied according to the source term information. For example, the fission product release fractions from the core vary according to the particular mode of release of the fission products, as shown in Table 12-2.

Containment spray parameters, filter flow rates, and decontamination factors associated with external leakage also are input values for the CORRAL code. The times for all events in the accident sequence, along with physical conditions as a function of time, were provided by a degraded core analysis. The physical conditions included pressure, temperature, and composition of the internal atmosphere, flow rates between compartments, leak rates to the external atmosphere, and the sizes of puff-type releases from the containment.

Table 12-2 *Total Core Fission Product Release Fractions*[a]

Fission product	Gap release fraction	Meltdown release fraction	Vaporization release fraction[e]	Steam explosion fraction[f]
Xe, Kr	0.030	0.870	0.100	$(X)(Y)$ 0.90
I, Br	0.017	0.883	0.100	$(X)(Y)$ 0.90
Cs, Rb	0.050	0.760	0.190	
Te[b]	0.0001	0.150	0.850	$(X)(Y)$ (0.60)
Sr, Ba	0.000001	0.100	0.010	
Ru[c]		0.030	0.050	$(X)(Y)$ (0.90)
La[d]		0.003	0.010	

[a] From Reactor Safety Study, Appendix VII, WASH-1400 (October 1975).
[b] Includes Se, Sb.
[c] Includes Mo, Pd, Rh, Tc.
[d] Includes Nd, Eu, Y, Ce, Pr, Pm, Sm, Np, Pu, Zr, Nb.
[e] Exponential loss over 2 hours with halftime of 30 minutes. If a steam explosion occurs prior to this, only the core fraction not involved in the steam explosion can experience vaporization.
[f] X = Fraction of core involved in the steam explosion. Y = Fraction of inventory remaining for release by oxidation.

After the input specifications have been completed and initial conditions have been set, the code uses the data to continuously compute changing properties and fission product removal rates as a function of time. These values are used in incremental solutions to the coupled set of differential equations to obtain the time-dependent fission product concentrations and accumulations in each compartment of the containment.

The output from CORRAL calculations consists of cumulative fractional releases from containment versus time for each of the seven fission product species. The releases are identified as occurring at ground level or at an elevation, and the temperature of the released gases is provided to assist subsequent estimations of plume buoyancy. The time of release is also specified in order that the radioactive decay corrections can be made for the plume.

12-4-2 Determination of Consequences

Introduction

To evaluate the consequences of the radioactive releases from each category, it is necessary to consider how the radioactivity is dispersed in the environment, the number of people and amount of property exposed, and the effects of radiation exposure on people and contamination of property. For this purpose, the Consequence (of) Reactor Accident (CRAC) code was developed [2].

A schematic outline of the consequence model calculations is given in Fig. 11-4. The starting point of the analysis is the quantity of radioactive material that could be released from the containment to the environment in the event of an accident. The spectrum of releases to the environment were discretized, in a manner shown in Fig. 11-3, into nine PWR and five BWR release categories, each with its associated point estimates of the frequency of occurrence and release magnitudes, as determined by event tree and fault tree analysis.

Dispersion Model

The meteorological model computes the dispersion of radioactive material in terms of its concentration in the air and on the ground as a function of time after the accident and distance from the reactor. As the plume from a reactor accident is carried away from the site by wind, atmospheric diffusion would be continually acting to disperse the contaminants at a rate depending on the wind speed, thermal stability, and underlying characteristics of the terrain. The average wind from ground level to a height that encompasses the plume top is the single most important parameter, as it determines the directions of transport and the initial volume of air into which the contaminant is diluted.

The accident consequence code uses the Gaussian plume model, with some modifications, to predict the way radioactivity is dispersed in the atmosphere. Such a model is a standard technique for calculating the dispersal of radioactivity in air [3, 4]. Computations of the plume dimension (vertical and lateral) were made for each of 34 downwind spatial intervals with midpoints varying from 0.4 km to 684 km.

The weather data used in the model was obtained from hour-by-hour meteorological records covering a 1-year period at 6 sites that would typify the locations of the first 100 large nuclear power plants. The characteristics of the 6 sites are shown in Table 12-3. The accident starting times were determined by systematic selection from the various sets of applicable meteorological data: one quarter of the data points were chosen from each season of the year and half from each group were taken in the daytime. The weather stability and wind velocity following an accident were assumed to change according to the weather data for the sites; in the RSS, 90 weather samples from the averaged 1-year records were taken, and each sample was thus assigned a probability of occurrence of 1/90.

Population Model

To determine the population that could be exposed to potential releases of radioactivity, 1970 census bureau data was used in the RSS to determine the number of people as a function of distance from the reactor in each of sixteen 22.5° sectors around each of the sites of Table 12-3.

In the case of a potentially serious accidental release, it was assumed

Table 12-3 *Number of Reactors Assigned to the Composite Sites*[a]

Characteristics	Number of sites	Number of reactors	
		BWR	PWR
Atlantic coastal site	10	5	9
Large river valley in Northeast	10	6	8
Great Lakes shore	4	3	2
Southeast river valley influenced by Bermuda High	17	7	23
Central Midwest plain	23	13	18
Pacific coastal site	4	0	6
Total	68	34	66

[a] From Reactor Safety Study, Appendix VI, WASH-1400 (October 1975).

that people living within 40 km of the plant, and located in the direction of the wind, would be evacuated in order to reduce their exposure to radioactivity. An evacuation model representing this process was developed based on a substantial number of actual evacuations that have occurred in the United States. The population was assumed to move radially outward, with 30% having an effective speed of 0 km/hr, 40% moving at 1.9 km/hr, and the remaining 30% traveling at 11.3 km/hr. The time available for evacuation of people took into account the time required for the plume to reach the individual.

Health Effects Model

Having computed the concentrations of radioactivity in the air and on the ground, the consequence model then computed the potential doses that could accrue from the following potential modes of exposure:

1. External irradiation from the passing cloud. This exposure would occur over a period of about one-half hour to a few hours.
2. Internal irradiation from inhaled radionuclides. While the inhalation would take place over the same time period as external irradiation from the passing cloud, the dose accumulated would be controlled by the residence time of the various radionuclides in the various parts of the body.
3. External irradiation from radionuclides deposited on the ground.
4. Internal irradiation from the inhalation of resuspended radionuclides that had been deposited on the ground. (This exposure mode does not contribute significantly to predicted doses.)
5. Ingestion of radionuclides from contaminated crops, water, and milk. (Since this type of exposure could be controlled by constraints placed on consumption until levels of radioactivity were below maximum permissible concentrations, it would not contribute significantly to predicted doses.)

The health effects that could be associated with a reactor accident are divided into three categories. Early and continuing somatic effects include the deaths, within days to weeks after exposure, due to large, acute doses of radiation; they also include illnesses and deaths within a year of exposure. In general, these early effects are primarily associated with individual total-body doses of 100 rads* or more and thus would be limited to persons within, at most, 80 km or so of the reactor.

The latent somatic effects include cancer fatalities as well as benign thyroid nodules; from radiation therapy experience, such effects are typically observed 2 to 30 years after irradiation. Finally, there are genetic

* The SI equivalent, the gray (Gy), is simply related by 100 rad = 1 Gy.

effects, which do not manifest themselves in the irradiated individuals, but rather in their descendants. In contrast to the early somatic effects, both latent cancer and genetic diseases are random phenomena whose probability of occurrence is some function of the dose magnitude. For this reason, both late somatic and genetic effects were calculated on the basis of the total dose to the population (cases per million man-rem†) rather than from individual doses. Late somatic and genetic effects may result from even very low doses, but with a very low incidence. Consequently, these effects may occur at long distances (e.g., 320 km) from the reactor, at which a small dose might still be received.

Property Damage Model

The property damage model provides a means to estimate the financial costs to the public that might occur as the result of a potential accident in a nuclear power plant. Although the damage that might occur is called property damage, it must be recognized that no property located off the reactor site would be physically damaged. Rather, the property would become sufficiently contaminated with radioactive material that its usefulness would be temporarily or permanently impaired. Thus, before the property could become useful again, the radioactivity would have to decay or weather away until it reached acceptable levels, or decontamination action would be needed to achieve these levels.

The property damage model considers the effects of both decay and decontamination. Decontamination would range from a simple washing (that may yield decontamination factors of 2) to more thorough procedures that yield a decontamination factor of about 20. The model assumed that, should an accident occur, such measures would be regarded as reasonable and would be implemented where appropriate, thus reducing the area that would not be inhabited. In the RSS an acceptable dose level after decontamination was chosen to be 25 rem in 30 years for urban areas and 10 rem in 30 years for areas in which the population density is low.

The major contributor to the overall cost would be from those areas in which reasonable decontamination procedures could not reduce levels of radioactivity to acceptable levels of dose. The dollar costs in the RSS for permanent relocation were $17,000 per capita to account for the value of property, land, and relocation costs. (Obviously this figure is now much too low, in view of inflation and escalating property values since 1975!) The cost for thorough decontamination was estimated to be about 10% of the value of the property. The model also includes the cost of managing a

† The SI equivalent of the rem is the sievert (Sv), which is related by 100 rem = 1 Sv; the man-rem unit implies that there is a linear relationship between radiation dose and increased probability of initiation of a cancer (see Section 18-6).

possible evacuation, the cost of temporary accommodation for the evacuees, and the decrease in value of interdicted property (such as impounded milk and crops).

12-5 QUANTIFICATION OF RADIOACTIVE RELEASES

12-5-1 Unavailabilities

The quantitative results of the PWR and BWR fault tree evaluations are collected in Tables 12-4 and 12-5, which list the unavailability according to system function.

The unavailability values can be categorized by hardware faults, test and maintenance outage, human errors, and common cause faults. Some of the major contributors are missing from the function unavailability tables because the contributor could not be simply classified when different failure contributions were combined (for example, a test and maintenance contribution combined with a hardware contribution).

The unavailabilities in Tables 12-4 and 12-5 are given in terms of median values, with the upper and lower values bounding the 90% probability range.

12-5-2 Key Accident Sequences

The event trees systematically identified about 1000 explicitly defined accident sequences potentially capable of causing significant releases of radioactive material from PWR and BWR nuclear power plants. Examination of both the Engineered Safety Feature (ESF) failures and the resulting physical processes involved in the various event tree sequences revealed patterns in timing or physical processes taking place that could be used to characterize the spectrum of releases of radioactivity from a reactor. Recognition of these patterns made it possible to select sets of key accident sequences to define the spectrum of releases. These key accident sequences are defined in terms of the accident symbols and containment failure modes defined in Tables 12-6 and 12-7; the key accident sequences for the PWR and BWR are given in Table 12-8.

PWR Release Categories

Each of the 38 PWR key accident sequences in Table 12-8 was analyzed by using the CORRAL code to obtain the magnitude of radioactivity released to the atmosphere. From these results it was found that the

Table 12-4 *PWR Function Unavailabilities*[a]

System	Unavailability (Q)			Major contributors (point estimates)			
	Q_{upper}	Q_{median}	Q_{lower}	Hardware	Test and maintenance	Human	Common cause
Electrical power	1.0×10^{-4}	1.0×10^{-5}	1.0×10^{-6}				
Reactor protection	1.0×10^{-4}	3.6×10^{-5}	1.3×10^{-5}	2.7×10^{-5}	1.2×10^{-5}		
Auxiliary feedwater							
0–8 hr after small LOCA	3.0×10^{-4}	3.7×10^{-5}	7.0×10^{-6}	2.0×10^{-6}	3.2×10^{-6}		3.0×10^{-5}[b]
8–24 hr after small LOCA	2.6×10^{-3}	1.2×10^{-3}	5.3×10^{-4}	1.1×10^{-3}			
0–8 hr without net	2.8×10^{-3}	4.5×10^{-4}	1.2×10^{-4}	2.0×10^{-6}	1.4×10^{-4}		1.1×10^{-4}[b]
Containment spray injection	7.8×10^{-3}	2.4×10^{-3}	1.0×10^{-3}	3.2×10^{-4}	1.5×10^{-4}		1.9×10^{-3}[b]
Emergency coolant injection							
Large LOCA	9.0×10^{-3}	5.6×10^{-3}	3.0×10^{-3}	NA[d]	NA	NA	NA
Small LOCA (5–15 cm diam.)	3.0×10^{-2}	9.5×10^{-3}	4.0×10^{-3}	NA	NA	NA	NA
Small–small LOCA (1–5 cm diam.)	2.7×10^{-2}	8.6×10^{-3}	4.4×10^{-3}	NA	NA	NA	NA
Containment spray recirculation	9.0×10^{-4}	1.0×10^{-4}	2.5×10^{-5}	5.2×10^{-6}	4.3×10^{-5}		2.8×10^{-5}[b]
Containment heat removal	3.0×10^{-4}	8.5×10^{-5}	3.0×10^{-5}	6.4×10^{-5}			1.0×10^{-5}[b]
Emergency coolant recirculation							
Large LOCA	3.1×10^{-2}	1.3×10^{-2}	4.4×10^{-3}	2.7×10^{-3}	1.0×10^{-4}	1.0×10^{-5}	6.0×10^{-3}[b]
Small LOCA	2.2×10^{-2}	9.0×10^{-3}	4.3×10^{-3}	2.0×10^{-3}			6.0×10^{-3}[b]
Small–small LOCA		6.0×10^{-3}[c]		2.0×10^{-3}			3.0×10^{-3}[b]
Containment leakage							
Pressure reduced to <0.0034 MPa (0.5 psi)	7.5×10^{-4}	2.0×10^{-4}	8.4×10^{-5}	1.3×10^{-4}			
Pressure not reduced to <0.0034 MPa	1.0×10^{-2}	2.0×10^{-3}	6.0×10^{-4}	8.8×10^{-4}			
Sodium hydroxide addition	1.1×10^{-2}	5.9×10^{-3}	3.6×10^{-3}	2.1×10^{-4}	4.8×10^{-3}		1.2×10^{-3}[b]

[a] From Reactor Safety Study, Appendix II, WASH-1400 (October 1975).
[b] Common cause resulting from human error.
[c] Point estimate value obtained only.
[d] NA—Not applicable to combined systems.

Table 12-5 *BWR Function Unavailabilities*[a]

System	Unavailability (Q)			Major contributors (point estimates)			
	Q_{upper}	Q_{median}	Q_{lower}	Hardware	Test and maintenance	Human	Common cause
Electrical power	1.0×10^{-5}	1.0×10^{-6}	1.0×10^{-7}				
Reactor protection	4.8×10^{-5}	1.3×10^{-5}	4.3×10^{-6}	5.8×10^{-6}	2.6×10^{-7}		1.9×10^{-6d}
Vapor suppression							
Large LOCA	4.5×10^{-4}	4.6×10^{-5}	6.8×10^{-6}	4.6×10^{-5}			
Small LOCA	2.3×10^{-3}	1.6×10^{-3}	1.2×10^{-3}	1.3×10^{-3}			
Emergency coolant injection[b]							
Large LOCA	2.1×10^{-3}	1.5×10^{-3}	1.0×10^{-3}	1.1×10^{-4}	1.1×10^{-3}		
Small LOCA (2.5–15 cm diam)	3.0×10^{-3}	6.5×10^{-4}	1.4×10^{-4}	1.7×10^{-6}	3.9×10^{-4}		6.5×10^{-5}
Small–small LOCA (1–2.5 cm diam)	3.9×10^{-4}	5.2×10^{-5}	7.8×10^{-6}	1.9×10^{-8}	9.0×10^{-6}		7.2×10^{-7}
Emergency coolant injection[c]							
Large LOCA	3.0×10^{-4}	2.0×10^{-4}	1.0×10^{-4}	1.2×10^{-5}	1.5×10^{-4}		
Small LOCA (2.5–15 cm diam)	2.3×10^{-3}	5.2×10^{-4}	1.2×10^{-4}	4.3×10^{-7}	3.8×10^{-4}		6.5×10^{-5}
Small–small LOCA (1–2.5 cm diam)	2.5×10^{-4}	4.2×10^{-5}	6.5×10^{-6}	4.7×10^{-9}	8.6×10^{-6}		7.2×10^{-7}
Containment leakage							
Drywell ($\geq$39 cm^2)	5.1×10^{-4}	1.8×10^{-4}	7.6×10^{-5}	8.0×10^{-5}			
Drywell (6–26 cm^2)	5.4×10^{-2}	5.5×10^{-3}	5.3×10^{-4}	2.1×10^{-5}			5.0×10^{-3e}
Wetwell ($\geq$39 cm^2)	7.5×10^{-4}	3.8×10^{-4}	2.1×10^{-4}	2.0×10^{-4}			
Wetwell (6–26 cm^2)	1.8×10^{-3}	8.8×10^{-4}	5.0×10^{-4}	3.4×10^{-4}			
Low-pressure coolant and core spray recirculation	3.6×10^{-4}	1.2×10^{-4}	4.8×10^{-5}	1.1×10^{-4}	7.0×10^{-9}		1.0×10^{-8}
High-pressure service water							
Required in 30 min	1.1×10^{-3}	4.3×10^{-4}	1.6×10^{-4}	6.4×10^{-6}	1.0×10^{-4}		1.2×10^{-4d}
Required in 25 hr	2.9×10^{-4}	1.1×10^{-4}	4.1×10^{-5}	6.4×10^{-6}	2.7×10^{-5}		3.0×10^{-5d}
Secondary containment							
Ground release	2.2×10^{-4}	3.5×10^{-5}	1.4×10^{-5}	1.8×10^{-5}			
Unfiltered elevated release	3.0×10^{-4}	5.6×10^{-5}	1.4×10^{-5}	4.1×10^{-5}			

[a] From Reactor Safety Study, Appendix II, WASH-1400 (October 1975).
[b] Minimum coolant requirement acceptable to AEC licensing.
[c] Minimum coolant requirement acceptable to others.
[d] Also human error.
[e] Common with large LOCA.

Table 12-6 *Key to PWR Accident Sequence Symbols*[a]

Symbol	Meaning
A	Intermediate to large LOCA
B	Failure of electric power to ESFs
B′	Failure to recover either on-site or off-site electric power within about 1 to 3 hr following an initiating transient which is a loss of off-site ac power
C	Failure of the containment spray injection system
D	Failure of the emergency core cooling injection system
F	Failure of the containment spray recirculation system
G	Failure of the containment heat removal system
H	Failure of the emergency core cooling recirculation system
K	Failure of the reactor protection system
L	Failure of the secondary system steam relief valves and the auxiliary feedwater system
M	Failure of the secondary system steam relief valves and the power conversion system
Q	Failure of the primary system safety relief valves to reclose after opening
R	Massive rupture of the reactor vessel
S_1	A small LOCA with an equivalent diameter of about 5 to 15 cm
S_2	A small LOCA with an equivalent diameter of about 1 to 5 cm
T	Transient event
V	Failure of low-pressure injection system check valve
α	Containment rupture due to a reactor vessel steam explosion
β	Containment failure resulting from inadequate isolation of containment openings and penetrations
γ	Containment failure due to hydrogen burning
δ	Containment failure due to overpressure
ϵ	Containment vessel melt-through

[a] From Reactor Safety Study, WASH-1400 (October 1975).

spectrum of releases could be well represented by a set of 9 different radioactive release categories.

The largest releases, category 1, are associated with a potential steam explosion in the reactor vessel. Such accidents would involve a large volume of molten UO_2 falling into a pool of water in the bottom of the reactor vessel, and becoming finely enough dispersed in the water to produce a steam explosion. This could potentially release large enough amounts of energy to rupture the vessel and, in some cases, even the containment as a result of missiles generated by the vessel rupture. Because of the heavy concrete shielding around the reactor vessel, a missile having sufficient energy to rupture the containment would almost certainly go up through the containment dome. Half of the molten core that was assumed finely dispersed in water could be ejected into the containment oxidizing atmosphere, thus producing a large release that would be discharged from the upper part of the containment.

Category 2 releases are also associated with core melt and basically involve failure of radioactivity removal systems to operate, followed by rupture of the containment caused by hydrogen burning and steam overpressure. Category 3 includes some of the cases that are similar to those in categories 1 and 2, but requires partial success of radioactivity removal systems. Category 4 involves core melt cases in which the containment is not fully isolated and the containment radioactivity removal systems have failed. Category 5 is similar to 4 except that the radioactivity removal

Table 12-7 *Key to BWR Accident Sequence Symbols*[a]

Symbol	Meaning
A	Rupture of reactor coolant boundary with an equivalent diameter of greater than 15 cm
B	Failure of electric power to ESFs
C	Failure of the reactor protection system
D	Failure of vapor suppression
E	Failure of emergency core cooling injection
F	Failure of emergency core cooling functionability
G	Failure of containment isolation to limit leakage to less than 100 vol % per day
H	Failure of core spray recirculation system
I	Failure of low-pressure recirculation system
J	Failure of high-pressure service water system
M	Failure of safety/relief valves to open
P	Failure of safety/relief valves to reclose after opening
Q	Failure of normal feedwater system to provide core make-up water
S_1	Small pipe break with an equivalent diameter of about 5 to 15 cm
S_2	Small pipe break with an equivalent diameter of about 1 to 5 cm
T	Transient event
U	Failure of high-pressure coolant injection system or reactor core isolation cooling system to provide core make-up water
V	Failure of low-pressure ECCS to provide core make-up water
W	Failure to remove residual core heat
α	Containment failure due to steam explosion in vessel
β	Containment failure due to steam explosion in containment
γ	Containment failure due to overpressure–release through reactor building
γ'	Containment failure due to overpressure–release direct to atmosphere
δ	Containment isolation failure in drywell
ϵ	Containment isolation failure in wetwell
ζ	Containment leakage greater than 2400 vol % per day
η	Reactor building isolation failure
θ	Standby gas treatment system failure

[a] From Reactor Safety Study, WASH-1400 (October 1975).

Table 12-8 *Key Accident Sequences from Event Trees*[a]

Sequence designation	Event tree failure(s)	Containment event tree failure
PWR sequences		
1. A	Large rupture only	None
2. A-β	Large rupture only	Containment leakage
3. AH-α	ECCS recirculation	Vessel steam explosion
4. AH-β	ECCS recirculation	Containment leakage
5. AH-ϵ	ECCS recirculation	Melt-through
6. AHI-α	ECCS recirculation plus sodium hydroxide	Vessel steam explosion
7. AHI-β	ECCS recirculation plus sodium hydroxide	Containment leakage
8. AHI-ϵ	ECCS recirculation plus sodium hydroxide	Melt-through
9. AG-δ	Containment heat removal	Overpressure
10. AHG-δ	ECCS recirculation plus containment heat removal	Overpressure
11. AHG-ϵ	ECCS recirculation plus containment heat removal	Melt-through
12. AHF-α	ECCS recirculation plus containment spray recirculation	Vessel steam explosion
13. AHF-β	ECCS recirculation plus containment spray recirculation	Containment leakage
14. AHF-δ	ECCS recirculation plus containment spray recirculation	Overpressure
15. AHF-ϵ	ECCS recirculation plus containment spray recirculation	Melt-through
16. AD-α	ECCS injection	Vessel steam explosion
17. AD-β	ECCS injection	Containment leakage
18. AD-ϵ	ECCS injection	Melt-through
19. ADI-α	ECCS injection plus sodium hydroxide	Vessel steam explosion
20. ADI-ϵ	ECCS injection plus sodium hydroxide	Melt-through
21. ADG-ϵ	ECCS injection plus containment heat removal	Melt-through
22. ADGI-ϵ	ECCS injection plus containment heat removal plus sodium hydroxide	Melt-through
23. ADF-β	ECCS injection plus containment spray recirculation	Containment leakage
24. ADF-ϵ	ECCS injection plus containment spray recirculation	Melt-through
25. ACD-β	Containment spray injection plus ECCS injection	Containment leakage
26. ACD-ϵ	Containment spray injection plus ECCS injection	Melt-through
27. ACDGI-α	All except electric power and containment spray recirculation	Vessel steam explosion
28. ACDGI-β	All except electric power and containment spray recirculation	Containment leakage
29. ACDGI-δ	All except electric power and containment spray recirculation	Overpressure

Table 12-8 (*Continued*)

Sequence designation	Event tree failure(s)	Containment event tree failure
30. ACDGI-ϵ	All except electric power and containment spray recirculation	Melt-through
31. AB-α	Electric power	Vessel steam explosion
32. AB-γ	Electric power	Hydrogen combustion
33. AB-ϵ	Electric power	Melt-through
34. S_2C-α	Small LOCA plus containment spray injection	Vessel steam explosion
35. S_2C-δ	Small LOCA plus containment spray injection	Overpressure
36. TMLB$'$-γ	Transient plus feedwater plus electric power	Hydrogen combustion
37. TMLB$'$-δ	Transient plus feedwater plus electric power	Overpressure
38. TMLB$'$-α	Transient plus feedwater plus electric power	Vessel steam explosion
BWR sequences		
1. AE-β (dry)	All emergency coolant injection	Containment steam explosion
2. AE-γ (dry)	All emergency coolant injection	Containment overpressure[b]
3. AF-β	ECCS functionability	Containment steam explosion
4. AF-γ	ECCS functionability	Containment overpressure
5. AF-α	ECCS functionability	Vessel steam explosion
6. AJ-γ	Post-accident heat removal	Containment overpressure
7. ADE-γ (dry)	All emergency coolant injection plus vapor suppression	Containment overpressure
8. ADF-γ	ECCS functionability plus vapor suppression	Containment overpressure
9. ADJ-γ	Post-accident heat removal plus vapor suppression	Containment overpressure
10. AEG-$\delta\eta$ (dry)	All emergency coolant injection plus containment leakage	Drywell leakage plus secondary containment isolation
11. AEG-δ (dry)	All emergency coolant injection plus containment leakage	Drywell leakage only
12. AGJ-$\delta\theta$	Post-accident heat removal plus containment leakage	Drywell leakage plus SGTS filters
13. AGJ-δ	Post-accident heat removal plus containment leakage	Drywell leakage only
14. AE-α	Partial emergency coolant injection	Vessel steam explosion
15. ADE-γ' (dry)	All emergency coolant injection plus vapor suppression	Containment overpressure[c]
16. ADF-γ'	ECCS functionability plus vapor suppression	Containment overpressure

(*Continued*)

Table 12-8 (*Continued*)

Sequence designation	Event tree failure(s)	Containment event tree failure
17. AJ-γ'	Post-accident heat removal	Containment overpressure
18. ADJ-γ'	Post-accident heat removal plus vapor suppression	Containment overpressure
19. A	Pipe break only	None
20. TC-α	Transient requiring shutdown plus reactor trip plus liquid poison injection	Vessel steam explosion
21. TC-γ	Transient requiring shutdown plus reactor trip plus liquid poison injection	Containment overpressure
22. TW-α	Transient shutdown plus post-accident heat removal	Vessel steam explosion
23. TW-γ	Transient shutdown plus post-accident heat removal	Containment overpressure
24. TW-γ'	Transient shutdown plus post-accident heat removal	Containment overpressure

[a] From Reactor Safety Study, Appendix V, WASH-1400 (October 1975).

[b] γ is used to indicate that containment failure occurs by overpressurization with a release path where a large amount of deposition of radioactivity occurs.

[c] γ' is used to indicate that containment failure occurs by overpressure, but with a release path where no deposition of radioactivity occurs. Examination of the containment and reactor building layout indicated that the probability for this release path was about 0.2 for all overpressure failures.

systems are operating. Categories 6 and 7 cover cases in which the molten core melts through the bottom of the containment, with and without radioactivity removal systems operating, respectively, but the above-ground part of the containment remains intact. In categories 8 and 9 the core does not melt, and only some of the activity in the gaps of the fuel rods is released. Category 8 involves gap releases with failure of the containment to isolate properly, while in category 9 the containment isolates correctly.

BWR Release Categories

The paths for release of radioactivity in a BWR are quite different from those for the PWR. Although the BWR has containment sprays, they are not designed as ESFs and were not credited in the RSS for removal of radioactivity. Further, the vapor suppression system that has some capability for removal of radioactivity is largely ineffective in a number of the core melt cases. Thus the principal mechanism for removal of radioactivity is natural deposition on the surface inside the containment and the reactor building.

From the CORRAL code calculations performed for each of the 24 key accident sequences defined in Table 12-8, 5 release categories were identified. Again, category 1 involves a steam explosion in the reactor vessel in which about half the core is ejected from the containment. The resulting exposure of the finely dispersed molten fuel to an oxidizing atmosphere results in a very large release of radioactive material to the atmosphere.

Category 2 involves a core meltdown after containment overpressure rupture caused by loss of decay heat removal systems. In this category, only a limited amount of deposition of the radioactive materials occurs and thus the release is made directly to the atmosphere. Category 3 covers overpressure ruptures of containment, similar to category 2, but the radioactive materials released from the core escape through the reactor building to the atmosphere. The radioactive release is smaller in magnitude than for category 2 since there is deposition and some retention of the radioactivity. Category 4 covers the case in which the containment fails to properly isolate and the leakage is enough to prevent containment overpressure rupture. However, the magnitude of the radioactivity release is significantly reduced by additional deposition in the containment because of the longer release times and by deposition in the reactor building. Category 5 covers the case in which the core does not melt and a small amount of activity may be released from the gap of the fuel rods; the small amount of radioactivity that may leak from containment is further reduced by processing through the reactor building gas treatment system.

12-6 PREDICTED FREQUENCIES FOR ACCIDENT SEQUENCES

12-6-1 Dominant Accident Sequences

The grouping of all accident sequences into release categories resulted in a restructuring of the problem into a form that showed clearly that, of all the postulated accidents leading to a given release, only a few dominant sequences made a significant contribution to the predicted frequency of occurrence of the release. Thus, the effort to estimate the magnitude of releases and frequencies was concentrated on those RSS sequences that were the main contributors to the risk. Tables 12-9 and 12-10 summarize the allocation of the dominant accident sequences from Table 12-8 to the release categories and include the frequencies of occurrence as calculated from event tree/fault tree analysis.

The frequencies for the individual sequences in the tables do not always add up to the median values shown there. There are two reasons for these

Table 12-9 *PWR Dominant Accident Sequences versus Release Category*[a]

Accident	Release categories								
	Core melt							No core melt	
	1	2	3	4	5	6	7	8	9
Large LOCA dominant accident sequences A	AB-α 1×10^{-11}	AB-γ 1×10^{-10}	AD-α 2×10^{-8}	ACD-β 1×10^{-11}	AD-β 4×10^{-9}	AB-ϵ 1×10^{-9}	AD-ϵ 2×10^{-6}	A-β 2×10^{-7}	A 1×10^{-4}
	AF-α 1×10^{-10}	AB-δ 4×10^{-11}	AH-α 1×10^{-8}		AH-β 3×10^{-9}	AHF-ϵ 1×10^{-10}	AH-ϵ 1×10^{-6}		
	ACD-α 5×10^{-11}	AHF-γ 2×10^{-11}	AF-δ 1×10^{-8}			ADF-ϵ 2×10^{-10}			
	AG-α 9×10^{-11}		AG-δ 9×10^{-9}						
A frequencies[b]	2×10^{-9}	1×10^{-8}	1×10^{-7}	1×10^{-8}	4×10^{-8}	3×10^{-7}	3×10^{-6}	1×10^{-5}	1×10^{-4}
Small LOCA dominant accident sequences S_1	S_1B-α 3×10^{-11}	S_1B-γ 4×10^{-10}	S_1D-α 3×10^{-8}	S_1CD-β 1×10^{-11}	S_1H-β 5×10^{-9}	S_1DF-ϵ 3×10^{-10}	S_1D-ϵ 3×10^{-6}	S_1-β 6×10^{-7}	S_1 3×10^{-4}
	S_1CD-α 7×10^{-11}	S_1B-δ 1×10^{-10}	S_1H-α 3×10^{-8}		S_1D-β 6×10^{-9}	S_1B-ϵ 2×10^{-9}	S_1H-ϵ 3×10^{-6}		
	S_1F-α 3×10^{-10}	S_1HF-γ 6×10^{-11}	S_1F-δ 3×10^{-8}			S_1HF-ϵ 4×10^{-10}			
	S_1G-α 3×10^{-10}		S_1G-δ 3×10^{-8}						
S_1 frequencies	3×10^{-9}	2×10^{-8}	2×10^{-7}	3×10^{-8}	8×10^{-8}	6×10^{-7}	6×10^{-6}	3×10^{-5}	3×10^{-4}
Small LOCA dominant accident sequences S_2	S_2B-α 1×10^{-10}	S_2B-γ 1×10^{-9}	S_2D-α 9×10^{-8}	S_2DG-β 1×10^{-12}	S_2D-β 2×10^{-8}	S_2B-ϵ 8×10^{-9}	S_2D-ϵ 9×10^{-6}		
	S_2F-α 1×10^{-9}	S_2HF-γ 2×10^{-10}	S_2H-α 6×10^{-8}		S_2H-β 1×10^{-8}	S_2CD-ϵ 2×10^{-8}	S_2H-ϵ 6×10^{-6}		
	S_2CD-α 2×10^{-10}	S_2B-δ 4×10^{-10}	S_2F-δ 1×10^{-7}			S_2HF-ϵ 1×10^{-9}			

	S_2G-α 9×10^{-10} S_2C-α 2×10^{-8}		S_2C-δ 2×10^{-6} S_2G-δ 9×10^{-8}						
S_2 frequencies	1×10^{-7}	3×10^{-7}	3×10^{-6}	3×10^{-7}	3×10^{-7}	2×10^{-6}	2×10^{-5}		
Reactor vessel rupture dominant accident sequences R	RC-α 2×10^{-12}	RC-γ 3×10^{-11} RF-δ 1×10^{-11} RC-δ 1×10^{-12}	R-α 1×10^{-9}				R-ϵ 1×10^{-7}		
R frequencies	2×10^{-11}	1×10^{-10}	1×10^{-9}	2×10^{-10}	1×10^{-9}	1×10^{-8}	1×10^{-7}		
Interfacing systems LOCA (check valve) V		V 4×10^{-6}							
V frequencies	4×10^{-7}	4×10^{-6}	4×10^{-7}	4×10^{-8}					
Transient event dominant accident sequencies T	TMLB′-α 3×10^{-8}	TMLB′-γ 7×10^{-7} TMLB′-δ 2×10^{-6}	TML-α 6×10^{-8} TKQ-α 3×10^{-8} TKMQ-α 1×10^{-8}		TML-β 3×10^{-10} TKQ-β 3×10^{-10}	TMLB′-ϵ 6×10^{-7}	TML-ϵ 6×10^{-6} TKQ-ϵ 3×10^{-6} TKMQ-ϵ 1×10^{-6}		
T frequencies	3×10^{-7}	3×10^{-6}	4×10^{-7}	7×10^{-8}	2×10^{-7}	2×10^{-6}	1×10^{-5}		
Summation of all accident sequences per release category									
Median (50% value)	9×10^{-7}	8×10^{-6}	4×10^{-6}	5×10^{-7}	7×10^{-7}	6×10^{-6}	4×10^{-5}	4×10^{-5}	4×10^{-4}
Lower bound (5% value)	9×10^{-8}	8×10^{-7}	6×10^{-7}	9×10^{-8}	2×10^{-7}	2×10^{-6}	1×10^{-5}	4×10^{-6}	4×10^{-5}
Upper bound (95% value)	9×10^{-6}	8×10^{-5}	4×10^{-5}	5×10^{-6}	4×10^{-6}	2×10^{-5}	2×10^{-4}	4×10^{-4}	4×10^{-3}

[a] From Reactor Safety Study, WASH-1400 (October 1975).
[b] Frequencies are yr.$^{-1}$

Table 12-10 *BWR Dominant Accident Sequences versus Release Category*[a]

	Release categories				
	Core melt				No core melt
Accident	1	2	3	4	5
Large LOCA dominant accident sequences A	AE-α 2×10^{-9}	AE-γ' 3×10^{-8}	AE-γ 1×10^{-7}	AGJ-δ 6×10^{-11}	A 1×10^{-4}
	AJ-α 1×10^{-10}	AE-β 1×10^{-8}	AJ-γ 1×10^{-8}	AEG-δ 7×10^{-10}	
	AHI-α 1×10^{-10}	AJ-γ' 2×10^{-9}	AI-γ 1×10^{-8}	AGHI-δ 6×10^{-11}	
	AI-α 1×10^{-10}	AI-γ' 2×10^{-9}	AHI-γ 1×10^{-8}		
		AHI-γ' 2×10^{-9}			
A frequencies[b]	8×10^{-9}	6×10^{-8}	2×10^{-7}	2×10^{-8}	1×10^{-4}
Small LOCA dominant accident sequences S_1	S_1E-α 2×10^{-9}	S_1E-γ' 4×10^{-8}	S_1E-γ 1×10^{-7}	S_1GJ-δ 2×10^{-10}	
	S_1J-α 3×10^{-10}	S_1E-β 1×10^{-8}	S_1J-γ 3×10^{-8}	S_1GE-δ 2×10^{-10}	
	S_1I-α 4×10^{-10}	S_1J-γ' 7×10^{-9}	S_1I-γ 4×10^{-8}	S_1EI-ϵ 1×10^{-10}	
	S_1HI-α 4×10^{-10}	S_1I-γ' 7×10^{-9}	S_1HI-γ 2×10^{-8}	S_1GHI-δ 2×10^{-10}	
		S_1HI-γ' 6×10^{-9}	S_1C-γ 3×10^{-9}		
S_1 frequencies	1×10^{-8}	9×10^{-8}	2×10^{-7}	2×10^{-8}	
Small LOCA dominant accident sequences S_2	S_2J-α 1×10^{-9}	S_2E-γ' 1×10^{-8}	S_2E-γ 4×10^{-8}	S_2CG-δ 6×10^{-11}	
	S_2I-α 1×10^{-9}	S_2E-β 4×10^{-9}	S_2J-γ 8×10^{-8}	S_2GHI-δ 6×10^{-10}	
	S_2HI-α 1×10^{-9}	S_2J-γ' 2×10^{-8}	S_2I-γ 9×10^{-8}	S_2EG-δ 3×10^{-10}	
	S_2E-α 5×10^{-10}	S_2I-γ' 2×10^{-8}	S_2HI-γ 9×10^{-8}	S_2GJ-δ 6×10^{-10}	
		S_2HI-γ' 2×10^{-8}	S_2C-γ 8×10^{-9}	S_2GI-δ 2×10^{-10}	
S_2 frequencies	2×10^{-8}	1×10^{-7}	4×10^{-7}	4×10^{-8}	
Transient dominant accident sequences T	TW-α 2×10^{-7}	TW-γ' 3×10^{-6}	TW-γ 1×10^{-5}		
	TC-α 1×10^{-7}	TQUV-γ' 8×10^{-8}	TC-γ 1×10^{-5}		
	TQUV-α 5×10^{-9}		TQUV-γ 4×10^{-7}		
T frequencies	1×10^{-6}	6×10^{-6}	2×10^{-5}	2×10^{-6}	

Table 12-10 (*Continued*)

	Release categories				
	Core melt				No core melt
Accident	1	2	3	4	5
Pressure vessel rupture accidents R		P.V. rupt. 1×10^{-8} Oxidizing atmosphere	P.V. rupt. 1×10^{-7} Nonoxidizing atmosphere		
R frequencies	2×10^{-9}	2×10^{-8}	1×10^{-7}	1×10^{-8}	
Summation of all accident sequences per release category					
Median (50% value)	1×10^{-6}	6×10^{-6}	2×10^{-5}	2×10^{-6}	1×10^{-4}
Lower bound (5% value)	1×10^{-7}	1×10^{-6}	5×10^{-6}	5×10^{-7}	1×10^{-5}
Upper bound (95% value)	8×10^{-6}	3×10^{-5}	8×10^{-5}	1×10^{-5}	1×10^{-3}

[a] From Reactor Safety Study, WASH-1400 (October 1975).

[b] Frequencies are $yr.^{-1}$

differences. First, the probability distribution of the final values was obtained by the SAMPLE-A program (see Chapter 10), which sampled from the distribution of possible values. The result of this sampling can make the median value slightly larger than an arithmetic summation when significant uncertainties exist.

The second reason for the differences is that an effort was made to account for the possible variation of release magnitudes. That is, because of the variability inherent in the parameters describing the physical processes that could occur during an accident, a given accident sequence will in reality have a spectrum of release magnitudes and hence a distribution of release fractions for each isotope.

To account for this distribution in possible release magnitudes, the Reactor Safety Study assigned a 10% chance that the sequence would lie in an adjacent release category, and a 1% chance that it would lie in the next adjacent category, etc., as shown in Fig. 12-1. That is, the original sequence frequency (initiating event frequency, system failure, and containment mode failure) was multiplied by the appropriate factor

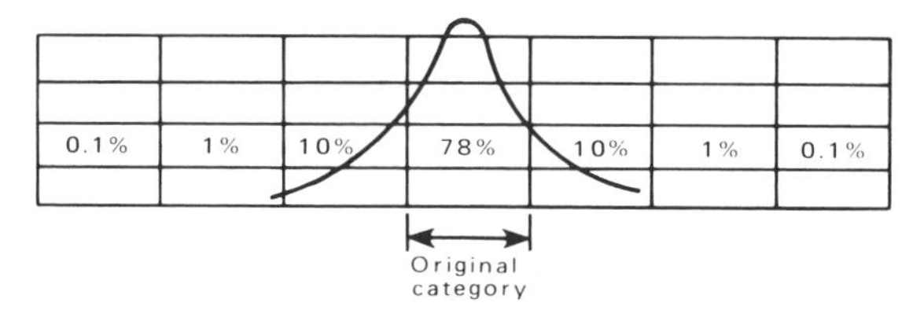

Fig. 12-1 *Allocation of frequencies of release to the adjacent release categories.* [*From Reactor Safety Study, Appendix* V, WASH-*1400* (*October 1975*).]

(0.78, 0.10, etc.). In this way, new sequences were effectively created with the additional frequency factors and were treated in the calculations as any other sequences; an example is the PWR check valve in Table 12-9. In order to be conservative, however, the factor of 0.78 was replaced by unity, so that the predicted frequency of each accident sequence was effectively multiplied by a factor of 1.22 (if no accounting was made of the release category in which the accident was placed).

We will now briefly examine the more important key accident sequences of Table 12-8 that contribute in a significant manner to the release categories in Tables 12-9 and 12-10.

Large LOCA

A large LOCA (labeled event A) is defined as a reactor coolant system (RCS) rupture having a size larger than the equivalent of about a 15 cm diameter hole. From Table 12-9 for the PWR, the median frequency of such an event is assessed at 1×10^{-4} per reactor-year. The dominant sequences that contribute to core melt are AD-ϵ and AH-ϵ in release category 7. The AD-ϵ is a large RCS rupture followed by failure of the emergency core cooling injection (D), which leads to core melt and failure of containment caused by the molten core melting through the bottom (ϵ). The AH-ϵ sequence is a large RCS rupture, followed by failure of the emergency core cooling recirculation system (H), and core melt, followed by the same mode of containment failure.

Although the predicted frequency of a large LOCA also was 1×10^{-4} per reactor-year for the BWR, the availability of the ECC system is better than for the PWR. Thus the frequency for LOCA-caused core melts in Table 12-10 for a BWR is less than 10^{-6} for all release categories and is not a significant risk contributor to any category. A core melt in a BWR would result in a larger radioactive release if it occurred, however. This is because the BWR containment volume is considerably smaller, and when the core melts the noncondensable gases generated by the zircalloy–water reaction and molten fuel interaction with the concrete will overpressurize and rupture the containment. Thus, the most likely path to containment failure is by overpressure rather than by melt-through.

Small LOCA

The small LOCA (S_1) is defined as an RCS rupture of between 5 cm and 15 cm equivalent diameter. From Table 12-9, it has a median frequency of about 3×10^{-4} per PWR reactor-year. The dominant sequences leading to core melt are $S_1D\text{-}\epsilon$ and $S_1H\text{-}\epsilon$, as in the large LOCA. Both sequences have a frequency of 3×10^{-6}, so the total frequency of core melt by S_1-type LOCA events is about 6×10^{-6} per year. As can be seen, however, S_1 events are not significant contributors to PWR release categories, other than category 7 for which there is a relatively small release, so S_1 events contribute very little to the risk from a PWR core melt.

For the BWR, the dominant accident sequence $S_1E\text{-}\gamma$ corresponding to failure of all emergency coolant injection systems leads to containment failure by overpressure; most other S_1 initiating events are nearly an order of magnitude less likely. The small LOCA (S_1) events do not contribute in a major way to risks of BWR operation.

Small LOCA

The small LOCA (S_2) is defined as an RCS rupture of between 1 cm and 5 cm equivalent diameter. From Table 12-9, the dominant sequences are $S_2D\text{-}\epsilon$ and $S_2H\text{-}\epsilon$, as in the A and S_1 sequences already discussed. The median frequency of core melt caused by S_2 events is about 2×10^{-5} per year, which is the largest contribution to a PWR core melt.

For the BWR, Table 12-10 shows that the predicted frequency of a core melt is relatively small for all S_2 accidents.

Reactor Vessel Rupture

The reactor vessel rupture event (R) is defined as a vessel rupture large enough to negate the effectiveness of the ECCS system required to prevent core melt. The median value of the frequency of such an event for either a PWR or BWR is about 10^{-7} per reactor year, as shown in Tables 12-9 and 12-10. Hence the rupture event is a negligible contributor to the frequency of release for either type of reactor.

Interfacing Systems LOCA

The interfacing systems LOCA (V) is caused by the failure of PWR check valves that isolate the low-pressure injection system from the reactor coolant system. This event requires the failure of two in-series check valves and, if this were to occur during reactor operation, it would produce a LOCA of about 15 cm effective diameter. The frequency of this event was calculated in the RSS to be 4×10^{-6} per year. (This value is somewhat large because a failure of one of the check valves was assumed to be undetectable; a monthly, independent test of these valves to assure their seating integrity, however, would lower the probability of their combined failure by a factor of about 20. It should be noted that this problem since has been corrected in PWRs.)

Transient Events

The transient events (T) refer to a wide range of conditions that require shutdown of the reactor and include all unplanned shutdowns, which have occurred at the rate of about ten per year [1]. Following a transient event, one of the shutdown systems must operate and, following that, operation of decay heat removal systems is required to prevent a core melt.

Transient sequences involving a failure to trip the reactor control rods are not important contributors to transient-produced core melt situations in a PWR. This is because of the relatively low failure probability of the shutdown systems and the additional protection against core melting provided by RCS safety valves and auxiliary feedwater; thus, sequences involving failure to trip (K) do not appear as major contributors in any category of Table 12-9. Rather, the dominant PWR transient contributors are the TML sequences in which the decay heat removal systems fail. The transient sequences give significant, although not dominant, contributions to the PWR release frequencies.

Transient events (T) dominate the predicted frequencies of release in categories 1 through 3 of the BWR core melt accidents. The probability of failure of the BWR decay heat removal system (W) was determined to be 1.6×10^{-6}; when combined with the expected ten transients that occur per reactor-year [1], a frequency of about 2×10^{-5} per year is obtained for the TW sequence. The release path for radioactivity can be through either the BWR containment drywell and the reactor building, where some deposition of the radioactive materials occur (γ), or more directly to the atmosphere (γ').

12-6-2 Histograms of Frequencies of Release

The median values for each class of accident (such as A, S_1, S_2, etc.) in Tables 12-9 and 12-10 were used to generate the PWR and BWR histograms in Figs. 12-2 and 12-3, with the error bars representing the 90% confidence limits. These figures help emphasize the uncertainties in the calculations.

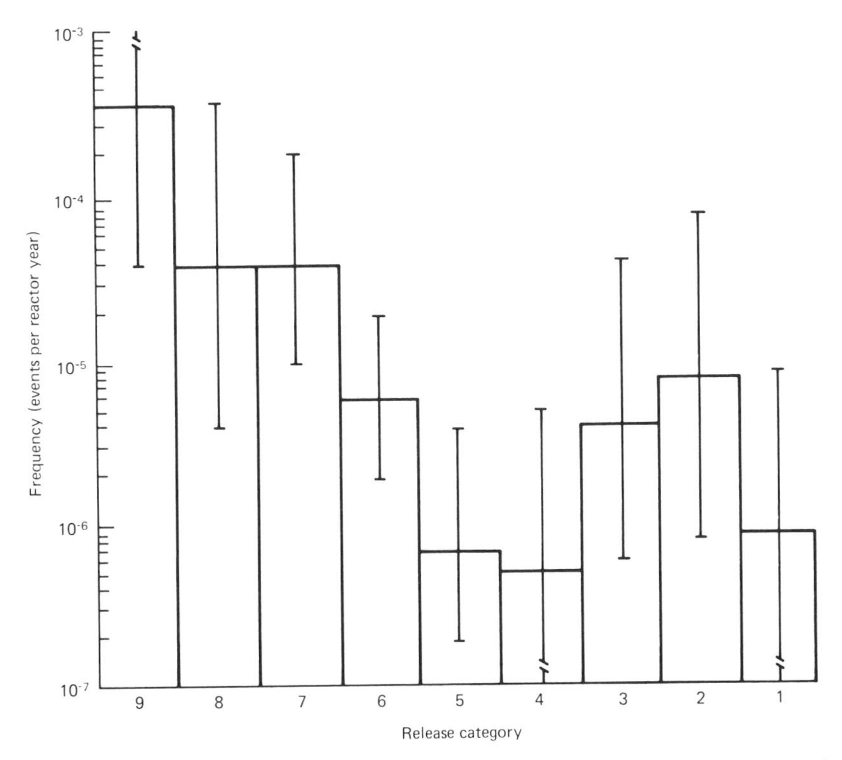

Fig. 12-2 *Histogram of* PWR *radioactive release frequencies. [From Reactor Safety Study,* WASH-*1400 (October 1975).]*

Figure 12-4 shows the result of the smoothing illustrated in Fig. 12-1 for the small LOCA sequences of the PWR. We can see that the net effect of the smoothing technique for several release categories was to increase by orders of magnitude their probability of occurrence. As a result of the smoothing, the cumulative probabilities for all PWR core melt release categories were principally determined by only six sequences: S_2D-ϵ, S_2H-ϵ, S_2C-δ, V, TML-ϵ, and $TMLB'$-δ.

12-6-3 Nonintrinsic Failures

The frequencies and releases just discussed are associated with a variety of intrinsic failures that potentially can occur within a nuclear power plant. The data on which the frequencies are based include a wide variety of causes such as design errors, failures in quality control procedures, operating errors, etc. However, they do not explicitly include considera-

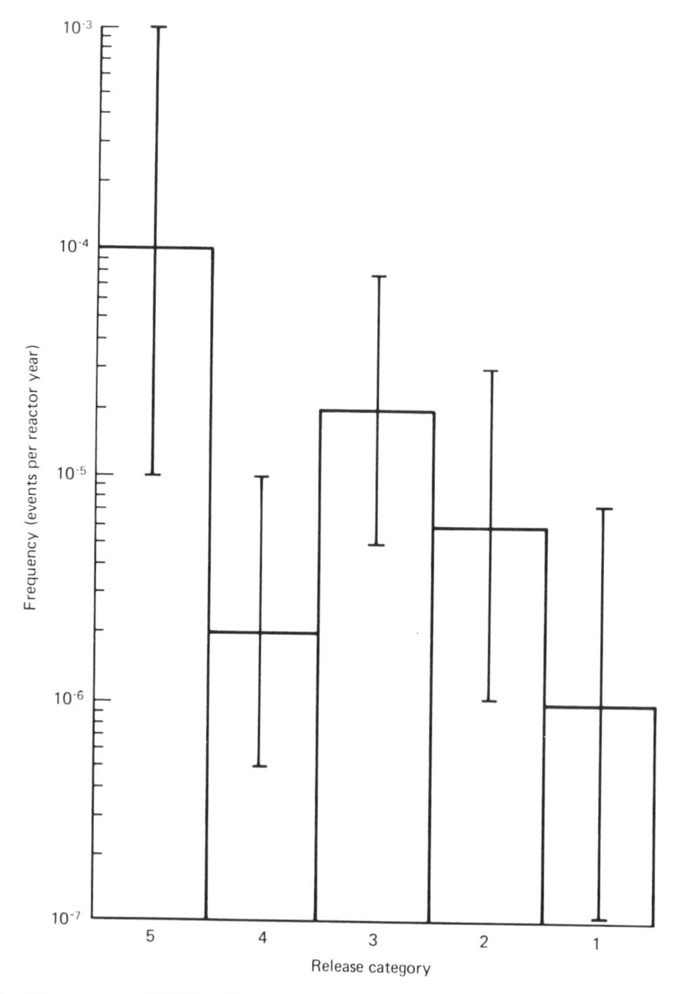

Fig. 12-3 *Histogram of* BWR *radioactive release frequencies.* [*From Reactor Safety Study,* WASH-*1400* (*October 1975*).]

tion of potential failures due to major external events that might affect the plant. If nonintrinsic events such as earthquakes, windstorms, and aircraft impacts were to cause an accident, they could produce a core melt by one of the paths described by the event trees and possibly could make the accident sequence frequencies higher than those due only to intrinsic failures.

Although it is difficult to predict with precision the frequency of potential accidents due to earthquake damage to a nuclear power plant (because

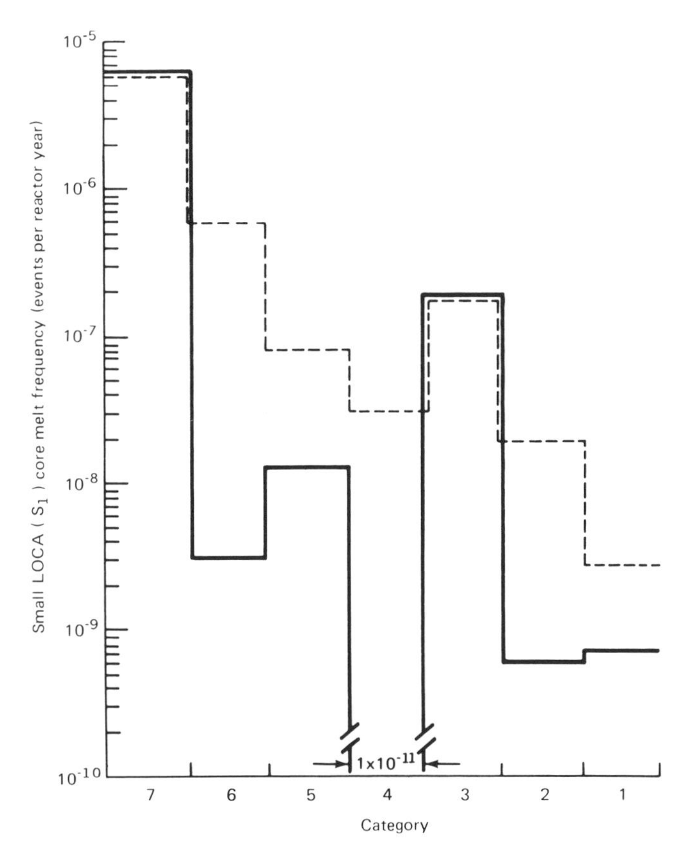

Fig. 12-4 *Application of frequency smoothing to small* LOCA (S_1) *core melt probabilities per* PWR *reactor-year;—original category frequencies; --- smoothed category frequencies.* [*From Reactor Safety Study, Appendix V,* WASH-*1400* (*October 1975*).]

of the general sparsity of quantitative data on the sizes and effects of earthquakes), it appears possible to make order of magnitude estimates that are useful. For example, since reactors are principally located on firm sites, a reasonable estimate for core melt initiated by an earthquake is [1] 10^{-7} per year with the true value lying between 10^{-6} to 10^{-8}. At this level, earthquake-induced accidents should not contribute significantly to reactor accident risks.

As another example, tornados may be considered. All United States power reactors are now being designed to withstand the effects of sizeable tornados. The reactor design basis tornado is usually assumed to have internal winds of about 500 km/hr, to move with a translational velocity of about 100 km/hr, and to develop pressure differentials of about 0.021 MPa (3 psi) in 3 sec. Since less than 1% of all tornados are expected to be as large as the design basis tornado, the frequency of such a tornado striking a nuclear power plant on the average would be less than 5×10^{-6} per year [1]. Furthermore, it was found that the effects on a reactor would be benign. Thus, a tornado-caused core melt sequence is considered to be very improbable compared to the other core melt sequences.

For nuclear power plants located within 8 km of airports, the frequency for a potentially damaging aircraft crash was calculated to be between 10^{-6} and 10^{-7} per year, based on a conservative calculation [1]. The majority of reactors are located farther than 8 km from airports, and the frequency for a crash at these plants would fall in the range of 10^{-6} to 10^{-8} per year. Because the containment is a fairly strong structure and the vital parts of the plant present quite a small area, any impact would have a small probability of producing a core melt sequence. It is therefore concluded that aircraft impacts are very much less likely to cause a core melt than the accident sequences already considered.

Finally, an investigation of the possibility of deliberate human acts to destroy the plant led to the following conclusions regarding sabotage [1]:

1. Nuclear power plants appear far less susceptible to sabotage efforts than most other civil or industrial targets.
2. Nuclear plants have inherent characteristics that provide built-in difficulties for successful sabotage efforts.
3. The worst consequences associated with acts of sabotage at reactors are not expected to lead to consequences more severe than the maximum consequences predicted by the Reactor Safety Study; the expected consequences of successful sabotage are but a small fraction of these maximum consequences.

With the implementation of current security measures, it appears that the probability of a successful serious sabotage of a reactor is very small.

12-6-4 Accidents Not Involving the Reactor Core

Potential releases of radioactivity from sources other than the core arise from the sources shown in Table 12-11. The frequency estimates summarized in the table were based largely on engineering judgments obtained from considering plant experience and the results of detailed analyses of other similar systems; they were not supported by fault tree analyses [1]. The releases from the accidents in the table would result in minor off-site consequences in comparison with other accidents analyzed in this section and represent only a very small contribution to overall risk.

Table 12-11 *Summary of Accidents Not Involving Core*[a,b]

Accident	Frequency, events per reactor-year
Loss of cooling in spent fuel storage pool	$< 10^{-6}$ [c]
Dropped shipping cask	6×10^{-7}
Refueling accident	10^{-3}
Waste gas storage tank rupture	10^{-2}
Liquid waste storage tank rupture	10^{-2}
Earthquake-induced loss of cooling in spent fuel storage pool with loss of air cooling system	3×10^{-8}

[a] Extracted from Reactor Safety Study, Appendix V, WASH-1400 (October 1975).

[b] PWR and BWR designs were examined, and the more severe accidents were selected as representative bounds for both.

[c] Estimated probability includes consideration of turbine- and tornado-generated missiles.

12-7 REACTOR SAFETY STUDY FINDINGS

12-7-1 Summary of Results

Table 12-12 shows the severity of various consequences versus point estimates of the frequency that they could occur. Similar results are available for latent cancer fatalities, thyroid illness, and genetic effects [1]. More extensive information is given, however, in risk curves for the individual consequence types of damages, as shown in Fig. 12-5, for example. Note that the lower frequency value in the figure is 10^{-9} per reactor-year, which is roughly the inverse of the age (5×10^9 yr) of the earth.

The results of the RSS are based on population and weather distributions applicable to the 68 sites at which the first 100 reactors will be located; they are not necessarily the correct curves for a given plant on a given site. (For example, a plant on a very low population site would have a different curve from a plant on a very high population site.) In averaging the two individual curves for the PWR and BWR, the results were weighted to account for the fact that there are about twice as many PWRs as BWRs in the 100 reactors covered by the RSS [1]. Also, it should be noted

Table 12-12 *Summary of Consequences of Reactor Accidents*[a]

	Consequences				
Chance per reactor-year	Early fatalities	Early illness	Total property damage $10^9	Decontamination area, km^2	Relocation area, km^2
One in 20,000[b]	<1.0	<1.0	<0.1	<0.3	<0.3
One in 1,000,000	<1.0	300	0.9	5200	340
One in 10,000,000	110	3000	3	8300	650
One in 100,000,000	900	14,000	8		750
One in 1,000,000,000	3300	45,000	14		

[a] From Reactor Safety Study, WASH-1400 (October 1975).
[b] This is the predicted chance of core melt per reactor year.

from the captions to the figures that the curves are only for the estimated frequencies, whereas there are uncertainties in both the magnitude of the damages and the frequencies.

Early Fatalities and Illnesses

The predicted frequency of accidents that produce fatalities versus the expected number of fatalities is shown in Fig. 12-5 for both the PWR and the BWR. The differences between the curves for the PWR and BWR are less than the uncertainties inherent in the calculations. The frequencies in the figure are essentially comprised of four contributing factors: the predicted frequency of core melt, the relative probability of various radioactive release categories following core melt, the probability of the existing weather conditions, and the probability that a particular population density will be exposed. The frequency of core melt affects the absolute value of the frequency scale but not the shape of the curve; the shape is principally determined by the other three factors.

The average frequency of an accident that results in more than 10 fatalities is predicted to be about 3×10^{-7} per reactor-year. Accidents involving 100 or more fatalities are predicted to have a frequency of about 10^{-7} per year and would be expected on the average to occur only once in 10 million reactor-years of operation. The largest number of fatalities is predicted to be about 3300 and to occur with a frequency of about 10^{-9} per reactor-year; this would happen with the simultaneous occurrence of a core melt accident, the worst weather, and the wind blowing in the direction of one of the high population density sectors.

Early illnesses are defined as those that require medical attention

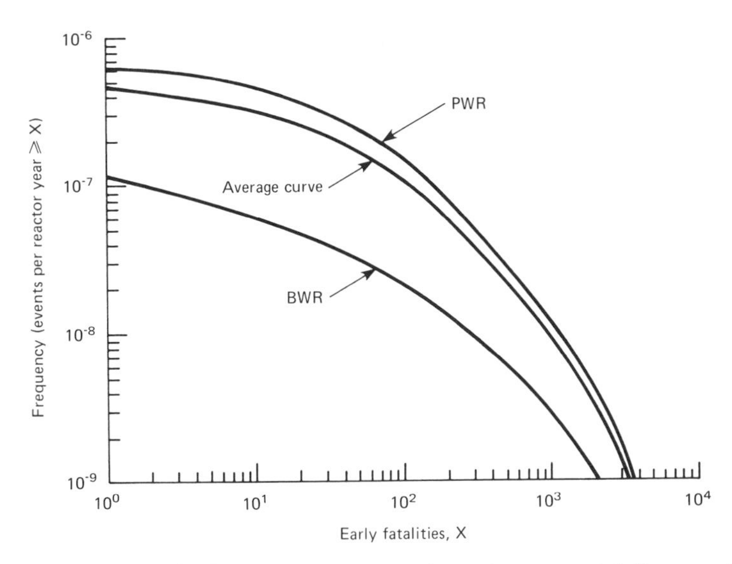

Fig. 12-5 *Risk of early fatalities per reactor-year. Approximate uncertainties are estimated to be represented by factors of $\frac{1}{4}$ and 4 on early fatalities and by factors of $\frac{1}{5}$ and 5 on frequencies.* [*From Reactor Safety Study,* WASH-*1400* (*October 1975*).]

shortly after an accident; the most important illness in this category is respiratory impairment. Some of these illnesses are expected to require continuing treatment. The predicted risk of early illness is shown in Fig. 12-6.

Long-Term Health Effects

Exposure to even low levels of radiation over the natural background of radiation that exists is generally believed to increase the likelihood of certain diseases and to increase certain genetic effects. Since these effects may be evidenced many years after the exposure, they are classified as long-term health effects. These include latent-cancer fatalities, thyroid illness, and genetic defects.

In the Reactor Safety Study, it was assumed that the total number of predicted cancers is equivalent to about 100 cancer fatalities per 10^6 man-rem, based on a whole body dose. Figure 12-7 is a plot of the frequency distribution of latent-cancer fatalities per year, most of which would occur over a period of 10 to 40 years following such an accident.

In the largest accident predicted in the study, the 1500 latent-cancer fatalities per year would be distributed over approximately 10 million people for whom the normal incidence rate of fatal cancer is about 17,000 per year. Thus the largest potential accident would represent an increase

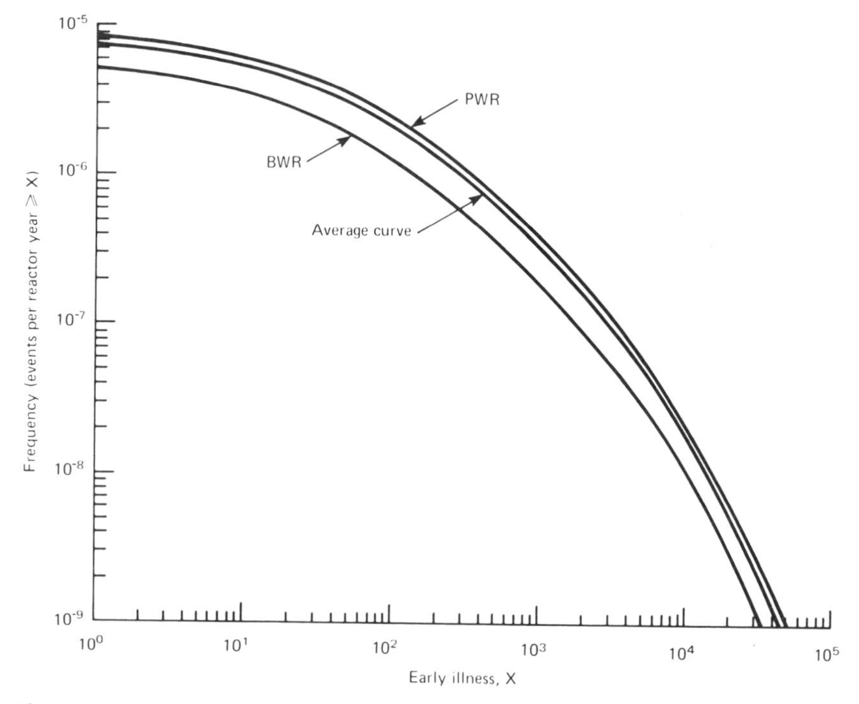

Fig. 12-6 *Risk of early illness per reactor-year. Approximate uncertainties are estimated to be represented by factors of* $\frac{1}{4}$ *and 4 on early illness and by factors of* $\frac{1}{5}$ *and 5 on frequencies.* [*From Reactor Safety Study,* WASH-*1400* (*October 1975*).]

over the normal rate of 1500/17000, or about 9%. This effect would probably not be measurable statistically because of the large variations in the normal rate.

A second effect of radiation is that exposure of the thyroid gland increases the likelihood of thyroid nodules. On the average, considering different age groups, about $\frac{1}{3}$ of all nodules would be malignant. Both types of nodules can be medically treated with good success. In the RSS, it was assumed that 10% of the malignant nodules would have a fatal outcome, and this number has been added to the number of latent cancer fatalities.

Figure 12-8 presents the incidence rate for all nodules. This rate would be expected to persist from about 10 to 40 years after the accident. In the largest accident predicted in the study, the 8000 cases per year would be distributed over about 10 million people; the normal annual incidence rate of nodules in this population is about 8000 per year. Thus the largest accident would approximately double the normal incidence; this effect would be detectable in the population subject to the risk.

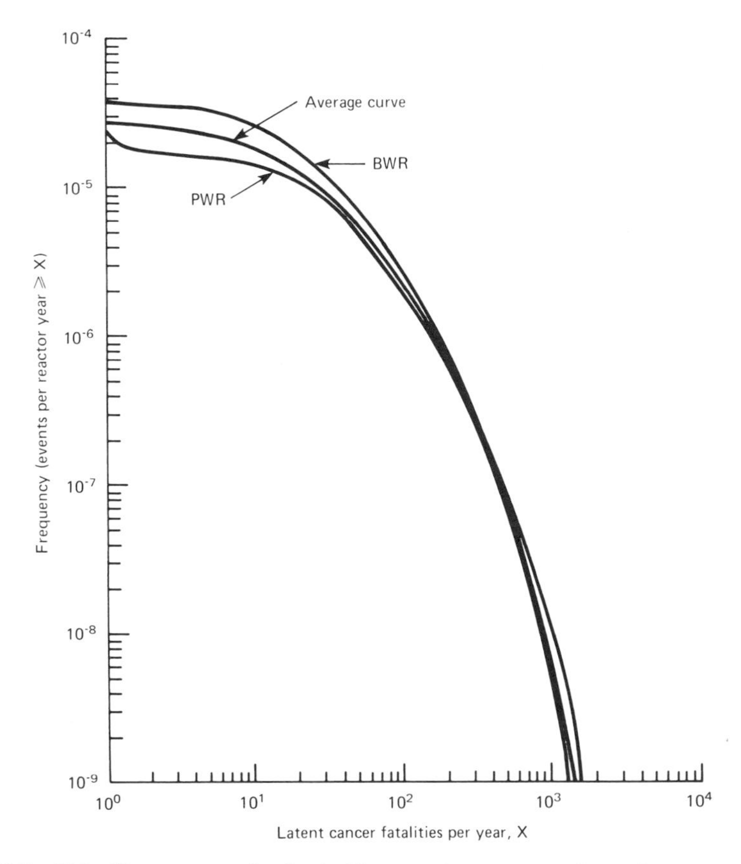

Fig. 12-7 *Risk of latent-cancer fatality incidence per reactor-year. Approximate uncertainties are estimated to be represented by factors of $\frac{1}{6}$ and 3 on latent-cancer fatalities and by factors of $\frac{1}{5}$ and 5 on frequencies.* [*From Reactor Safety Study,* WASH-*1400 (October 1975).*]

As a third long-term health effect, genetic mutations can occur spontaneously, from unknown causes, or can be induced by a variety of physical or chemical agents, one of which is ionizing radiation. The effect of radiation is to increase the mutation rate, but genetic disorders that would arise from radiation-induced mutation would not differ from those that have been occurring naturally for as long as people have existed.

The frequency distribution for the number of genetic effects that might occur is given in Fig. 12-9. In this curve the genetic effects per year apply to the first generation and would occur over about a 30-year period. Additional genetic effects could also occur in later generations. The number of

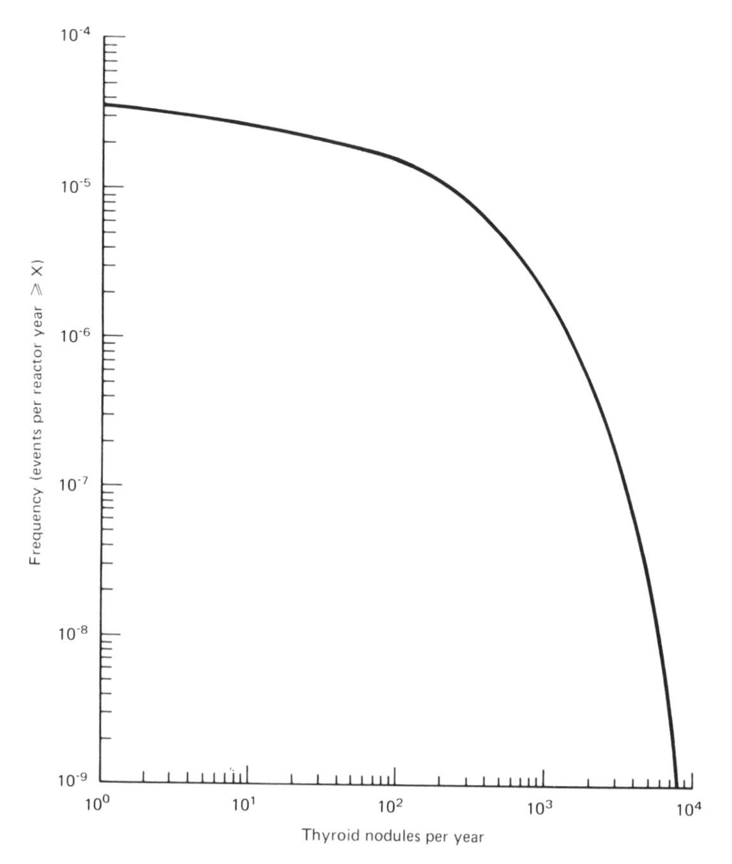

Fig. 12-8 *Risk of thyroid nodule incidence per reactor-year. Approximate uncertainties are estimated to be represented by factors of* $\frac{1}{3}$ *and 3 on thyroid nodules and by factors of* $\frac{1}{5}$ *and 5 on frequencies.* PWR *and* BWR *are nearly identical.* [*From Reactor Safety Study,* WASH-*1400* (*October 1975*).[

cases of genetic defects that could be produced by the largest accident predicted in the RSS is 190 per year. Since the normal incidence rate of genetic effects in the approximately 10 million people affected is approximately 8000 per year, the 190 cases per year would represent an increase in the normal rate of approximately 2%.

Property Damage

The predicted property damage costs are shown in Fig. 12-10. The curve shows that 80% of all core melt cases would have damage costs of less than $300 million and that 99% would have costs less than $4 billion.

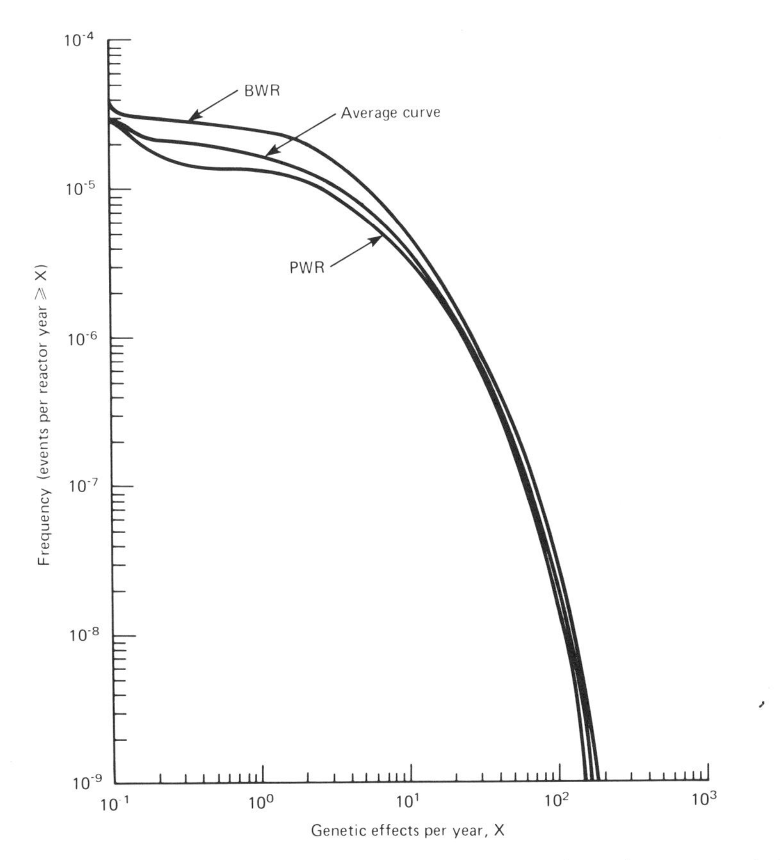

Fig. 12-9 *Risk of incidence of genetic effects per reactor-year. Approximate uncertainties are estimated to be represented by factors of* $\frac{1}{3}$ *and 6 on genetic effects and by factors of* $\frac{1}{5}$ *and 5 on frequencies.* [*From Reactor Safety Study,* WASH-*1400* (*October 1975*).]

These curves were considered to represent a conservative estimate of the costs in 1975 dollars [1]; of course, inflation and the escalation in property values since 1975 means the dollar magnitudes are considerably underestimated.

Figure 12-11 shows the frequency distribution of land area affected for the case in which people would not be relocated, but in which decontamination would be required, and an area from which people would have to be relocated. Although a portion of this latter area would become useful after decontamination, the dollar damage estimates incorporated the total value of structures and land within this area. In 80% of all potential core

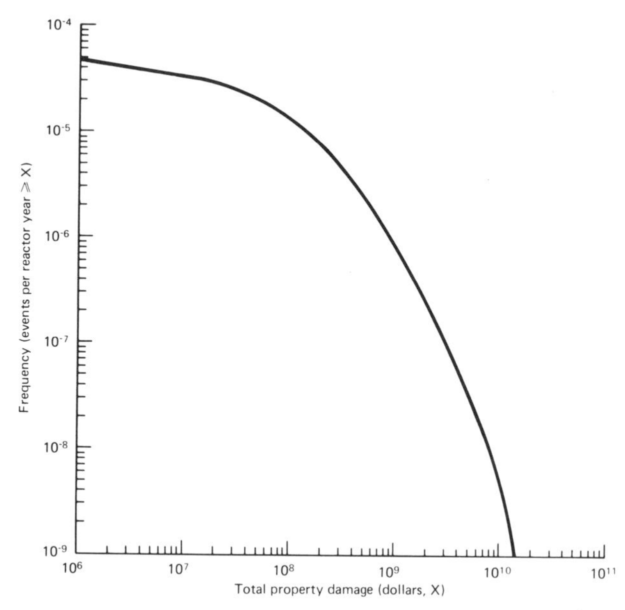

Fig. 12-10 *Risk of property damage per reactor-year. Approximate uncertainties are estimated to be represented by factors of $\frac{1}{5}$ and 2 on property damage and by factors of $\frac{1}{5}$ and 5 on frequencies.* [*From Reactor Safety Study,* WASH-*1400* (*October 1975*).]

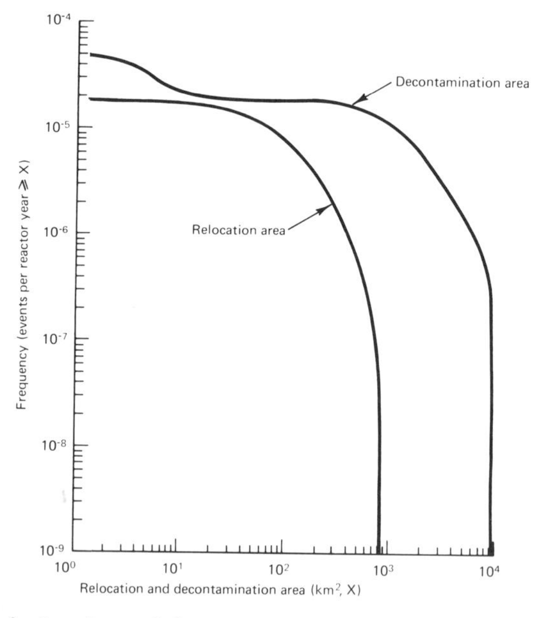

Fig. 12-11 *Risk of relocation and decontamination area per reactor-year. Approximate uncertainties are estimated to be represented by factors of $\frac{1}{5}$ and 2 on areas and by factors of $\frac{1}{5}$ and 5 on frequencies.* [*From Reactor Safety Study,* WASH-*1400* (*October 1975*).]

melt accidents, the area that might require relocation is less than about 50 km^2 and the area requiring decontamination is less than about 1000 km^2.

12-7-2 Conclusions

The Reactor Safety Study was a risk assessment with a more realistic approach than had previously been used in the licensing process. Factors that contributed to this realism include [5]:

1. The definition of dependencies between and among safety functions and the engineered safety systems.
2. Relating core melt to containment failure modes.
3. Consideration and application of human error, test and maintenance, and common cause contributions to system failures.
4. A consequence model that incorporates probability distributions for population and weather conditions and the effect of both plume rise and evacuation.
5. Realistic failure definitions for safety functions where possible.
6. Realistic values for radioactivity removal from containment by various means.

An important result of the study was the determination that large LOCAs are not the only significant accidents for either reactor type. In fact, the frequency of core melt due to large LOCAs was predicted to be a factor of ten less than the frequency of core melt from other causes.

Small LOCA sequences and non-LOCA transients were determined to be significant contributors to the predicted frequency of core melt for the PWR. The major contributor to the core melt frequency for the BWR was failure of the decay heat removal systems following a transient. It is noteworthy that reactor vessel and steam generator failures are not significant contributors to overall risk.

An important result in Fig. 12-12 is a comparison of the predicted frequency of fatalities from 100 nuclear reactors compared to the risks from other hazards. The curves for the nonnuclear hazards have been obtained from historical records, whereas the risk for the nuclear power plants was obtained by multiplying by 100 the frequencies predicted for the average reactor (Fig. 12-5). Thus the curve for 100 reactors assumes there is no common cause, such as an earthquake, that will damage more than one reactor at a multiple-reactor site.

The orders-of-magnitude differences in Fig. 12-12 between nuclear and nonnuclear risks is clearly evident. Other comparisons of risks will be made in Chapter 16, while Chapter 18 contains arguments about why such comparisons between nuclear and nonnuclear risks are often criticized.

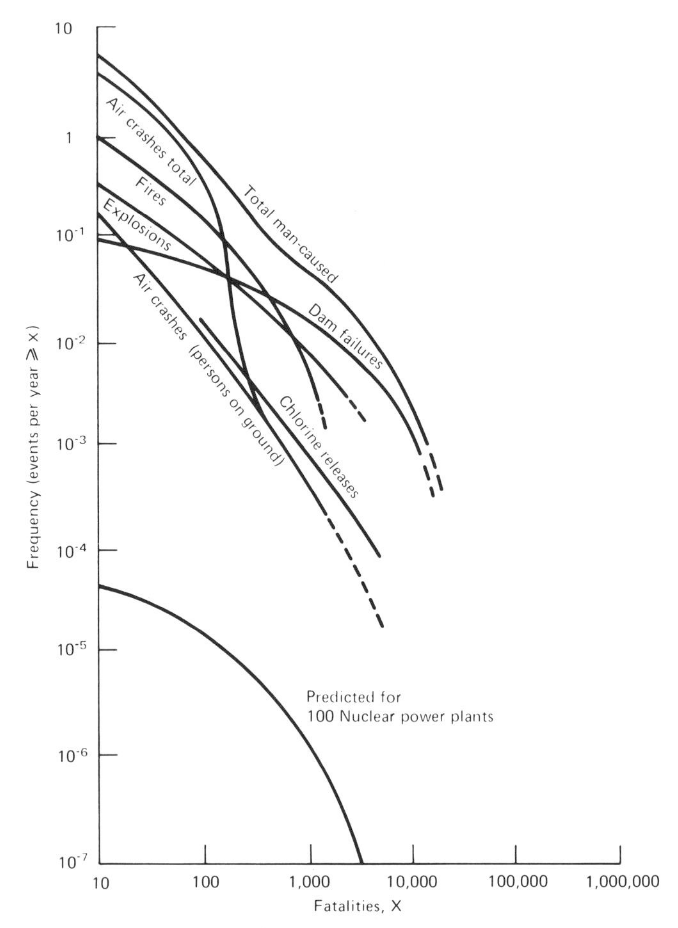

Fig. 12-12 *Comparisons of risks for fatalities.* [*From Reactor Safety Study,* WASH-*1400* (*October 1975*).]

12-8 LIMITATIONS OF THE REACTOR SAFETY STUDY

As pointed out in a review of the RSS [5]:

1. The study analyzed only spontaneously generated accidents at nuclear reactor sites. For example, no attempt was made to quantify the frequency or consequences of sabotage. A separate study of this subject has been completed [6]. Likewise, no attempt was made to treat risks from other portions of the nuclear fuel cycle. (Some of these other risks will be considered in Chapters 14 through 16.)

2. The study assumed that a partial core melt always led to complete core melt. Because of lack of direct experience , it was further assumed that once a core lost its initial configuration, it would be difficult to contain; in such a situation, however, it might melt downward but remain contained from below by the foundation because of adequate upward heat transfer. For cases in which the molten core would penetrate the containment building through the bottom, its interaction with the surrounding soil and water table was not well-defined in the study.

3. The study did not specifically address the effect of initiating events at multi-unit sites. It is expected that such sites and site events can be treated in the same manner as single unit sites; however, it is unclear just how such multi-unit sites may modify the risk estimates presented.

4. The study treated common cause concerns in a largely macroscopic manner, essentially ignoring them in the microscopic sense. Although the overview technique that was used has merit, a method that also locates components common to more than one system is better.

5. The use of the "smoothing" technique, which was applied to the consequence categories (whereby 10% of each category was added to adjacent categories and 1% to the second adjacent categories), is conservative but may mask significant results.

6. The study determined source terms only for the key accident sequences resulting from a large LOCA. These were then used to establish release categories. Source terms for other important accident sequences arising from different initiating events were estimated by comparing the phenomena involved in each sequence with the phenomena involved in the large LOCA key sequences. Thus, source terms for events other than the large LOCA were not independently determined by means of the CORRAL code.

This approach was expected to provide a conservative bound on the release magnitudes (and therefore consequences) resulting from these other initiating events. If the study's assumption of conservative bounding was, in fact, correct, the conservatism injected into the analysis of accident initiators other than the large LOCA may have resulted in an overemphasis of the importance of other initiating events. Thus, the study's conclusion that the small LOCA and transient event initiators are more important than the large LOCA may, in fact, partially result from the use of the source terms from the large LOCA key accident sequences for all initiators.

7. The study does not make clear how to break down risk contributors. For example, it is important to know whether the major contributors to risk were the result of mechanical malfunctions of plant equipment due to intrinsic failure rates, external events, or human interaction (either

operator or test and maintenance crew error). This type of information would be valuable in improving future reactor designs; it would also have an impact on maintenance operations and possibly on certain common cause or common mode problems.

8. The study did not attempt to quantify the conservatism claimed for its analysis. Indeed, little or no sensitivity work was performed to show the effect of variations in significant parameters on the results.

12-9 CRITIQUES OF THE REACTOR SAFETY STUDY

Following publication of the Reactor Safety Study (WASH-1400) report, the Nuclear Regulatory Commission in 1977 organized a Risk Assessment Review Group, in order to [7, 8]:

1. Clarify the achievements and limitations of WASH-1400.
2. Assess the peer comments thereon and responses to those comments.
3. Study the present state of such risk assessment methodology.
4. Recommend to the Commission how (and whether) such methodology can be used in the regulatory and licensing process.

The Group was formed to represent a wide spectrum of views about nuclear safety, although each member was chosen for his or her technical expertise. The Group's report is sometimes called the "Lewis Committee Report," named after the man who served as chairman.

It was concluded that the methodology used in WASH-1400 is an important advance over earlier techniques applied to reactor risks, is sound, and should be developed and used more widely under circumstances in which there is an adequate data base or sufficient technical expertise to insert credible subjective probabilities into the calculations. Proper application of the methodology can provide a tool to make the licensing and regulatory process for nuclear reactors more rational and to match available resources (research, quality assurance, inspection, licensing regulations) to the risks.

The Group also found that the Executive Summary of the RSS (a 12-page overview of the study prepared for the layperson) is a poor description of the contents of the report and could be misused in a discussion of reactor risks. For this reason, the Nuclear Regulatory Commission in early 1979 withdrew any explicit or implicit past endorsement of the Executive Summary (but not of the rest of the RSS).

The Group stated there were a number of sources of both conservatism and nonconservatism in the risk calculations in WASH-1400, but it was unable to determine whether there was any net bias in one direction or the other. Among the sources of conservatism were the inability to quantify human adaptability during the course of an accident, and the effects from a pervasive regulatory influence in the choice of uncertain parameters. Sources of nonconservatism were issues about completeness of the accident sequences and an inadequate treatment of common cause failures. Thus the Group was unable to define whether the overall probability of a core melt given in WASH-1400 is high or low, but it concluded that certain error bands were understated by an unknown amount. Reasons for this understatement included an inadequate data base, a poor statistical treatment, and an inconsistent propagation of uncertainties throughout the calculation.

The Nuclear Regulatory Commission accepted the Group's conclusion that absolute values of the risks presented in WASH-1400 should not be used uncritically either in the regulatory process or for public policy purposes, and agreed that the numerical estimate of the overall risk of reactor accident reported in WASH-1400 is not reliable. This conclusion echoed that of an independent review of nuclear reactor safety conducted by the American Physical Society, which concluded from the (draft) WASH-1400 report that [9]:

> the event-tree and fault-tree approach can have merit in highlighting *relative* strengths and weaknesses of reactor systems, particularly through comparison of different sequences of reactor behavior. However . . . we do not now have confidence in the presently calculated absolute values of the probabilities of the various branches.

In still another critique of nuclear power, the 1977 Ford Foundation–MITRE Corporation study [10] considered that the WASH-1400 may have seriously underestimated the chances of reactor accidents (or conversely may have overestimated them); similar doubts were expressed in a study sponsored by the National Academy of Sciences [11].

The American Nuclear Society (ANS) has reviewed both reports and [12] "finds no major discrepancies between the conclusions of the Rasmussen Report and the Lewis Report." Furthermore the ANS "supports the development and application of probabilistic risk assessment to improve safety in design, to assist in safety analysis, and to help put into perspective the relative risks associated with all forms of energy production and use."

Perhaps the most detailed and quantitative analysis of the Rasmussen (WASH-1400) and Lewis Reports has been published by the Electric Power Research Institute (EPRI) [5, 13–15], which considered the same issues as did the Lewis Group. Although there was a general agreement on the findings of the two reports, the EPRI study concluded that the maximum degree of nonconservatism in the WASH-1400 numerical results was less than a factor of 4 in both the median and upper bound estimates of the

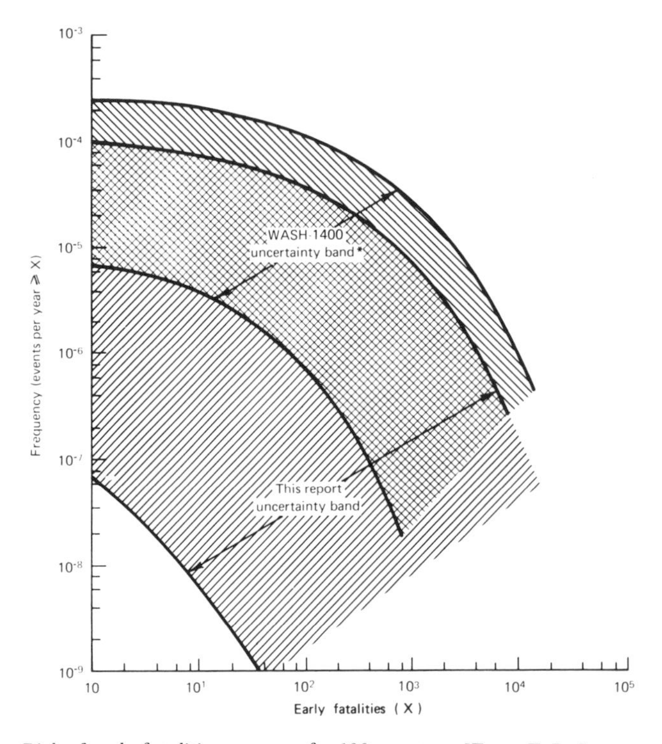

Fig. 12-13 *Risk of early fatalities per year for 100 reactors. [From F. L. Leverenz and R. C. Erdmann, Electric Power Research Institute Rep.* EPRI NP-*1130* (*July 1979*).]

core melt probability; the more likely result is that WASH-1400 is overly conservative by a factor of 12 on the median [15]. Another conclusion of the EPRI study was that the 90% confidence limits for the values in WASH-1400 were too small by a factor of 20; the effect of this larger uncertainty band on the risk curve for early fatalities is shown in Fig. 12-13.

12-10 FOLLOW-UP REACTOR SAFETY STUDIES

Risk analyses of light water reactors are continually being performed [16]. For example, the Nuclear Regulatory Commission is in the process of an Integrated Reliability Evaluation Program (IREP). This program is being conducted to help ensure that specific differences between any United States nuclear power plant and those portrayed in the RSS do not lead to potential risks greater than those given in the study. Studies are also underway to assess the level of potential accident risks for nuclear plants located near large populations, again to see that the risks do not exceed those in WASH-1400 by a unacceptable amount. It also may be mentioned that the Department of Energy is performing a detailed probabilistic risk analysis of a baseline version of each reactor manufactured by the four United States vendors; the purpose of these analyses is to improve the safety for each particular type of plant.

Some of the post-RSS analyses already completed have involved a reexamination of the safety systems studied in the RSS [14, 17–19]. The sensitivity of the RSS results to changes in component failure rates also has been examined [20], as well as the potential impact upon the conclusions of the RSS because of inadvertently overlooked accident sequences [21].

Other investigations have focused on improving the methods used to evaluate the probability of pressure-vessel failure and the unavailability of safety systems [22–25]. Improvements also have been made in the calculation of the potential effects of radioactive releases [26].

The impact upon the safety of nuclear plants from specific natural hazards, such as fires [27], earthquakes [28], or tornadoes [29], has been investigated in detail.

Another type of risk study is an examination of the relative reduction in risk obtained from different reactor safety systems, such as the containment, engineered safety features, and the emergency core cooling system (ECCS). In such a case, the containment functions are the most important for reducing the risk to the general public in the event of an accident [30], whereas the ECCS is most important for minimizing damage to the nuclear power plant itself. This will be discussed in more detail in Chapter 17.

Still another extension of the RSS-type analysis is to the remainder of the nuclear fuel cycle, including mining and milling, transportation, fuel fabrication, and waste disposal. Results of an Electric Power Research Institute study indicate that the supporting fuel cycle contributes about 1% of the risk to the public of an LWR, and thus the results of the RSS reasonably approximate the full risk of nuclear power [31]. Of the different portions of the remainder of the cycle, mining and milling contribute health hazards from the routine emissions of low-level radiation, while transportation (discussed in Chapter 14) contributes the greatest risk to the public from accidents. The risks of radiation releases from accidents at nuclear waste disposal facilities is orders of magnitude smaller than even the risks from transportation accidents [31], but waste disposal risks are discussed in Chapter 15 because of the interest by the general public in the past few years.

12-11 THREE MILE ISLAND ACCIDENT

On March 28, 1979, there was an accident at the Three Mile Island Unit 2 reactor near Middletown, Pennsylvania, that led to the dispersal of a small amount of radioactivity. The accident in the Babcock and Wilcox PWR began as a small loss of coolant accident caused by the failure of a 5-cm valve in the pressurizer to reseat. Because of a series of maintenance and control room human errors, control room design deficiencies, and nuclear regulatory lapses in not providing better training for the operators, the chain of events ultimately led to a release of radioactivity from the turbine auxiliary building [32–34].

The overall release of radioactivity from the plant was such that if any member of the public had been at the site of the highest level of off-site radiation for the first 11 days after the accident, he or she would have received a dose of 83 millirem [35]. By way of comparison, had the individual resided in Denver, Colorado, instead of in Middletown for the year preceding the accident, he or she would have received a comparable dose. Thus, the additional radioactivity burden on the residents of Pennsylvania was not excessive.

One of the questions of interest to risk analysts is: Were the events of the Three Mile Island accident included as a possible scenario in the Reactor Safety Study? The answer is: yes and no. The small LOCA due to failure of the valve to reseat in the pressurizer had been considered, but many of the following (incorrect) operator actions, such as the shutoff of the emergency coolant injection system in the midst of a loss of coolant accident, were not. The consequences from the uncontrolled small LOCA considered in the Reactor Safety Study were conservative in the case of the Three Mile Island accident.

In retrospect it is probably safe to say that the major damage done during and after the accident was some loss of public acceptance of nuclear power. The principal gain was an increased understanding of the way to better design the reactor control room and the need for better training of reactor operators under conditions simulating the simultaneous failure of several plant components [36–39].

EXERCISES

12-1 Summarize the analysis methods used in WASH-1400. Your exposition should be aimed at an engineer or physicist who has some familiarity with nuclear reactors and how accidents can occur, but who has not studied any of WASH-1400. A length of 1000 to 1500 words, exclusive of figures or tables, is preferred.

12-2 Summarize the results of WASH-1400. Your exposition should be aimed at an engineer or physicist who has some familiarity with nuclear reactors and how accidents can occur, but who has not studied any of WASH-1400. A length of 1000 to 1500 words, exclusive of figures or tables, is preferred.

REFERENCES

1. Reactor Safety Study—An Assessment of Accident Risks in U.S. Commercial Nuclear Power Plants. U.S. Nuclear Regulatory Commission Rep. WASH-1400, NUREG-75/014 (October 1975). [This report is commonly called the "Rasmussen Report" after N. C. Rasmussen who directed the study.]
2. I. B. Wall *et al.*, Overview of the reactor safety study consequence model, *in* "Nuclear Systems Reliability Engineering and Risk Assessment" (J. B. Fussell and G. R. Burdick, eds.), p. 87. Soc. Industrial and Applied Math., Philadelphia, Pennsylvania, 1977.
3. J. Lamarsh, "Introduction to Nuclear Engineering," Chapter 11. Addison-Wesley, Reading, Massachusetts, 1975.
4. R. H. Clarke and H. F. Macdonald, Radioactive releases from nuclear installations: Evaluation of accidental atmospheric discharges, *Prog. Nucl. Energy* **2,** 77 (1978).
5. Summary of the AEC Reactor Safety Study (WASH-1400). Electric Power Research Institute Rep. EPRI-217-2-1 (April 1975).
6. Safety and security of nuclear power reactors to acts of sabotage, *Nucl. Safety* **17,** 665 (1976).
7. Risk Assessment Review Group Report to the U.S. Nuclear Regulatory Commission. U.S. Nuclear Regulatory Commission Rep. NUREG/CR-0400 (September 1978). [This report is commonly called the "Lewis Committee Report" after H. W. Lewis who chaired the Risk Assessment Review Group.]
8. Report of the NRC risk assessment review group on the reactor safety study, *Nucl. Safety* **20,** 24 (1979).
9. Report to the American Physical Society by the study group on light-water nuclear reactor safety, *Rev. Mod. Phys.* **47,** Suppl. No. 1 (Summer 1975).
10. Report of the Nuclear Energy Policy Study Group, Ford Foundation and MITRE Corporation, "Nuclear Power Issues and Choices." Ballinger, Cambridge, Massachusetts, 1977.
11. Risks Associated with Nuclear Power: A Critical Review of the Literature, Summary and Synthesis Chapter. National Academy of Sciences (April 1979).
12. An Evaluation of the Rasmussen and Lewis Reports and of the January 1979 NRC Policy Statement on Risk Assessment, (ANS Public Policy Statement), *Nuclear News,* p. 177 (June 1980).
13. F. L. Leverenz and R. C. Erdmann, Critique of the AEC Reactor Safety Study (WASH-1400). Electric Power Research Institute Rep. EPRI 217-2-3 (June 1975).
14. R. C. Erdmann *et al.*, Probabilistic Safety Analysis, Electric Power Research Institute Rep. 217-2-4 (July 1975) and NP-424 (April 1977).

15. F. L. Leverenz and R. C. Erdmann, Comparison of the EPRI and Lewis Committee Review of the Reactor Safety Study. Electric Power Research Institute Rep. NP-1130 (July 1979).
16. S. Levine, The role of risk assessment in the nuclear regulatory process, *Ann. Nucl. Energy* **6,** 281 (1979).
17. R. C. Erdmann *et al.,* ATWS: A Reappraisal Part II Evaluation of Societal Risks Due to Reactor Protection System Failure, Vols. 1 and 2, BWR Risk Analysis; Vol. 3 PWR Risk Analysis. Electric Power Research Institute Rep. NP-265 (August 1976).
18. J. E. Kelly and R. C. Erdmann, BWR Decay Heat Removal System Appraisal. Electric Power Research Institute Rep. NP-861 (August 1978).
19. S. V. Asselin, D. D. Carlson, J. W. Hickman, and M. A. Fedele, System event tree analyses for determining accident sequences that dominate risk in LWR power plants, "Probabilistic Analysis of Nuclear Reactor Safety," Vol. 3, p. XII.2-1. American Nuclear Society, LaGrange Park, Illinois, 1978.
20. J. E. Kelly, F. L. Leverenz, N. J. McCormick, and R. C. Erdmann, Sensitivity assessments in reactor safety analyses, *Nucl. Technol.* **32,** 155 (1977).
21. W. B. Loewenstein and G. S. Lellouche, The potential impact of probabilistic safety analysis, *Proc. Topical Meeting Thermal Reactor Safety,* p. 1–43. Dept. of Energy Rep. CONF-770708 (1977).
22. S. Levine and W. E. Vesely, Prospects and problems in risk analysis: Some viewpoints, *in* "Nuclear Systems Reliability Engineering and Risk Assessment" (J. B. Fussell and G. R. Burdick, eds.), p. 5. Soc. Industrial Applied Mathematics, Philadelphia, Pennsylvania, 1977.
23. S. Levine, The role of risk assessment in the nuclear regulatory process, *Nucl. Safety* **19,** 556 (1978).
24. W. E. Vesely, F. F. Goldberg, and E. K. Lynn, OCTAVIA—Computer Code: PWR Reactor Pressure Vessel Failure Probabilities Due to Operationally Caused Pressure Transients. U. S. Nuclear Regulatory Commission Rep. NUREG-0258 (March 1978).
25. W. E. Vesely and F. F. Goldberg, FRANTIC—A Computer Code for Time-Dependent Unavailability Analysis. U. S. Nuclear Regulatory Commission Rep. NUREG-0193 (October 1978).
26. R. C. Erdmann *et al.,* Probabilistic Safety Analysis IV. Electric Power Research Institute Rep. NP-1039 (April 1979).
27. M. Kazarians and G. E. Apostolakis, On the Fire Hazard in Nuclear Power Plants, *Nucl. Eng. Design* **47,** 157 (1978).
28. Analysis of the Risk to the Public from Possible Damage to the Diablo Canyon Nuclear Power Station from Seismic Events. U. S. Nuclear Regulatory Commission Docket Numbers 50-275 and 50-323 (August 1977).
29. L. A. Twisdale *et al.,* Tornado Missile Risk Analysis. Electric Power Research Institute Rep. NP-768 and NP-769 (May 1978).
30. E. P. O'Donnell and J. J. Mauro, A cost-benefit comparison of nuclear and nonnuclear health and safety protective measures and regulations, *Nucl. Safety* **20,** 525 (1979).
31. R. C. Erdmann *et al.,* Status Report on the EPRI Fuel Cycle Accident Risk Assessment. Electric Power Research Institute Rep. NP-1128 (1979).
32. The President's Commission on the Accident at Three Mile Island, "The Need for Change: The Legacy of TMI." Pergamon, Oxford, 1979; summarized in *Nucl. Safety* **21,** 234 (1980).
33. H. W. Lewis, The safety of fission reactors, *Sci. Am.* **242,** 53 (1980).
34. Three Mile Island: A Report to the Commissioners and to the Public. Rogovin, Stern & Huge law firm report (January 1980); summarized in *Nucl. Safety* **21,** 389 (1980).

35. Ad Hoc Population Dose Assessment Group, Summary and Discussion of Findings From: Population Dose and Health Impact of the Accident at Three Mile Island Nuclear Station (A Preliminary Assessment for the Period March 28 to April 7, 1979). Bureau of Radiological Health, Rockville, Maryland (May 10, 1979).
36. TMI-2 Lessons Learned Task Force Status Report and Short Term Recommendations. U. S. Nuclear Regulatory Commission Rep. NUREG-0578 (July 1979).
37. TMI-2 Lessons Learned Task Force Final Report. U. S. Nuclear Regulatory Commission Rep. NUREG-0585 (October 1979).
38. Investigation into the March 28, 1979 Three Mile Island Accident by Office of Inspection and Enforcement. U. S. Nuclear Regulatory Commission Rep. NUREG-0600 (August 1979).
39. Human Factors Evaluation of Control Room Design and Operator Performance at Three Mile Island-2. U. S. Nuclear Regulatory Commission Rep. NUREG/CR-1270 (January 1980).

CHAPTER 13

Risks for Liquid Metal Fast Breeder and High Temperature Gas Reactors

13-1 LIQUID METAL FAST BREEDER REACTOR

13-1-1 Comparison of Safety Characteristics of LMFBRs and LWRs

Probabilistic methods have been used to assist the planning of research and development goals for the liquid metal fast breeder reactor (LMFBR) [1, 2], and an extensive probabilistic study of risks has been performed for the Clinch River Breeder Reactor Plant (CRBRP) [3–5]. This LMFBR may eventually be constructed in Tennessee and is a reactor designed to produce 380 MW_e.

Although the CRBRP and light water reactors (LWRs) have many similarities in inherent characteristics and safety design features, the following key differences substantially affect the important accident sequences and calculated radionuclide releases [3]:

1. The low stored energy of the low-pressure sodium primary coolant is the principal reason that coolant piping ruptures are significantly lower-risk accidents than for LWRs. In addition, the provision of a second piping enclosure, called the guard piping and guard vessel, provides assurance that the sodium level in the reactor vessel is maintained above the core in the event of a pipe break.
2. The chemical composition, temperatures, and pressure of the containment atmosphere are considerably different following significant accidents in the CRBRP than they are in LWRs. Those differences lead to lower pressures in the containment, and an inherent depletion mechanism in the CRBRP containment resulting from sodium oxide aerosol agglomeration and settling.
3. The radionuclide releases from containment to the atmosphere following a core accident are expected to be lower for the CRBRP than for LWRs. Also, the chemical affinity between sodium and iodine and the

aerosol settling phenomenon make the release of iodine significantly lower for the CRBRP.

4. The CRBRP design includes two independent, diverse, fast-acting shutdown systems. Thus, although anticipated transients without scrams have more severe consequences for the CRBRP than for LWRs, especially if the reactor reactivity exceeds prompt-critical, the probabilities of such events have been reduced through this additional shutdown mechanism.

5. The thermal, hydraulic, nuclear, and geometric characteristics of the CRBRP core do not exclude events leading to fuel vaporization, as in the case of LWR cores. These characteristics result in energetic core-disruptive accidents that are significant risk contributors for the CRBRP.

6. The inventory of plutonium per megawatt of electrical output is approximately nine times greater for the CRBRP than for a typical LWR. Approximately one-third of the plutonium in the CRBRP is contained in the radial blankets, however, which are assumed not to be involved in core accidents. The greater plutonium content of the CRBRP, together with the potential that core accidents could lead to vaporization of the fuel, makes plutonium a more important contributor to risks for the CRBRP than for LWRs.

7. During full-power operation, the core of an LMFBR is not in its most reactive configuration. Thus certain mechanisms that can rapidly move fuel material toward the center of the core or move sodium away from the center of the core may produce an increase in power. If the power increased rapidly enough, the fuel could heat up sufficiently to be vaporized. Expansion of the vaporized fuel might do mechanical damage to the reactor primary enclosure.

As we can see, characteristics 1 through 4 tend to lessen the risks from an LMFBR accident relative to an LWR accident, whereas characteristics 5 through 7 enhance those risks.

13-1-2 Risk Assessment Analysis

The Clinch River Breeder Reactor Plant Safety Study [3–5] was conducted to evaluate the risk to the public associated with the operation of the CRBRP. Because this work focused on risk to the public health, the only events investigated in the CRBRP Safety Study were those leading to the release of radioactivity to the environment, i.e., those accidents sufficient to cause a core disruptive accident (CDA) in which the core coolable geometry is lost. Such accidents include loss of heat removal capability at power, loss of piping integrity, loss of flow with failure to shutdown, and loss of decay heat removal capability following reactor shutdown.

The approach employed in the CRBRP study paralleled that used in the Reactor Safety Study [6]. The similarity of approach facilitated a comparison between the calculated risk of operating the CRBRP at its own site and the corresponding risk for a typical LWR located on a composite of sites of the first 100 light water plants.

Although the design of the CRBRP was not yet finalized at the time of the study, it was felt that the systems significant to plant safety were sufficiently well designed to allow a meaningful risk assessment [3]. However, the study could not be completed to the depth or precision possible for operating reactors.

There are 12 causal categories with the potential to cause a CDA [3]:

Inadequate Heat Removal

1. Inadequate heat transfer between the fuel and the primary sodium,
2. Blocked or ruptured primary heat transport system,
3. Inadequate heat transfer between the primary and the intermediate heat transport systems,
4. Blocked or ruptured intermediate heat transport system,
5. Inadequate heat transfer between the intermediate heat transport system and the steam system,
6. Blocked or ruptured steam system,
7. Inadequate heat transfer between the steam system and the cooling tower circulating water,
8. Inadequate cooling by the cooling tower circulating water system.

Excess Reactor Power

9. Fuel reactivity effects (temperature, density, enrichment, and core movement),
10. Control-rod reactivity effects (rod motion and incorrect loading),
11. Sodium reactivity effects (temperature and density),
12. Steel reactivity effects (density).

For most classes of CDAs, the predicted result is relatively slow melting of several fuel assemblies, possibly followed by collection of the molten fuel and steel in the bottom of the reactor vessel. If a sufficient fraction of the core is involved, penetration by the molten fuel material through the bottom of the reactor vessel and guard vessel may occur; this is termed *thermal damage*. In addition to thermal damage, mechanical damage resulting from the CDA can occur. The degree of mechanical damage to the vessel can be related to the amount of fuel vapor formed during the accident: the larger the fraction of the core vaporized, the greater will be the energy transmitted to the reactor vessel and head and the greater will be the potential for mechanical damage.

The causal categories already given ultimately led to a list of 33 accident initiators that were potentially significant contributors to risk. The 33 accident initiators are itemized in Table 13-1 along with their mean frequency of occurrence.

The procedures used for the core accident analysis and evaluation differed from those used for the light water Reactor Safety Study. Specifically, the CORRAL code was replaced by the SAS-3A code [7] to analyze the initiating phase of each accident, and the VENUS-II code [8] was used to analyze the hydrodynamic disassemblies that were predicted to occur. No computer code existed for a mechanistic analysis of the transition phase between the initiation and disassembly phases, so the approach was to examine the key phenomena as separate effects, perform out-of-pile experiments where possible, and employ engineering judgment to construct the scenario [3].

The SAS-3A code models the reactor core by combining groups of subassemblies into "channels." The channel itself is represented in the code by a two-dimensional average fuel pin surrounded by an annular sodium gap and an annular structural shell. Each channel has an average pin (fuel, gap, and cladding described by one-dimensional radial heat conduction models) and an axial sodium coolant flow gap, which includes a boiling model. Examples of other important phenomena included in SAS-3A are those that describe fission gas release and restructuring within the fuel, cladding and fuel motion, molten-fuel-coolant interactions due to bursting of the cladding during rapid power transients, and the hydraulic response for the primary loop and core. The VENUS-II code is a hydrodynamic core-disruption model that assumes that the core has reached a state in which the fuel and core structures are largely molten and a large increase in reactor power has commenced, as established by the SAS analysis output. The disassembly of the core is then described by use of a two-dimensional (r–z) hydrodynamics-point kinetics algorithm.

Two general types of core accident terminations were identified [3]: melting of fuel in a nonenergetic core disruptive accident (CDA) and vaporization of fuel (energetic CDA). In a typical accident scenario, an appropriate accident initiator would cause some of the core to vaporize so that a high-pressure fuel vapor bubble would form. This bubble would expand, driving the sodium in the reactor vessel above the core upward as a slug toward the reactor head. If the energy expended in expansion of the vapor bubble were large enough, some damage might occur to the seals in the reactor head. The damaged seals would provide one pathway through which sodium and radioactive core material could be released from the reactor vessel to the reactor containment building.

Following this initial hypothesized expansion of the fuel vapor bubble, some portion of the reactor core also could melt and accumulate in the

Table 13-1 *Summary of Accident Initiators and Their Probability of Occurrence*[a]

Initiator	Median frequency (per year)	Uncertainty factors[c]
1.1 Fuel failure caused by the formation of coolant blockages of fuel subassemblies due to crud, fuel particles, or external objects (1)[b]	10^{-4}	10
1.2 Fuel failure caused by fuel-pin enrichment errors (9)	10^{-5}	10
1.3 Failure of the core to be supported because of either reactor vessel or core support failure (1)	10^{-8}	100
1.4 Single control-rod assembly withdrawal at maximum design speed (i.e., 24 cm/min) (ramp insertion of <10¢/sec) (10)	3×10^{-1}	3
1.5 Single control-rod assembly withdrawal at maximum speed capability (i.e., 183 cm/min for primary rods) (ramp insertion of 10 to 50¢/sec) (10)	10^{-6}	10
1.6 Control assembly group (7 assemblies) withdrawal at maximum design speed (i.e., 24 cm/min) (ramp insertion of 10 to 50¢/sec) (10)	3×10^{-2}	3
1.7 Control assembly group (7 assemblies) withdrawal at maximum speed capability (i.e., 183 cm/min for primary rods) ramp insertion of >50¢/sec) (10)	3×10^{-7}	10
1.8 Sudden core radial movement (≤60¢ step) (9)	unlikely; not considered further	
1.9 Core, radial blanket, and control-rod movement due to an SSE (60¢ step) (9,10)	1.5×10^{-4}	3
1.10 Core, radial blanket, and control-rod movement due to an OBE (30¢ step) (9,10)	1.4×10^{-3}	3
1.11 Voiding or gas bubbles in core (1,11)	10^{-5}	10
1.12 Moderator in the coolant (11)	10^{-5}	10
1.13 Loss of hydraulic holddown (19¢ step for 30 assemblies moving) (9)	10^{-4}	10
1.14 Severe damage to control-rod system preventing scram (1,3)	no credible initiators identified; not considered further	

bottom of the reactor vessel. If more than 4% of the fuel in the core were to accumulate in the bottom of the CRBRP reactor vessel, the molten fuel would melt through the reactor vessel and guard vessel as a result of fission product decay heating. Following this melt-through, a significant fraction of the sodium coolant in the primary loops would flow out of the vessel and into the bottom of the reactor cavity.

After a period of several hours, the heat generated by radioactive decay of fission products in the fuel would be sufficient to heat the sodium in the reactor cavity to its boiling point. Sodium boiling would produce a source of radioactive material and sodium vapor that would be released to the

Table 13-1 (*Continued*)

Initiator	Median frequency (per year)	Uncertainty factors[c]
2.1 Rupture of upper part of reactor vessel or of 1 primary-loop hot leg (2)	10^{-7}	100
2.2 Rupture of lower part of reactor vessel or of 1 primary-loop cold leg (2)	10^{-8}	100
2.3 Loss of one heat transport loop (2,3)	2	3
2.4 Rupture of an intermediate loop in the steam generator bay (4)	10^{-7}	100
2.5 Internal failure of a steam generator module (5)	10^{-1}	3
2.6 Loss of flow in 3 primary loops (1,3)	10^{-2}	3
2.7 Pump trip in 3 primary loops due to PPS signal (1,3)	7	3
3.1 Loss of the main condenser system (7)	3×10^{-1}	2
3.2 Loss of the main feedwater supply (5)	2	2
3.3 Loss of 1 steam generator loop (5)	10^{-1}	3
3.4 Loss of 3 steam generator loops (5)	1.4×10^{-5}	10
4.1 Loss of off-site power (1,3)	1×10^{-1}	3
4.2 Turbine-generator trip (1,3)	1	2
4.3 Loss of dc power (1,3)	10^{-8}	100
4.4 Loss of emergency ac power (1,3)	10^{-6}	100
4.5 Loss of instrumentation and control for critical plant components (1,3)	1.2×10^{-6}	100
5.1 Normal shutdown with all SHRS configurations available	3	2
5.2 Normal shutdown with one heat transport loop unavailable	2	2
5.3 Normal shutdown with the DHRS unavailable	5×10^{-1}	3

[a] From Rep. CRBRP-1 (1977).

[b] The number in parentheses after each initiator refers to the causal categories discussed on page 292.

[c] The product of the uncertainty factor and the median frequency represents the 95% confidence level for the initiator frequency.

reactor containment building. Once inside this building, radioactive material that is not decaying and not removed from the containment atmosphere by naturally occurring depletion mechanisms could be released to the environment.

13-1-3 Risks for the CRBRP

The consequences of a CRBRP reactor core accident were analyzed by using the same accident consequence analysis methods as those for light water reactors (see Section 12-4). A total of 43 different meteorological conditions were considered, each weighted according to its relative frequency of occurrence. The first 42 conditions consisted of the 6 Pasquill stability classifications [9–11], each for 7 wind speeds, and the last condition was for rain. Meteorological conditions were derived from 3 years of data taken at the CRBRP site. The population distribution was well known from census data completed every 10 years.

Important conclusions from the CRBRP risk assessment are [3]:

1. The accident sequences dominating the risk from the CRBRP are those involving loss of flow with failure to shut down, loss of off-site and on-site power, and seismically initiated sequences. The latter two types of sequences led to failure to adequately cool the core following shutdown.
2. The frequency of nonenergetic core disruptive accidents (CDAs) in the CRBRP is estimated to be 1 in 45,000 per year of reactor operation; most of these CDAs would have an insignificant effect on public health.
3. The frequency of a highly energetic CDA that could simultaneously fail the primary system boundary and the reactor containment building is estimated to be 1 in 200 million per year of reactor operation.
4. Sensitivity studies indicate that major variations in important input data and assumptions do not alter the conclusion that the CRBRP contributes negligibly to overall societal risk.

Figure 13-1 is the frequency for early fatalities caused by accidental releases of radioactivity from the CRBRP. The fatalities are those occurring within one year of the accident. The median curve represents the 50% confidence level (i.e., there is a 50% probability that the risk is above the curve and a 50% probability that the risk is below the curve). The upper bound represents a 95% confidence level (i.e., there is a 95% probability that the risk is below this curve) and the lower bound represents a 5% confidence level. Similar uncertainties exist for latent-cancer fatalities shown in Fig. 13-2.

The CRBRP early-fatality curve is actually made up of two curves representing two fundamentally different accident events. One curve dominates the low-consequence end, and the other the high-consequence end. The portions of these two curves that contribute insignificantly to frequencies of lower or higher consequences are shown dashed in the figures.

Figure 13-3 compares the early-fatality risk from the CRBR with that for an LWR considered in the Reactor Safety Study. Figure 13-4 shows a similar comparison for latent fatalities. Potential consequences associated with the CRBR accidents are comparable to those of the LWR; however,

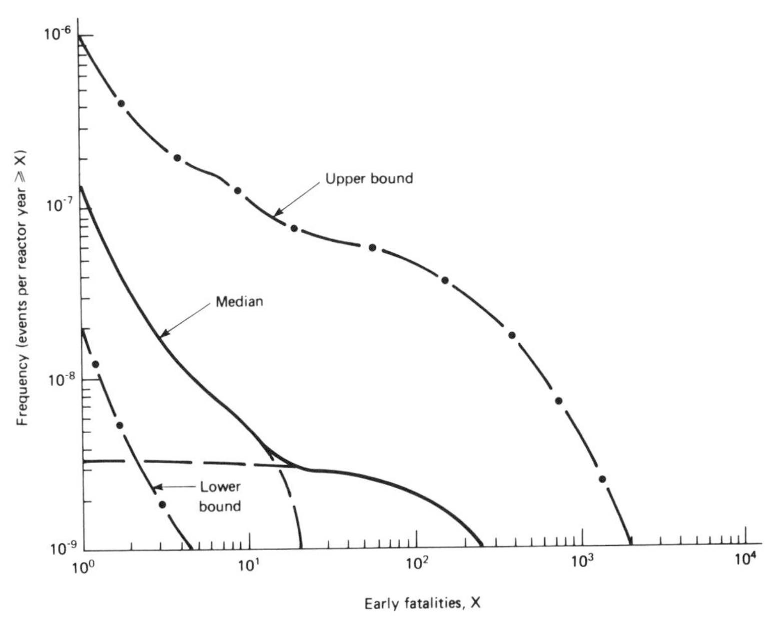

Fig. 13-1 *Risk of early fatalities per reactor-year.* (*From Rep.* CRBRP-*1, 1977.*)

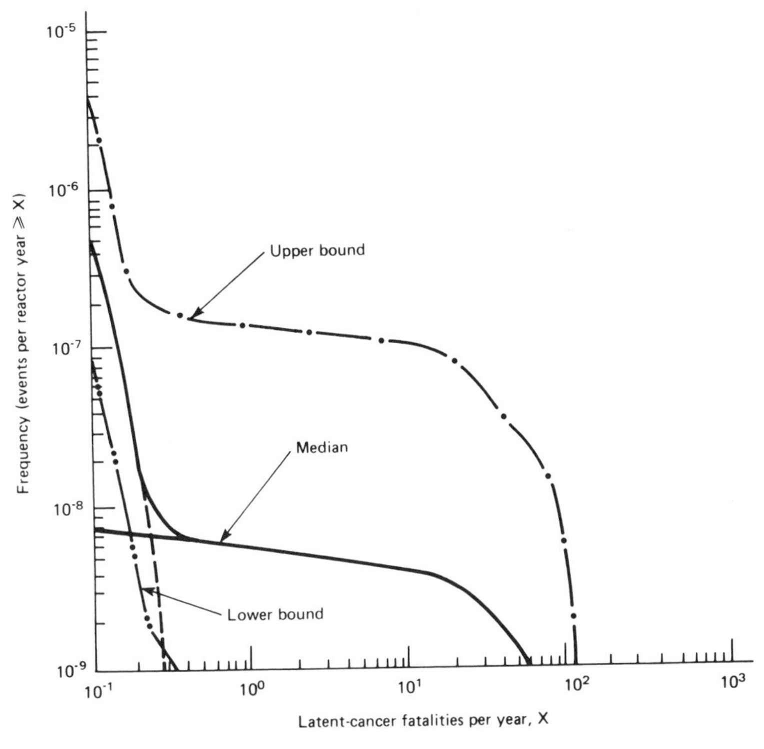

Fig. 13-2 *Risk of latent-cancer fatalities per reactor-year.* (*From Rep.* CRBRP-*1, 1977.*)

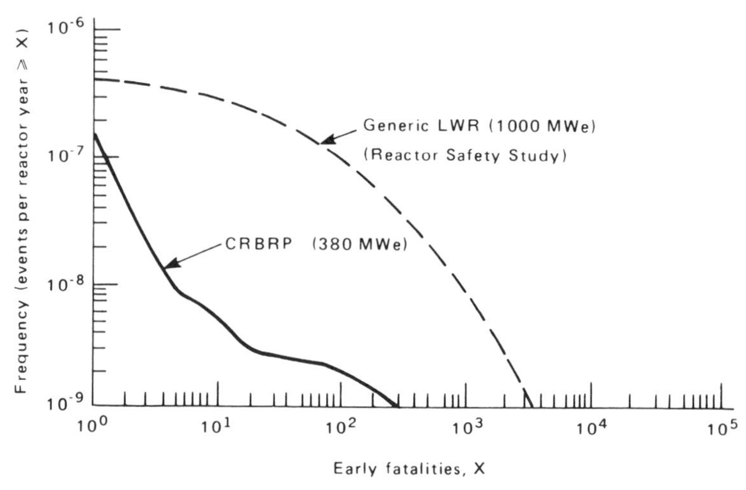

Fig. 13-3 *Comparison of risks for early fatalities for the* CRBRP *and* LWR. (*From Rep.* CRBRP-*1*, *1977*.)

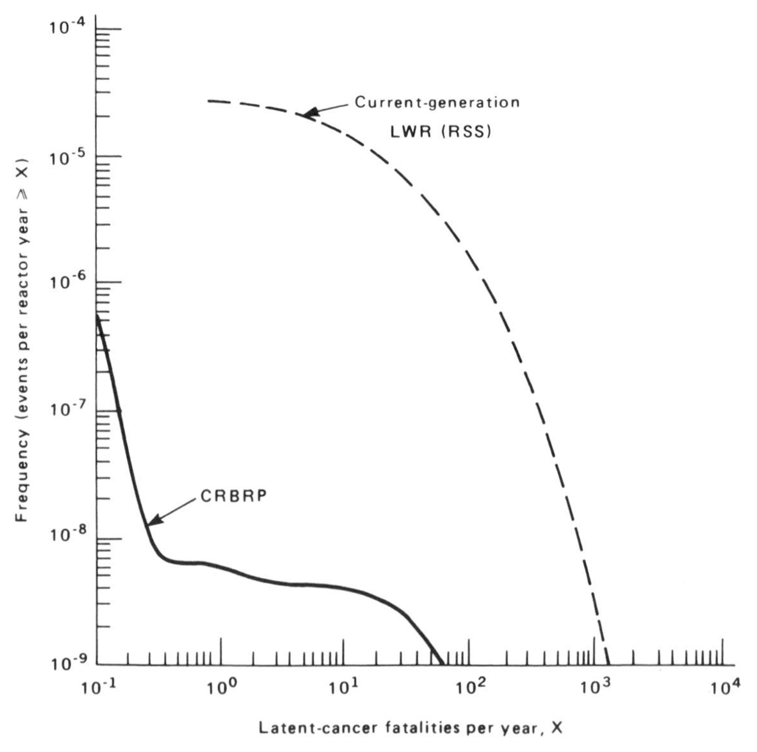

Fig. 13-4 *Comparison of risks for latent-cancer fatalities for the* CRBRP *and* LWR (*From Rep.* CRBRP-*1*, *1977*.)

care must be exercised in drawing direct comparisons from these curves because [3]:

1. Even though the CRBR consequences appear to be significantly less than those for an LWR, the larger uncertainties associated with the CRBR curve do not allow for a point-by-point comparison. Note that if the upper

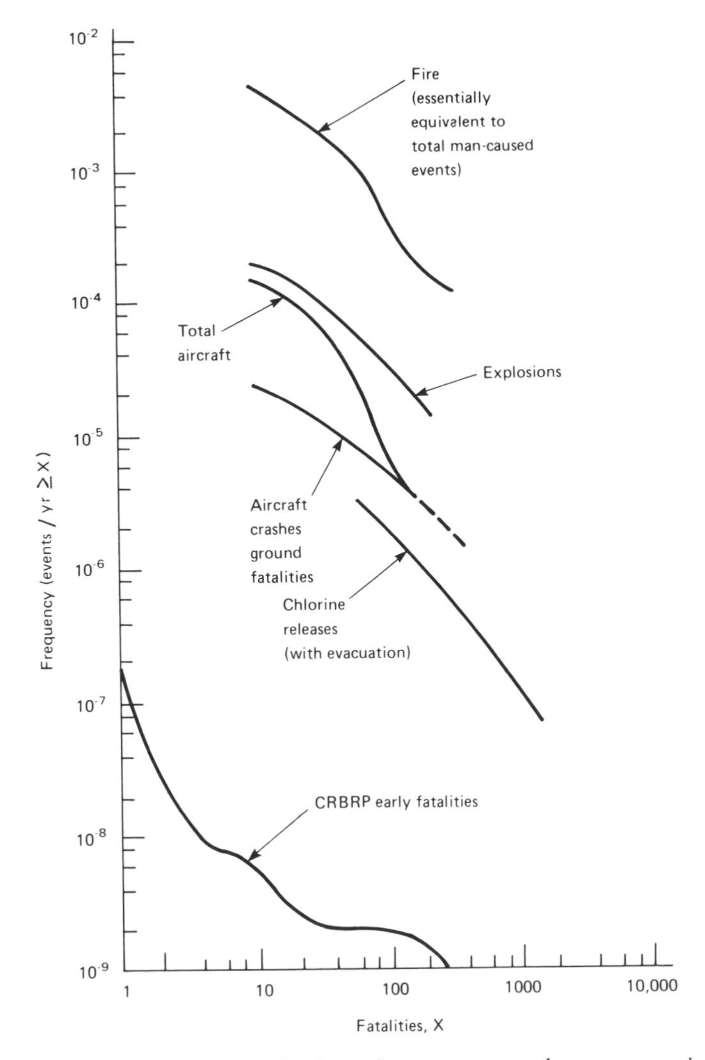

Fig. 13-5 *Comparison of risks for fatalities due to man-caused events occurring within 10 miles of the* CRBRP *site*. (*From Rep*. CRBRP-*1, 1977*.)

bound values of the CRBR curve were plotted from Fig. 13-1, the curve would be close to but still lower than the LWR median-value curve.

2. The LWR curve is for a 1000-MW_e plant, whereas the CRBR power is 380 MW_e; this means that the fission product inventory in the CRBR, at an equivalent fuel burnup, is approximately one-third that in a current-generation LWR.

3. The LWR curve is the average of LWR sites throughout the United States, whereas the CRBR curve is specific to the Clinch River site in Tennessee.

Figure 13-5 shows the CRBR risk due to early fatalities compared to the estimated frequency for other low-frequency high-consequence events. As we can see, the predicted risk is orders of magnitude smaller than that from other hazards in the vicinity of the CRBRP site.

13-2 HIGH TEMPERATURE GAS REACTOR

A probabilistic risk assessment of a reference design High Temperature Gas Reactor (HTGR) has been performed during the Accident Initiation and Progression Analysis (AIPA) Study [12–15]. The initial objective of the AIPA study was to provide guidance for safety research and development programs for HTGRs. Another objective of the study was to consider alternative design options related to HTGR safety. More recently, the objective has been expanded to include an assessment of risk to the public, in order to support the licensing of HTGRs.

The reference 1975 General Atomic 3000 MW_t HTGR was assumed to be located on a representative site in the United States, which is defined by an exclusion area boundary at a radius of 500 m and a low-population zone boundary at a radius of 2500 m. The site was derived from demographic and meteorological data provided in WASH-1400 (Section 12-4).

The HTGR has several advantages over a light water reactor. For example, the large heat capacity of the graphite moderator would slow down any fuel melting in the event there were a complete failure of all normal and emergency systems for providing forced convection cooling. Hence the probability of a meltdown is lower than for an LWR. The slow temperature transient in the HTGR would cause delayed release of radioactive material from the core so that the concentration of radioactive material would be reduced by natural decay. Furthermore, the geometry of the primary coolant system allows for significant deposition of fission products on surfaces within the concrete pressure vessel during the heatup, and this would also reduce their availability for release from the containment.

Table 13-2 *Dominant Failure Modes of the HTGR Main Loop Cooling System*[a]

Description	Assessed frequency (events/reactor-yr)
Loss of secondary coolant heat sink (including faults in condensate and circulating water systems)	2×10^{-1}
Spurious control signal isolates all main loops	4×10^{-2}
Loss of off-site power followed by main turbine generator trip	1×10^{-2}
Total loss of steam generator feedwater	7×10^{-3}
Loss of bearing water to all main loop cooling system circulators	6×10^{-3}
Turbine trip followed by failure of steam by-pass valves (loss of secondary coolant)	5×10^{-3}
Reactor trip followed by unavailability of (redundant) auxiliary boiler system	5×10^{-3}
Rupture of main steam header (loss of secondary coolant)	1×10^{-3}

[a] From V. Joksimovic and W. J. Houghton, *in* "Probabilistic Analysis of Nuclear Reactor Safety," Vol. 2, p. V. 6-1. American Nuclear Society, LaGrange Park, Illinois, 1978.

A second advantage is that rupture of the prestressed concrete reactor vessel is not possible from a single propagating crack [16]. Also, there are two diverse systems (rod insertion and boron carbide pellets) for reactor shutdown.

Some of the disadvantages of the HTGR, from the point of view of safety to the public, are itemized in Table 13-2. The estimated frequency

Table 13-3 *Health Effects and Property Damage for HTGR Accidents*[a]

	Frequency (events/reactor-yr)	
Accident consequences	10^{-6}	10^{-7}
Early fatalities	<1	<1
Early illnesses	<1	<1
Property damage	<$1 million	$1 million
Relocation area (sq km)	0	0
Decontamination area (sq km)	0	0.5
Latent-cancer fatalities[b]	1	8
Thyroid nodules[c]	10	100
Genetic effects	<1	1

[a] From V. Joksimovic and W. J. Houghton, *in* "Probabilistic Analysis of Nuclear Reactor Safety," Vol. 2, p. V.6-1, American Nuclear Society, LaGrange Park, Illinois, 1978.

[b] BEIR commission recommendations used; linear hypothesis used with no dose rate effectiveness factor reduction allowed.

[c] Sum of benign and cancerous nodules.

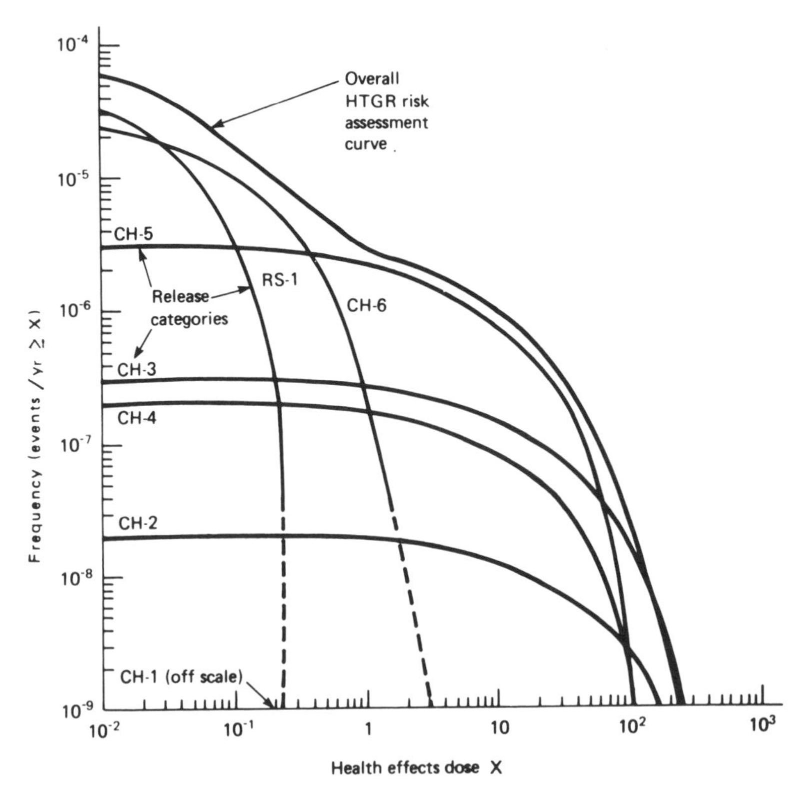

Fig. 13-6 *Risk of latent-cancer fatalities as expressed in health effects dose. (From V. Joksimovic and W. J. Houghton, in "Probabilistic Analysis of Nuclear Reactor Safety," Vol. 2, p. V.6-1, American Nuclear Society, LaGrange Park, Illinois, 1978.)*

of a core heatup occurring because of accidents of all types is 3×10^{-5} per reactor-year but, in about 90% of these cases, the pressure vessel and the containment are protected and very little radioactivity is released to the environment (even though the core is damaged beyond repair).

The core heatup (CH) analyses of the accidents were grouped into six categories ranked according to the severity of release of radioactivity, with category CH-1 the worst and CH-6 the most benign [15]. The results for the risk are plotted in Fig. 13-6 for the frequency versus health effects dose, which is a linear combination of the whole body gamma and inhalation dose. This health effects dose is proportional to the total number of latent-cancer fatalities in such a way that for the representative United States site assumed in the study, one health effects dose unit equals approximately 0.13 fatalities. The dominant core heating mode (CH-5 in Fig.

13-6) is caused primarily by a loss of forced circulation of the helium cooling.

Point estimates for risks from all the accident sequences were evaluated for a number of consequences [15]. The risks at frequencies of 10^{-6} and 10^{-7} per reactor-year are listed in Table 13-3. The 10^{-7} per reactor-year frequency level is of interest since it has sometimes been considered as an acceptable goal by regulatory authorities.

EXERCISE

13-1 Compare the calculated risks from operation of PWRs, BWRs, LMFBRs, and HTGRs. Your exposition should be aimed at an engineer or physicist who has some familiarity with nuclear reactors, and how accidents can occur, but who has not studied any of the published risk studies. A length of about 1500 words, exclusive of figures and tables, is preferred.

REFERENCES

1. D. E. Simpson, LMFBR accident consequences assessment, *in* "Nuclear Systems Reliability Engineering and Risk Assessment" (J. B. Fussell and G. R. Burdick, eds.), p. 128. Soc. Industrial Applied Math., Philadelphia, Pennsylvania, 1977.
2. F. X. Gavigan and J. D. Griffith, The application of probabilistic methods to safety R&D and design choices, *in* "Nuclear Systems Reliability Engineering and Risk Assessment" (J. B. Fussell and G. R. Burdick, eds.), p. 22. Soc. Industrial Applied Math., Philadelphia, Pennsylvania, 1977.
3. Clinch River Breeder Reactor Plant (CRBRP) Safety Study—Assessment of Accident Risks in CRBRP, Rep. CRBRP-1, Vols. 1 and 2 (1977).
4. H. B. Piper, L. L. Conradi, A. R. Buhl, P. J. Wood, and D. E. W. Leaver, Clinch River Breeder Reactor Plant Safety Study, *Nucl. Safety* **19**, 316 (1978).
5. P. J. Wood *et al.*, The Clinch River breeder reactor plant safety study: Summary and results, "Probabilistic Analysis of Nuclear Reactor Safety," Vol. 2, p. V.1-1. American Nuclear Society, LaGrange Park, Illinois, 1978.
6. Reactor Safety Study—An Assessment of Accident Risks in U. S. Commercial Nuclear Power Plants. U. S. Nuclear Regulatory Commission Rep. WASH-1400, NUREG-75/014 (October 1975).
7. M. G. Stevenson *et al.*, Current Status and Experimental Basis of SAS LMFBR Accident Code. Atomic Energy Commission Rep. CONF-740401-P3, 1303 (1974).
8. J. F. Jackson and R. B. Nicholson, VENUS-II: An LMFBR Disassembly Program. Argonne National Laboratory Rep. ANL-7951 (1972).
9. J. Lamarsh, "Introduction to Nuclear Engineering," Chapter 11. Addison-Wesley, Reading, Massachusetts, 1975.
10. G. G. Eichholz, "Environmental Aspects of Nuclear Power," Chapter 3. Ann Arbor Science, Ann Arbor, Michigan, 1976.
11. R. H. Clarke and H. F. Macdonald, Radioactive releases from nuclear installations: Evaluation of accidental atmospheric discharges, *Progr. Nucl. Energy* **2**, 77 (1978).

12. General Atomic Company, HTGR Accidental Initiation and Progression Analysis Status Rep., ERDA Rep. GA-A 13617, Vol. 1, Introduction and Summary (1976); Vol. II, AIPA Risk Assessment Methodology (1975); Vol. III, Preliminary Results (including design options) (1975); Vol. IV, Phase I Analyses and R&D Recommendations (1975); Vol. V, AIPA Fission Product Source Terms (1976); Vol. VI, Event Consequences and Uncertainties Demonstrating Safety R&D Importance of Fission Product Transport Mechanisms (1976); Vol. VII, Occupational Radiation Exposure from Gasborne and Plateout Activities (1976); Vol. VIII, Responses to Comments on AIPA Status Rep. (1977).
13. V. Joksimovic and K. N. Fleming, Applications of probabilistic risk assessment in the development of HTGR technology, *in* Proc. Topical Meeting on Thermal Reactor Safety CONF-770708, p. 1–53 (1977).
14. A. W. Barsell, V. Joksimovic, and F. A. Silady, An assessment of HTGR accident consequences, *Nucl. Safety* **18,** 761 (1977).
15. V. Joksimovic and W. J. Houghton, An update of HTGR risk assessment study, "Probabilistic Analysis of Nuclear Reactor Safety," Vol. 2, p. V.6-1. American Nuclear Society, LaGrange Park, Illinois, 1978.
16. Risks Associated With Nuclear Power: A Critical Review of the Literature, Summary and Synthesis Chapter. National Academy of Sciences (April 1979).

CHAPTER 14

Risks for Nuclear Materials Transportation

14-1 ANALYSIS PROCEDURE

Fixed-site facilities, such as nuclear reactors, have a well-defined population distribution, and the population in the immediate vicinity of the plant (the exclusion area) is controlled by the facility operator; furthermore, evacuation procedures can be developed for each particular site. The population distribution in the vicinity of a transportation accident, however, is highly variable: transportation accidents may occur in areas with very low-population densities or in urban areas, and generally the evacuation plans are less well developed. Since transportation accidents may occur at virtually any location along the shipping route, the variability in the geography and meteorology for transportation accidents also adds a degree of complexity not found in risk assessments of fixed sites. However, the analysis of possible accident releases from casks used in materials transportation, which have *passive* safety systems, generally is simpler than that of nuclear reactors, which have *active* safety systems (like the emergency core cooling system) in addition to the (passive) containment system.

The maximum damage from a transportation accident generally is lower than from a nuclear reactor because there is a much smaller source of radioactive material in a cask than in a reactor, and because the potential energy available for its release and dispersion is much lower. A good discussion of the amounts of radioactive material transported and the casks in which it is shipped is covered in the text by Eichholz [1].

A series of probabilistic studies of risks incurred with the transport of materials associated with light-water-reactor nuclear fuel operations has been performed by Battelle Pacific Northwest Laboratories [2–9]. The studies provide a determination of the risks of transporting materials such as uranium hexafluoride and uranium oxide, calculation of risk curves, identification of the most important contributors to the risks, and identification and evaluation of possible methods to reduce the transportation risks. The risk assessment technique also can be used to predict the changes in risks incurred following changes in transportation regulations, shipping routes, transportation modes, or shipping containers.

The risk from any accidental release sequence is expressed as the product of the amount of material present in a shipment, the fraction of that material lost to the environment in that release sequence, the predicted frequency with which the release sequence can occur during transport, and the consequences of release of a unit source of material. The consequences of such releases are dependent on the type of release, the population distribution near the accident, the weather conditions at the time of release, and assumptions about the health effects caused by the material that is released. The weather conditions and population must be weighted according to their probabilities of occurrence along the shipping route. The risks for all the individual routes are weighted according to the amounts being shipped along each route, and the risks are finally summed to obtain the overall risk [7].

To perform an analysis of the risk, the system is first described according to the amount of material in a shipment, its origin and destination, and its physical and chemical characteristics, which are obtained from projected shipments; also, information is needed on the vehicles used. Finally, the shipping route is divided into segments, and each segment is described in terms of population and weather characteristics.

The release sequences can be identified by fault tree and event tree analysis. In addition to releases caused by forces produced in transportation accidents, one must consider releases resulting from package closure errors, substandard packaging construction, and deterioration in the condition of the packaging that is a result of the normal transportation environment.

The determination of the release fraction is an area in which uncertainty exists. A data base on which estimates can be made is generally not available; therefore, conservative estimates are required and sensitivity studies are made to determine the effect of these estimates on the overall risk. These estimates have been supported by mathematical modeling that has been used to approximate the container failure thresholds, and physical tests have been conducted to verify some calculated values [10].

14-2 SPENT NUCLEAR FUEL TRANSPORT

The risk of shipping spent nuclear fuel assemblies by truck and train from nuclear reactors to storage facilities away from the reactor has been analyzed assuming the existence of one hundred 1000 MW_e light water reactors [7, 8]. The truck shipments were assumed to be made in a water-filled, legal-weight (about 23,000 kg) cask with a capacity of 1 PWR or 2 BWR spent-fuel assemblies; rail shipments were assumed to be made in a water-filled cask with a capacity of 7 PWR or 18 BWR assemblies. It was estimated that 885 truck shipments and 580 rail shipments of spent fuel would be made to transport 2680 PWR fuel assemblies and 3275 BWR fuel assemblies per year. Twenty percent of the shipments were assumed to be made by truck, with the remaining shipments to be made by train. The fuel was assumed to be shipped 180 days after discharge from the reactor. The average shipping distance was calculated to be 690 km, corresponding to four away-from-reactor fuel storage facilities assumed to be operating. These facilities are located at Morris (Illinois), Barnwell (South Carolina), Oak Ridge (Tennessee), and Hanford (Washington).

The accident environment was categorized by five accident stresses that were studied at Sandia Laboratories: impact, crush, puncture, fire, and immersion [11, 12]. Impact forces act over periods of a few milliseconds and are from approximately one direction, whereas crush forces can exist for several seconds following an accident and are applied from several directions; puncture stresses occur when a cask is struck by an object, such as the coupler of a railcar. Impact forces, for example, are produced in 80% of all truck accidents and fires occur in 1.6% of the accidents; immersion accidents occur only infrequently (in about 1 out of every 3000 truck accidents), while the crushing of casks from static forces and punctures are improbable in truck accidents.

The population distribution along the spent nuclear fuel shipping routes was characterized by dividing the Continental United States into four zones based roughly on population density and degree of urbanization. A representative state for each of the zones was selected:

I. High urbanization: New Jersey
II. Densely populated: Massachusetts
III. Moderately populated: Missouri
IV. Low population: Washington

For the risk analysis of truck shipments, 43 specific basic events were identified to describe failure of the cask lid, 33 for failure of the lid closure seal, 20 for failure of the cask wall, and 41 events each for the pressure relief device, the drain valve, and the vent valve [7]. Probabilities of occurrence for each of these events then were identified and nonzero probabilities for a total of 198 release sequences were obtained by means of fault tree analysis.

Accidents involving truck shipments of spent fuel to interim storage facilities were calculated to occur at a rate of 1.5×10^{-6} per shipment kilometer; the expected frequency of accidents is about 1 in 935 shipments [7]. At this frequency, a truck accident involving spent fuel would be expected to occur about once every 1.1 years. Based on the release sequence probabilities determined in the analysis, 1 out of 3.6×10^4 spent fuel shipments to interim storage would be estimated to release a small amount of radioactive material. An example of this type of release would be a very small leak of cask coolant, which would not result in any consequence to the general public.

The probability per year of a release resulting in one or more latent-cancer fatalities was estimated to be 2.2×10^{-5}. The risk curve for the estimated frequency of latent-cancer fatalities from truck shipments from one hundred 1000 MW_e light water reactors is presented in Fig. 14-1. Curves for transportation of other radioactive materials [2–4] and for

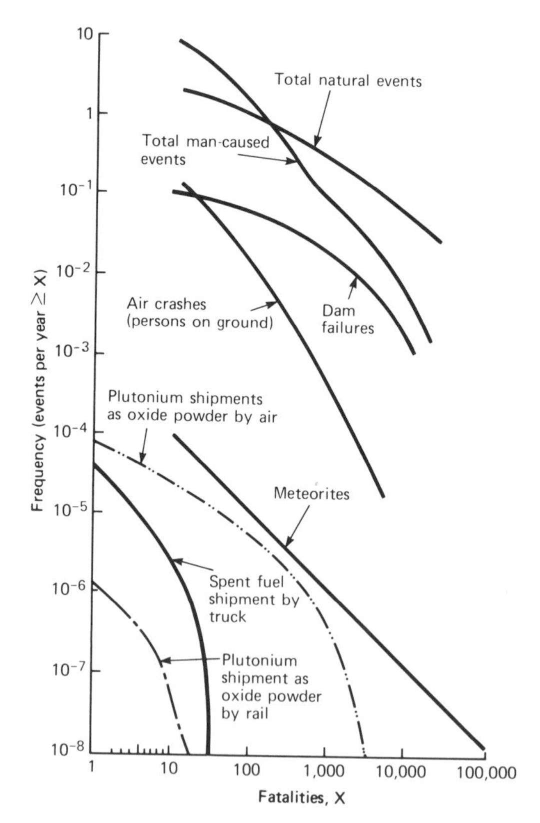

Fig. 14-1 *Risk of latent-cancer fatalities from truck shipments of spent fuel from one hundred 1000* MW_e LWRs *to storage.* [From *H. K. Elder,* Trans. Amer. Nucl. Soc. *30, 319 (1978).*]

other risks in society are included for comparison [13]. The risk of shipping spent fuel is well below the spectrum presented for nonnuclear man-caused events.

Two potential methods of reducing the risk level of truck shipments also were investigated [7, 8]. By replacing the rupture disk in the reference truck design with a pressure-relief valve, the risk level could be reduced by 24%. Second, by storing the spent fuel for 2 years rather than 180 days before shipment, fuel failures from overheating with no cavity coolant could be eliminated and the risk level could be reduced by 82%.

14-3 URANIUM HEXAFLUORIDE TRANSPORT

Uranium hexafluoride (UF_6) is transported from UF_6 conversion facilities to enrichment plants and from the enrichment plants to fuel fabricators. Shipments are also made to and from the enrichment plants for foreign customers purchasing enrichment services. The UF_6 is shipped in cylinders of various sizes by truck and train.

When UF_6 is released to the atmosphere, it rapidly combines with moisture in the air to produce hydrogen fluoride (HF) and uranyl fluoride (UO_2F_2). The chemical hazard to humans from the HF far outweighs any radiological hazard from the uranium. For example, sufficient concentrations of HF in air (40 mg/m^3 for exposures of several hours or 400 mg/m^3 for short exposures) can cause death; however, concentrations well below the lethal concentrations can be detected by smell (less than 1 mg/m^3) and can cause severe lung irritation (25 mg/m^3).

Table 14-1 summarizes the system considered for transporting UF_6 for one hundred 1000 MW_e nuclear plants [5, 6, 9]. Four separate transportation systems were considered in the analysis: 9-metric ton cylinders by truck, 9-metric ton cylinders with overpack by train, 12.7-metric ton cylinders by train, and 2.3-metric ton cylinders with overpack by truck. The risk for each of these systems was assessed individually and summed to give the overall risk of transporting UF_6.

The specific techniques that were used in the analysis of the risks followed the procedure outlined in Section 14-1 and were similar to those for the transport of spent fuel. The number of people exposed to lethal concentrations of HF is quite sensitive to assumptions about the effectiveness of evacuation efforts following the accident, however. Since sublethal concentrations of HF are detectable by smell and can cause lung irritation, most people exposed to the cloud of HF released from an accident involving a UF_6 cylinder would move to an area that was not hazardous. The available data did not permit a probabilistic treatment of evacuation [5, 6].

Table 14-1 *Uranium Hexafluoride Model Shipping System for One-Hundred 1000 MW_e LWRs*[a]

Shipping container	Transportation mode	Containers per shipment	Shipment origin/destination	Material shipped per year, metric tons	Average shipment distance, km	Rate of accidents per kilometer
9-metric ton cylinder (Model 48x)	Truck	2	Conversion/enrichment Reprocessing/enrichment	17,000	700	1.5×10^{-6}
9-metric ton cylinder with overpack (Model 48x)	Rail	4	Enrichment/enrichment	1600	400	8.1×10^{-7}
12.7 metric ton cylinder (Model 48y)	Rail	4	Import/enrichment Reprocessing/enrichment Conversion/enrichment	37,000	700	8.1×10^{-7}
2.3-metric ton cylinder with overpack (Model 30B)	Truck	5	Enrichment/fabrication Enrichment/export	9000	1000	1.5×10^{-6}

[a] From R. E. Rhoads and J. F. Johnson, *Nucl. Safety* **19**, 135 (1978).

The risk curve for the estimated frequency of latent-cancer fatalities from transporting UF_6 for one hundred 1000 MW_e plants is shown in Fig. 14-2. The UF_6 transportation risk spectrum presented assumes that 95% of the people in the areas of lethal concentration are evacuated; this was believed to be conservative [5, 6]. The effect of assuming that no evacuation occurs also is shown in Fig. 14-2.

14-4 PLUTONIUM TRANSPORT

The risks from the transportation by truck, train, and cargo aircraft of 18 metric tons of plutonium from commercial reprocessing facilities to fuel fabrication plants has been studied [2–4, 9]. This amount would be the annual shipping level for reprocessing fuel from one hundred 1000 MW_e light water nuclear reactors. The shipment of plutonium in both the

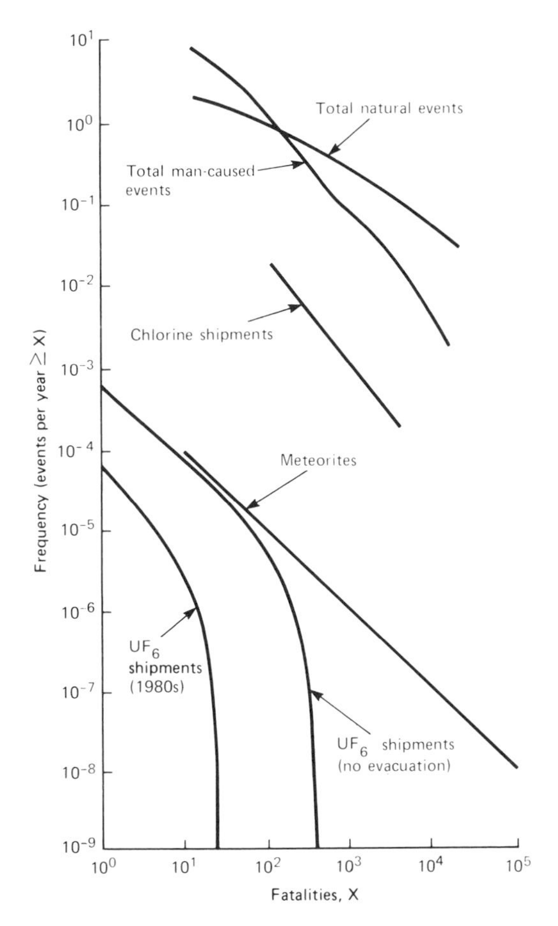

Fig. 14-2 *Risk of latent-cancer fatalities from transporting UF_6 for one hundred 1000* MW_e LWRs. [*From R. E. Rhoads and J. F. Johnson,* Nucl. Safety ***19,*** *135* (*1978*).]

liquid nitrate and powdered dioxide forms was considered in the truck and train risk assessments. Only plutonium dioxide shipments were considered in assessing the risks of transporting plutonium by cargo aircraft since nitrate shipments were prohibited at the time of the studies. (Now plutonium nitrate shipments by any carrier also have been discontinued.) For air transport, the risk of transporting the plutonium to and from the air terminal by truck also was included.

The system descriptions for the three plutonium studies are summarized in Table 14-2. The risk analysis was done by means of the proce-

Table 14-2 *System Description for Shipping Plutonium for One Hundred 1000 MW_e LWRs*[a]

Feature	Transport mode				
	Truck	Truck	Rail	Rail	Aircraft
Shipping container	6M	L-10	6M	L-10	6M
Containers per shipment	39	50	90	68	39
Quantity of plutonium per container, kg	2.55	2.0	2.55	2.0	2.55
Average shipping distance, km	2500	2500	2500	2500	2500
Total plutonium shipped, metric tons	18	18	18	18	18
Total number of shipments	180	180	90	144	180
Accidents per kilometer	1.5×10^{-6}	1.5×10^{-6}	8.1×10^{-7}	8.1×10^{-7}	5.9×10^{-9}

[a] From R. E. Rhoads and J. F. Johnson, *Nucl. Safety* **19**, 135 (1978).

dure outlined in Section 14-1 in the manner used to study spent fuel and UF_6 transport.

The risk curves for latent-cancer fatalities from transporting plutonium for one hundred 1000 MW_e plants by truck, train, and cargo aircraft are presented in Fig. 14-3. Each of the curves is calculated on the assumption that all 18 metric tons of plutonium are shipped by the mode and in the physical form (dioxide or nitrate) indicated [2–4]. This assumption permits comparisons to be made among the three transport modes and between shipments in the solid and liquid forms. Other societal risks are also included in Fig. 14-3 for comparison.

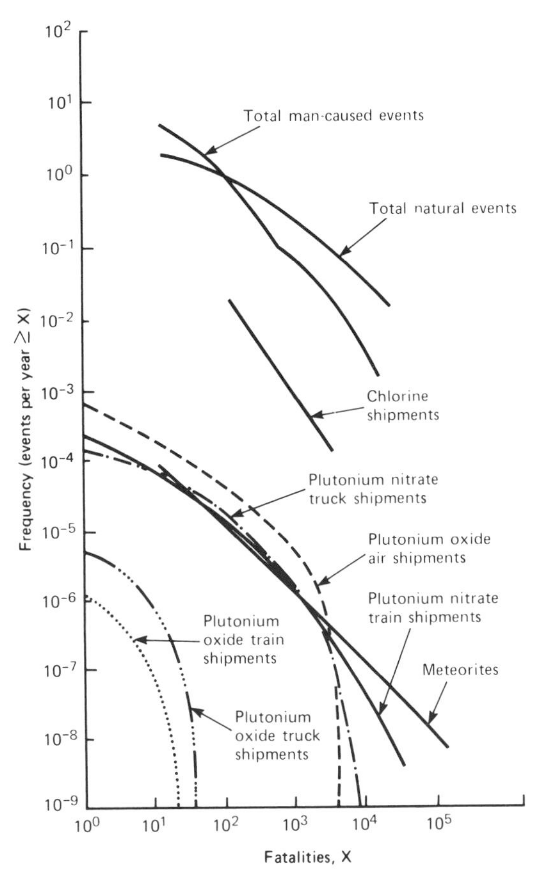

Fig. 14-3 *Risk of latent-cancer fatalities from transporting plutonium from one hundred 1000* MW_e LWRs. [*From R. E. Rhoads and J. F. Johnson,* Nucl. Safety ***19***, *135* (*1978*).]

The curves show that the risk of shipping plutonium in either the solid or liquid form by truck or train is similar [2–4]. Although the risk of transporting plutonium oxide in the 6M containers by cargo aircraft is greater than the risk of transporting plutonium oxide by truck or train, it is still relatively low. The increased risk for plutonium dioxide shipments by air, as compared to truck and train shipments, is because a large percentage of aircraft accidents result in a release and the larger accident forces present in an aircraft accident are expected to produce significantly larger release fractions.

Another way of presenting the risks from transportation of plutonium dioxide is to put the latent-cancer fatalities on the basis of a reactor-year. For example, from Fig. 14-3, the predicted frequency of one or more fatalities, about 5×10^{-6} per year, is for 39 gigawatts of electrical power (GW_e) or a little over 1×10^{-7} fatalities per GW_e-yr, which is the value obtained in an independent study by the Electric Power Research Institute (EPRI) [14].

14-5 NUCLEAR WASTES TRANSPORT

The risks from transporting vitrified high level wastes and cladding hulls from a reprocessing plant to a waste repository also have been examined as a part of the EPRI fuel cycle accident risk assessment project [14]. The methods of analysis were the same as those used for the Reactor Safety Study described in Chapter 12. Rail transport of high level wastes was predicted to cause about 1×10^{-8} latent-cancer fatalities per GW_e-yr, while the risks from either truck or train transport of cladding hulls was even smaller.

The risks for transporting non-high-level wastes in 55-gallon drums (such as rags, sludge, and various housekeeping wastes) from a power plant have been predicted to be about 5×10^{-7} latent-cancer fatalities per GW_e-yr, which exceeds all other transportation risks in the nuclear fuel cycle. The risks from all transportation activities was estimated to be 3×10^{-6} latent-cancer fatalities per GW_e-yr, which is still four orders of magnitude smaller than the comparable figure for the nuclear reactor itself [14].

EXERCISE

14-1 Compare the calculated risks from operation of LWRs and the transportation of UF_6 and spent nuclear fuel assemblies. Your exposition should be more than just an enumeration of risk results and should be aimed at an engineer or physicist who has some familiarity with nuclear power systems and how accidents can occur, but who has not studied any of the published risk studies. A length of 1000 to 1500 words, exclusive of figures or tables, is preferred.

REFERENCES

1. G. G. Eichholz, "Environmental Aspects of Nuclear Power," Chapter 10. Ann Arbor Science, Ann Arbor, Michigan, 1976.
2. T. I. McSweeney *et al.*, An Assessment of the Risk of Transporting Plutonium Oxide and Liquid Plutonium Nitrate by Truck. Battelle Pacific Northwest Laboratories Rep. BNWL-1846 (1975).
3. R. J. Hall *et al.*, An Assessment of the Risk of Transporting Plutonium Dioxide and Liquid Plutonium Nitrate by Train. Battelle Pacific Northwest Laboratories Rep. BNWL-1996 (1977).
4. T. I. McSweeney and J. F. Johnson, An Assessment of the Risk of Transporting Plutonium Oxide by Cargo Aircraft. Battelle Pacific Northwest Laboratories Rep. BNWL-2030 (1977).
5. C. A. Geffen *et al.*, An Assessment of the Risk of Transporting Uranium Hexafluoride by Truck and Train. Battelle Pacific Northwest Laboratories Rep. BNWL-2211 (1978).
6. C. A. Geffen, Risks of shipping uranium hexafluoride by truck and train, *Trans. Am. Nucl. Soc.* **30,** 321 (1978).
7. H. K. Elder *et al.*, An Assessment of the Risk of Transporting Spent Nuclear Fuel by Truck. Battelle Pacific Northwest Laboratories Rep. PNL-2588 (1978).
8. H. K. Elder, Risk of shipping spent fuel in the U.S., *Trans. Am. Nucl. Soc.* **30,** 319 (1978).
9. R. E. Rhoads and J. F. Johnson, Risks in transporting materials for various energy industries, *Nucl. Safety* **19,** 135 (1978).
10. L. L. Bonzon and M. McWhirter, Special tests of plutonium shipping containers, *in* "Transport Packaging for Radioactive Materials" (*Int. At. Energy Agency, Symp. Proc., Vienna*), pp. 335–349; also issued as Sandia Laboratories Rep. SAND-76-5370 (CONF-760813-8) (1976).
11. A. W. Dennis, J. T. Foley, W. F. Hartman, and D. W. Larson, Severities of Transportation Accidents Involving Large Packages. Sandia Laboratories Rep. SAND-77-0001 (1978).
12. A. W. Dennis, "Predicted occurrence rates of severe transportation accidents involving large casks, *Int. Symp. Packaging Transport. Radioactive Mater., 5th, Las Vegas, Nevada* (May 1978).
13. Reactor Safety Study—An Assessment of the Accident Risks in U. S. Commercial Nuclear Power Plants. U. S. Nuclear Regulatory Commission Rep. WASH-1400, NUREG-75/014 (October 1975).
14. R. C. Erdmann *et al.*, Status Report on the EPRI Fuel Cycle Accident Risk Assessment. Electric Power Research Institute Rep. NP-1128 (1979).

CHAPTER 15

Risks for Nuclear Waste Disposal

15-1 RISKS FROM PRECLOSURE ACCIDENTS

Risks from nuclear waste disposal can be classified into two broad categories: preclosure and postclosure. The preclosure risks may be analyzed with the methods used to study the rest of the risks from nuclear power production; risks from postclosure accidents will be considered in the following sections of this chapter.

The dominant preclosure accident that contributes to risk from a conceptual waste repository located at a bedded-salt western site is failure of a final high-efficiency particulate (HEPA) air filter [1]. The resulting release of the accumulated radioactivity was predicted to occur with a probability of 3×10^{-3} per reactor-year and to cause a risk of 1×10^{-9} latent-cancer fatalities per GW_e-yr. The risk from a truck crash into a high-level waste loading area is about the same, but only because the consequences from a larger predicted release are offset by a smaller probability of occurrence.

All accidents from preclosure of the waste repository were calculated to lead to risks of 2×10^{-9} fatalities per GW_e-yr (with an uncertainty factor of 10), but this value is more than three orders of magnitude smaller than the risks from transporting the wastes (see Chapter 14) and seven orders of magnitude smaller than risks of latent-cancer fatalities from operation of a nuclear plant. For this reason, risks from preclosure accidents are felt to be a negligible contributor to risks from the nuclear fuel cycle [1].

15-2 INTRODUCTION TO RISKS FROM POSTCLOSURE ACCIDENTS

The successful isolation of radioactive wastes from the biosphere appears technically feasible for periods of thousands of years, provided that a systems approach is used for the selection of the geologic environment, the repository site, the medium chosen for waste emplacement, the form of the waste, and the engineered barriers [2]. The principle behind the selection of an isolation system is to provide multiple and, to some extent, independent, natural and engineered barriers.

A variety of terrestrial and nonterrestrial methods has been proposed for disposal of nuclear wastes. Terrestrial methods include, in addition to conventional geologic storage, the techniques of chemical resynthesis, partitioning and transmutation (i.e., nuclear resynthesis), rockmelting, reverse well, and very deep hole disposal; nonterrestrial methods include seabed, island, ice-sheet, and space [3]. A procedure for overall evaluation of these different disposal methods has been proposed [4]. A good general introduction to nuclear waste disposal methods is provided by Eichholz [5], Cohen [6], and Pigford [7].

To characterize the future state of the natural and engineered barriers surrounding the waste, we must understand the behavior of the materials over longer periods of time than in other risk evaluations and be able to predict the geologic conditions of the site. To assess the frequencies of release from the long-term storage of waste, we must attempt to predict the barrier failure modes. These failure modes are quite different for the various disposal schemes since different initiating events are possible for many of the methods, and the event trees for failure of the engineered containment systems are usually different. For the conventional geologic storage method, the engineered barriers for a nuclear waste repository in basalt, for example, include [8]:

- The waste form itself,
- The waste canister and canister fill material,
- Grout/liner material surrounding the canister,
- Host rock (basalt),
- Backfill of the placement room, and
- Plugging of the shafts and other penetrations.

The chief problem in a risk assessment of waste disposal is predicting the integrity of the engineered containment systems and the geological conditions of the site as a function of time. Especially when the time periods are on the order of centuries, the uncertainties in the predictions can be very large. The evaluation of risks from geologic waste disposal also requires knowledge of the mass transport coefficients (i.e., leaching rates) for the radionuclides that escape the engineered barriers.

Another major difficulty with the evaluation of nuclear waste disposal risks is the uncertainty of the expected population that is subject to exposure at the time of any release. Generally, it is assumed that the waste disposal site would be permanently isolated from long-term occupancy within some exclusion area, although the long-term future use of land is a sociological question as well as an engineering one. For this reason, it is usually simpler to study only the mechanisms for release of radionuclides into the soil and their propagation. Hence we expect that a quantitative geologic waste disposal study would have final results, at most, of the radioactivity level of different nuclides (in curies) as a function of distance from the original site of disposal.

Four major classes of stored nuclear wastes are [2]:

- Uranium mine and mill tailings, which are the residues from uranium mining and milling operations that contain low concentrations of naturally occurring radioactive materials; tailings are generated in large volumes and are stored at the site of mining and milling operations.
- Low-level wastes (LLW), which contain less than ten nanocuries of transuranic contaminants per gram of material; low-level wastes are generated in almost all activities involving radioactive materials and are presently being disposed of by shallow land burial.
- Transuranic (TRU) wastes, which are materials containing more than ten nanocuries of transuranic activity per gram of material; the wastes result predominantly from the fabrication of plutonium to produce nuclear weapons and, when it occurs, spent fuel reprocessing.
- High-level wastes (HLW), which are either intact spent fuel assemblies or the portion of the wastes generated in the reprocessing of spent fuel that contain virtually all of the fission products and most of the actinides not separated out during reprocessing.

The volume and activity of these wastes is summarized in Table 15-1 [9].

The modest activity level of tailings and low-level wastes is such that the general public would incur no early fatalities or illnesses. The principal risk from these wastes is from latent cancers, but this is small compared to risks from other waste storage accidents. The risks from transuranic wastes arise from the reprocessing of spent fuel for the recycling of plutonium, but in the past this has not been part of the back-end of the fuel cycle, so we will focus our attention on high-level wastes.

The radioactivity in high-level waste that is generated in one year of operation of a 1000 MW_e is shown in Fig. 15-1 as a function of time after discharge of fuel irradiated for about three years [7]. After the first few hundred years, the activity occurs because of the α-emission of the actinides Am, Cm, Pu, and of ^{99}Tc and ^{129}I; the β-decays of ^{137}Cs and ^{90}Sr dominate prior to that. After about 10^4 years, the actinides become less important than the ^{99}Tc, which has a half-life of 2.12×10^5 yr. From Fig. 15-1, it is obvious that the time after initial storage at which a radioactive

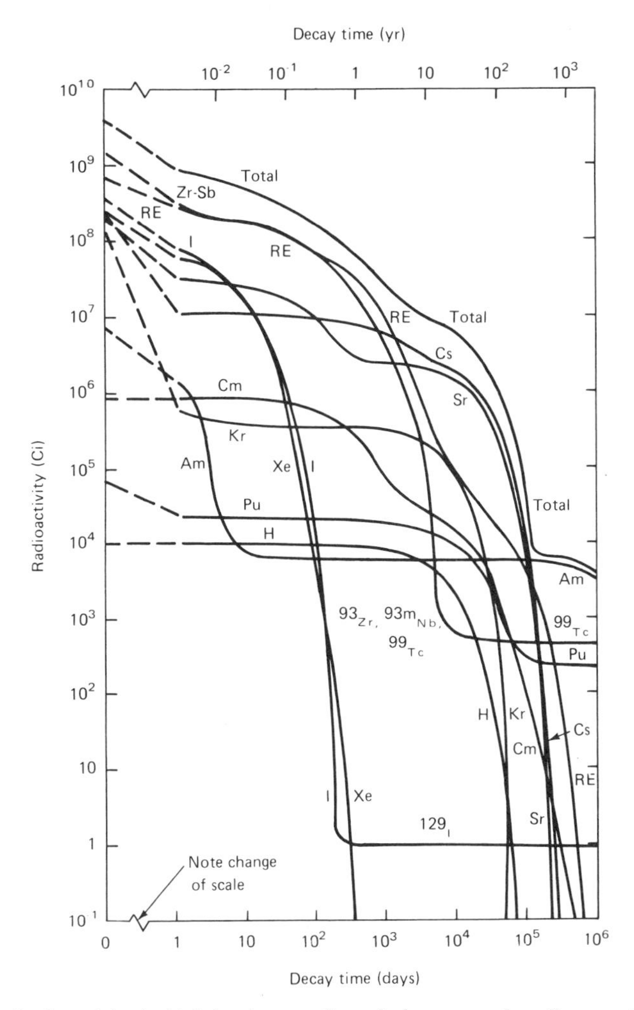

Fig. 15-1 *Radioactivity in high-level waste from fuel reprocessing. Rare earths* (RE) *include* La, Ce, Pr, Nd, Pm, Sm, Eu, Gd, Tb, Dy, Ho. Zr − Sb = Zr, Nb, Mo, Tc, Ru, Rh, Pd, Ag, Cd, In, Sn, Sb. *Pu assumed to remain with waste is 0.5% of total Pu present in spent fuel. Based on 30* MW/*ton, 33,000* MW *days/ton, 3125* MW (*thermal*), *100% load factor.* [*Reproduced with permission from the* Annual Review of Nuclear Science, **24,** *515.* © *1974 by Annual Reviews, Inc.*]

release sequence begins is important when determining the consequences of an initiating event.

In assuming the risks from permanent high-level radioactive waste storage, one should keep in mind that there is a "time horizon" beyond which the risks can be considered to be benign. For example, it has been pointed

Table 15-1 *Characteristics of the Various Kinds of Radioactive Wastes*[a]

Type of waste	Volume of material (m^3)	Activity in curies (and ingestion hazard[b] in m^3 water) per GW_e year of electricity produced by U.S. reactors[c] up to 2000 A.D.			Remarks on disposal
		2010 A.D.	300 years later	10^5 years later	
High-level wastes	2.1	$5.6 \times 10^6 (3.7 \times 10^{12})$	$1.9 \times 10^4 (4.8 \times 10^9)$	$4.5 \times 10^2 (8.5 \times 10^7)$[d]	Presumably to be solidified after 5 years, and ultimately emplaced somewhere remote from the surface environment
Cladding hulls (remains of metal tubes in which fuel is contained)	1.7	$1.3 \times 10^5 (4 \times 10^9)$	$1.1 \times 10^3 (4.8 \times 10^7)$	$2 \times 10 (8.3 \times 10^5)$	Similar disposal

TRU wastes (low- and intermediate-level wastes with a sizable actinide component)	~36. (estimates vary at least severalfold)	$1.6 \times 10^4 (1.4 \times 10^9)$	$7.3 \times 10^2 (1.7 \times 10^8)$	$6(1.1 \times 10^7)$	Surface mausolea becoming favored over deep-ocean dumping or near-surface burial
Other low-level wastes	$\sim 4.2 \times 10^2$ (estimates vary at least severalfold)	$4.3 \times 10^2 (6 \times 10^6)$	$5(1.3 \times 10^5)$	(5×10^2)	Near-surface burial
Iodine 129	2.2×10^{-3}	$0.8(1.3 \times 10^7)$	$0.8(1.3 \times 10^7)$	$0.8(1.3 \times 10^7)$	Not yet being isolated
Tritium, krypton-85, and carbon-14				Negligible	Not yet being isolated
Tailings piles	1.5×10^5	$10^3 (3 \times 10^9)$	$10^3 (3 \times 10^9)$	$5 \times 10^9 (1.5 \times 10^9)$	Now required to be covered

[a] Reproduced from Risks Associated with Nuclear Power: A Critical Review of the Literature, Summary and Synthesis Chapter (1979), with the permission of National Academy of Sciences, Washington, D.C.

[b] Defined as the volume of water that would be required to dilute an aqueous solution of the material in question to the "maximum permissible concentration" for human ingestion.

[c] The figures shown were computed for a particular set of assumptions regarding the growth of the number of United States reactors of the light water, high temperature, gas-cooled, and liquid-metal fast-breeder types, but are dominated by the light-water component. Total United States wastes created up to 2000 A.D. can be approximately estimated by multiplying the table entries by the projected United States nuclear-energy production in this period, for which recent estimates have ranged from about 4000 to 2400 GW_e-yr.

[d] Whereas the entries for the earlier times are not very sensitive to the type of reactor fuel used, the radioactivity of wastes after long times, dominated by actinide elements, may be up to ten or so times larger than the values shown, per unit of energy production, for some proposed plutonium enriched fuels.

out that "on any long time scale, nuclear power must be viewed as a means of cleansing the earth of radioactivity," since nuclear fission is a means of eliminating the radon gas resulting from the α-decay of uranium [6]. For a geologic waste repository, a natural time cut-off can be inferred by comparing the ingestion hazard from the waste to that from naturally occurring uranium in equilibrium with its daughter products. The comparison in Fig. 15-2 shows that the region of intersection is between 10^3 and 10^4 years. For this reason, it has been suggested that safety studies of waste disposal focus on the period up to 10^3 years [10, 11].

Even the thought of isolating wastes for hundreds of years causes concern for many people who argue that human social institutions and political systems rarely last that long. Such a response is based on the time constants encountered on the earth's surface, whereas characteristic time

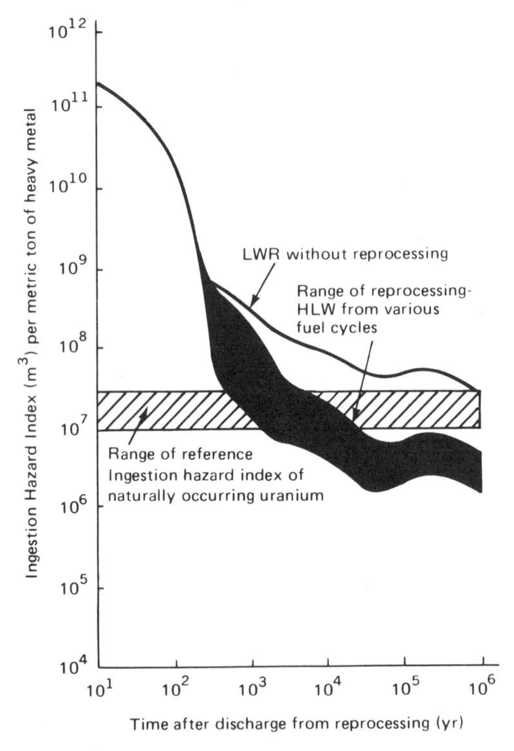

Fig. 15-2 *Comparison of ingestion hazard from high-level wastes with that from naturally occurring uranium. [From H. Levi, in* Int. Symp. Underground Disposal Radioactive Wastes, *Int. Atomic Energy Agency, Vienna. Vol. II, 437 (1980).]*

intervals for events deep beneath the surface are millions of years. If one takes into account the natural delays, such as the time required for circulating ground water to contact the waste and the travel time for dissolved waste to reach a discharge location, it can be argued that very little waste would escape to the human environment during the first few hundred years [6].

15-3 INTERIM STORAGE

Tanks for storage of high level liquid wastes (HLLW) range in size up to 4900 m^3 (1.3×10^6 gal) capacity and are normally located below ground in concrete vaults. There have been several reported cases of neutralized HLLW leaking to the ground from nonstress-relieved carbon steel storage tanks storing caustic defense waste, but the leaks have resulted in no significant impact on the environment [12]. Modern tank designs provide secondary containment (with a steel liner or second wall) and include the capability to detect leaks rapidly and transfer the tank's contents to a spare tank if a leak should occur; no leaks in the stainless steel tanks storing acidic HLLW have yet been reported [12].

A list of the predicted accidents arising from HLLW tank storage is given in Table 15-2, and a brief summary of the accidents from a sealed storage cask for solidified high-level waste (SHLW) storage appears in Table 15-3 [13]. Since the SHLW canister is placed in a metal overpack, which is contained in a reinforced-concrete radiation shield, the consequences are benign, especially when the waste is encased in borosilicate glass.

The potentially most severe interim storage accidents arise for water basin storage of unpackaged spent fuel (which is the method currently in

Table 15-2 *Accidents Causing Radioactivity Releases from a High-Level Liquid Waste Tank Storage Facility*[a]

Accident description	Radioactivity release	Predicted frequency
Off-gas by-passes high-efficiency particulate air filter	<0.1 g HLW released to atmospheric protection system	0.06/tank-yr
Filter fire from condensed organics ignited by a spark	Same	~0.05/tank-yr

[a] Excerpted from Technology for Commercial Radioactive Waste Management, Vol. 3, Interim Storage Alternatives, Rep. DOE/ET-0028 (1979).

Table 15-3 *Accidents Causing Radioactivity Releases from a Sealed-Storage-Cask Solidified High-Level Waste Storage Facility*[a]

Accident description	Radioactivity release	Predicted frequency
Canister failure by impact in receiving cell	0.048 μg release to the environment if encased in borosilicate glass, 0.38 μg if encased in calcine	2×10^{-6}/handling operation
Canister failure in weld and test cell	Same	Same

[a] Excerpted from Technology for Commercial Radioactive Waste Management, Vol. 3, Interim Storage Alternatives, Rep. DOE/ET-0028 (1979).

Table 15-4 *Accidents Causing Radioactivity Releases from a Water Basin Storage Facility for Unpackaged Spent Fuel*[a]

Accident description	Radioactivity release	Predicted frequency
Rough handling which causes cladding rupture of 1% of rods in a PWR assembly	Through stack: 13 Ci ^{85}Kr; 1.5×10^{-9} Ci ^{129}I	0.7/yr
Same as above, except no confinement protection	Through stack: 13 Ci ^{85}Kr; 1.5×10^{-7} Ci ^{129}I	5×10^{-4}/yr
Dropped fuel assembly and rupture of all rods	Through stack: 20 Ci of ^{85}Kr; 2.4×10^{-10} Ci of ^{129}I; 3.2×10^{-2} Ci of ^{3}H; 1.6×10^{-6} Ci of ^{14}C	0.06/yr
Criticality from storage array disruption or mishandling	Through stack: 13 Ci of ^{85}Kr; 9×10^{-3} Ci of ^{83}Br and ^{84}Br; 4.3×10^{-2} Ci of ^{131}I, ^{133}I, ^{134}I, ^{135}I	1×10^{-5}/yr
Design basis tornado, causing failure of roof and transport of basin water by tornado	At ground level: 16.5 MT water assumed entrained, which releases 3.3×10^{-3} Ci of mixed fission products (primarily ^{134}Cs and ^{137}Cs)	1×10^{-5}/yr
Rail cask venting, with 0.25% failed fuel due to improper off-gas equipment operation	Through stack in 0.25 hr: 5 Ci ^{85}Kr; 6.9×10^{-9} Ci of ^{129}I; 8.0×10^{-3} Ci of ^{3}H; 4.0×10^{-5} Ci of ^{14}C	4×10^{-5}/yr

[a] Excerpted from Technology for Commerical Radioactive Waste Management, Vol. 3, Interim Storage Alternatives Rep. DOE/ET-0028 (1979).

vogue). Although accidents have occurred in spent fuel receiving and storage or similar facilities, a review and analysis of incidents occurring in government and commercial nuclear facilities reports no instance of injury to a member of the general public [14]. Table 15-4 provides a list of possible accidents and the predicted probabilities for their occurrence.

An obvious conclusion is that risks from interim storage of high-level wastes lead to relatively benign consequences compared to nuclear reactor accidents, although the predicted probability of occurrence for some accidents is somewhat higher. A similar conclusion was reached after study of the interim waste storage program at Hanford [15, 16].

15-4 PERMANENT WASTE DISPOSAL

Potential mechanisms for the failure of a waste repository located in bedded salt in New Mexico were examined as early as 1974. A salt bed maximizes the integrity of the barrier separating circulating ground water from any waste that escapes through all the engineered barriers [17]. The worst possible consequences were predicted to arise from a large meteorite striking the site of the depository, but this was predicted to occur with a frequency of less than one event during the history of the earth (only 1.6×10^{-13} per year); even the burst of a 50-megaton nuclear weapon was shown to be incapable of breaching the containment system [17]. Other possible causes of containment failure were examined, including sabotage, inadvertant drilling, and geologic faulting, but all were found to lead to benign consequences.

Several natural geologic formations, besides the New Mexico site, are known to have been stable for many millions of years [18]. Some of these, such as bedded and domed salt formations, are also known to have been free from water intrusion over these same time periods, while others have had very low water intrusion. If high-level wastes were stored there, these formations would be disturbed only by the creation of potential pathways from the surface (i.e., shafts or boreholes) that afterwards must be sealed, and by the heat source from the waste.

Both of these disturbances can be accommodated with current technology through the use of appropriate design conservatisms and engineered barrier systems; this is because the sealed shafts and boreholes represent a vertical barrier of at least 610 m (2000 ft) to the surface. In the extremely unlikely event that water were to leak into the repository, it is even more improbable that a mechanism would exist to force the waste out [18]. The heat rate of emplaced waste can be controlled by placing less waste in each canister, providing wider spacing between canisters, or allowing the waste to cool for a longer period before emplacement. Such techniques have been utilized in conceptual designs to maintain peak repository temperatures of less than 200°C, and even lower temperatures are possible [18].

The use of a mined geologic repository for the permanent disposal of high-level waste also has been examined and found to be a safe, acceptable approach [19–22]. A similar conclusion was reached after study of the long-term disposal of wastes at the Idaho National Engineering Laboratory [23].

An interesting approach to the assessment of risks from storage of wastes is based on a comparison between an atom of waste buried at a depth of 600 m and a typical atom of radium in the rock and soil above the waste canister [24]. The key assumption is that the waste atom is no more likely to escape and find its way to a human than is the radium atom. By taking into account the natural time delays for ground water to travel to a discharge location, the probability that people would ingest the waste from the repository was predicted to be 4×10^{-13} per year [24], although this could be increased significantly by withdrawal of containment ground water from a well in the vicinity of the repository.

Another point worth mentioning about the public risk from radioactive waste disposal is that the lethal dose is substantially less than that from other toxic materials such as arsenic and barium; indeed, arsenic is used as a herbicide and hence is routinely scattered around on the ground in regions where food is grown [6]. More about the risks from such nonnuclear materials is discussed in Chapter 17.

EXERCISES

15-1 Assume that water is the only transport medium for waste nuclei, and consider a simplified waste repository model with only three barriers between the waste and the biosphere:

(a) the host rock and back-filling material, which prevents the penetration of surface or groundwater to the waste containers (barrier 1),

(b) the waste container walls and waste matrix (such as borosilicate glass), which delays the leaching of waste nuclides if water reaches the container (barrier 2),

(c) the geologic structure between the repository and biosphere, which delays the transport to the biosphere of water containing radionuclides (barrier 3).

Assume that each barrier fails randomly with a constant instantaneous failure rate λ_i, $i = 1$ to 3, and that all rates are different. You can ignore the radioactive decay of the waste.

(a) Derive the equation for the reliability of the 3-barrier system. [Hint: See an analogous mathematical problem in Exercise 6-14.]

(b) Assuming that $\lambda_1 = 10^{-3}$/yr, $\lambda_2 = 3.6 \times 10^{-3}$/yr, and $\lambda_3 = 5 \times 10^{-5}$/yr, calculate the probability of failure during the first 100 years.

(c) Repeat part (b) for the first 10^3 years.

15-2 The radioactive waste in a container is characterized by two species, a mother nuclide with decay rate $\hat{\lambda}_m$ and a daughter nuclide with decay rate $\hat{\lambda}_d$. The initial activity of the mother nuclide at the time of disposal is $\hat{\lambda}_m N_m(0)$, where $N_m(0)$ is the initial inventory of radionuclides; the probability per unit time that the engineered and natural radioactive waste confinement system will fail is $f(t)$ for both nuclides.

(a) Verify that the probability of release of mother nuclides to the biosphere is $\hat{\lambda}_m N_m(0) f(t) \exp(-\hat{\lambda}_m t)$.

(b) If there are initially no daughter nuclides present, verify that the probability of release of daughter nuclides to the biosphere is

$$\frac{\hat{\lambda}_d \hat{\lambda}_m N_m(0)\, f(t)}{\hat{\lambda}_d - \hat{\lambda}_m} [\exp(-\hat{\lambda}_m t) - \exp(-\hat{\lambda}_d t)].$$

(c) For the 3-barrier waste confinement system of Exercise 15-1, derive $f(t)$.

15-3 Exercises 15-1 and 15-2 are simplified approaches to a risk assessment model for nuclear waste repositories developed by A. Pritzker and J. Gassmann, *Nucl. Technol.* **48,** 289 (1980). Read this paper to see numerical examples for the probable discharge rate of important nuclides such as ^{99}Tc, ^{135}Cs, and ^{90}Sr.

REFERENCES

1. R. C. Erdmann *et al.*, Status Report on the EPRI Fuel Cycle Accident Risk Assessment. Electric Power Research Institute Rep. NP-1128 (July 1979).
2. Report to the President by the Interagency Review Group on Nuclear Waste Management. National Technical Information Service Rep. TID-29442 (March 1979).
3. Environmental Impact Statement on Management of Commercially Generated Radioactive Waste. Department of Energy Rep. DOE/EIS-0046-F (October 1980).
4. D. E. Deonigi, Evaluation methodology of waste management concepts, *Nucl. Technol.* **24,** 331 (1974).
5. G. G. Eichholz, "Environmental Aspects of Nuclear Power," Chapter 12. Ann Arbor Science, Ann Arbor, Michigan, 1976.

6. B. L. Cohen, The disposal of radioactive wastes from fission reactors, *Sci. Am.* **236,** No. 6, 21, (1977).
7. T. H. Pigford, Environmental aspects of nuclear energy production, *Ann. Rev. Nucl. Sci.* **24,** 545–559 (1974).
8. Basalt Waste Isolation Project Annual Report—Fiscal Year 1979. Rockwell Hanford Operations Rep. RHO-BWI-79-100 (November 1979).
9. Risks Associated With Nuclear Power: A Critical Review of the Literature, Summary and Synthesis Chapter. National Academy of Sciences, Washington, D.C. (April 1979).
10. H. Levi, The project-safety-studies Entsorgung. *Int. Symp. Underground Disposal Radioactive Wastes,* Int. Atomic Energy Agency, Vienna, Vol. II, p. 437 (1980).
11. J. Hamstra, Radiotoxic hazard measure for buried solid radioactive material, *Nucl. Safety* **16,** 180 (1975).
12. Alternatives for Managing Wastes from Reactors and Post-Fission Operations in the LWR Fuel Cycle. Energy Research and Development Administration Rep. ERDA 76-43, Vol. 3, Section 19 (May 1976).
13. Technology for Commercial Radioactive Waste Management, Vol. 3, Interim Storage Alternatives. Department of Energy Rep. DOE/ET-0028 (May 1979).
14. Safety Evaluation Report Related to Operation of Barnwell Fuel Receiving and Storage Station. U.S. Nuclear Regulatory Commission Rep. NUREG 0009 (1976).
15. Radioactive Wastes at the Hanford Reservation: A Technical Review. National Academy of Sciences, Washington, D.C. (1978).
16. Radioactive waste management at the Hanford Reservation, *Nucl. Safety* **20,** 434 (1979).
17. H. C. Claibourne and F. Gera, Potential Containment Failure Mechanisms and Their Consequences at a Radioactive Waste Repository in Bedded Salt in New Mexico. Oak Ridge National Laboratory Rep. ORNL-TM-4639 (October 1974).
18. High-level radioactive waste disposal, public policy statement by the American Nuclear Society, *Nucl. News,* 190 (November 1979).
19. K. J. Schneider and A. M. Platt (eds.), High-Level Radioactive Waste Management Alternatives. Battelle Pacific Northwest Laboratories Rep. BNWL-1900, Vol. 1 (May 1974).
20. Report to the American Physical Society on nuclear fuel cycles and waste management, *Rev. Mod. Phys.* **50,** No. 1, Part II (January 1978).
21. The nuclear fuel cycle: An appraisal, *Phys. Today* **30,** No. 10, 32 (1977).
22. Geologic Criteria for Repositories for High-Level Radioactive Wastes. National Academy of Sciences, Washington, D.C. (1978).
23. Environmental and Other Evaluations of Alternatives for Long-Term Management of Stored INEL Transuranic Waste. Department of Energy Rep. DOE/ET-0081 (February 1979).
24. B. L. Cohen, High-level radioactive waste from light water reactors, *Rev. Mod. Phys.* **49,** 1 (1977).

PART III

Other Risk Assessments

CHAPTER 16

Comparison of Risks

16-1 CONVENTIONAL ENERGY SOURCES

The occupational risks of producing electricity from coal, oil, natural gas, and nuclear are considered in this section. Since different technologies are used to generate the electricity, this type of risk assessment can be called a "competing risk comparison."

In Chapters 12 through 15, the public risks from nuclear power were considered. Since the risks from uranium mining and milling operations are primarily occupational, it was not necessary to consider the entire nuclear fuel cycle. In an assessment of competing occupational risks, however, the potential damages (i.e., magnitude of consequences) must be estimated for the entire fuel cycle. This means that the calculations should proceed from mining of the fuel or raw materials to the management of the resulting wastes, and must include not just the electrical generation by the power plant.

In a competing risk assessment, only the probability and damages are considered; there is no accounting for the costs of producing a product such as electricity, or for the benefits to society from that production. A comparative risk–cost–benefit study for the previously mentioned four sources of generating electricity has been published in WASH-1224 [1, 2], from which we will focus only on the risk analysis and a few of the environmental effects that can cause latent consequences.

The comparative risk study was restricted to impacts under *normal* operating conditions, including routine accidents whose frequencies can be established from historical data. Thus, the light water reactor (LWR) risks are "statistically observed risks" rather than "predicted risks" for large accidents, as in the Reactor Safety Study of Chapter 12. The risks reported in the study are those for a 1000 MW_e power plant operating at 75% capacity factor and generating 6.57×10^9 kW · hr of electricity per year; the WASH-1224 study was based on aggregate national data, and hence is a study of a generic plant rather than a specific plant.

Table 16-1 *Comparison of Annual Routine Health and Safety Risks For a 1000 MW_e Plant at 75% Capacity Factor*[a]

Health and safety risks	Coal	Oil	Gas	LWR
Occupational health and safety				
Occupational health (MDL/year)[b]	600	U[c]	U	480
Occupational safety				
Fatalities (deaths/year)	1.1	0.17	0.08	0.1
Nonfatal injuries (number/year)	46.8	13.1	5.3	6.0–7.0
Total man-days lost (MDL/year)	9250	1725	780	900–1000
Public health and safety				
Public health				
Routine pollutant release (MDL/year)	U	U	U	180–210
Public safety				
Transportation injuries				
Fatalities (deaths/year)	0.55	U	U	0.009
Nonfatal (injuries/year)	1.2	U	U	0.08
Total man-days lost (MDL/year)	3500	U	U	60

[a] From Comparative Risk-Cost Benefit Study of Alternative Sources of Electrical Energy, WASH-1224 (December 1974).
[b] MDL = man-days lost.
[c] U = undetermined.

Table 16-1 illustrates the routine occupational health risks from the four sources of electricity generation. The dominant occupational health effect in the coal fuel cycle is coal workers' pneumoconiosis (CWP), i.e. "black lung" disease, which results from the accumulation of coal dust in underground miners' lungs; an advanced stage of this disease is progressive massive fibrosis (PMF). It was estimated that, with future mine improvements, about 0.6 cases of CWP and 0.006 cases of PMF can be attributed annually to the mining requirements associated with a 1000 MW_e coal-fired plant, assuming 50% of the coal comes from underground mines. In Table 16-1, each predicted cancer malignancy from PMF was assigned a value of 6000 man-days lost (MDL), corresponding to 24 years of lost working time, while 1000 MDL was assigned to each case of simple CWP: thus $0.6(1000) + 0.006(6000) \approx 600$ MDL is the value for coal.

It is worth noting that prior to the passage of the United States Federal Coal Mine Health and Safety Act of 1969, the actual number of cases of PMF that occurred per United States plant per year was 0.6, a factor of 100 higher than the value estimated in the study. The small value used in Table 16-1 was based on British statistics in which the allowable dust concentrations were lower, like those of the 1969 Act.

Table 16-1 also contains data on occupational injuries as a result of routine industrial accidents that occur throughout the different fuel cycles. About one occupational fatality per year can be attributed to each coal-fired plant, primarily because of injuries in underground mining. This value assumes that 50% of the coal production is from underground mines, and 50% from surface mines (where the accident rate is much lower because fewer people are involved and working conditions are safer).

The injury rates in the oil, gas, and LWR fuel cycles are roughly equivalent, and they are an order of magnitude below that of coal. Mining in the nuclear fuel cycle accounts for most of the injuries. Because of the high-energy content of nuclear fuels, nuclear fuel mining injury rates are much lower than those in coal mining.

We now turn to a consideration of the public health risks reported in Table 16-1. The estimated 180 to 210 MDL per year for public health risks for nuclear power is based upon an analysis of the amount of expected release of radioactive gases from a 1000 MW_e plant (primarily ^{85}Kr and ^{3}H); the transport of the radionuclides through air, water, and biological pathways; the dose to humans; and the use of the linear dose-response function to correlate the incidence of cancer with exposure. The linear model predicts a response per unit population dose (in man-rem) independent of the magnitude of the dose to particular individuals and independent of the rate at which individuals receive the dose (see Section 18-6). The results from this theory are complicated by large uncertainties so the numerical result of 180 to 210 MDL per year for the population surrounding a nuclear plant may be in error by an order of magnitude [1].

Unfortunately, human health hazards from fossil fuel pollutants are not as well understood and quantified as the health hazards from radiation. Correlations between various measures of human respiratory impairment, including death, and levels of air pollution have been observed during and following episodes of exceptionally high concentration of air pollutants. However, there is no information on the effect of individual exposure to specific pollutants during such episodes, so no dose–response relationship has been formulated. Thus, although a quantitative comparison between radiation health effect and fossil air pollution health effect is an essential part of the comparison of nuclear and fossil fuel cycles, no quantitative estimate of fossil air pollution health effect was made in Table 16-1.

In another study, however, the National Academy of Sciences has estimated that the number of deaths per 1000 MW_e-yr of electricity produced by a coal plant ranges from 0.007 to 17 for new fossil plants with lime scrubbers, or from 3 to 170 for old plants using 3% sulfur coal [3].

Table 16-2 *Comparison of Annual Routine Environmental Degradation Causing Latent Consequences for a 1000 MW_e Plant at 75% Capacity Factor*[a]

Environmental	Coal	Oil	Gas	LWR
Air				
SO_2 release, without abatement (tons/year)	120,000	38,600	20	3,600
SO_2 release, with abatement (tons/year)	24,000	21,000	0	720
NO_x release, without abatement (tons/year)	27,000	26,000	13,400	810
Particulate releases, without abatement (tons/year)	270,000	26,000	518	8,000
Particulate releases, with abatement (tons/year)	2,000	150	4	60
Trace metals releases (tons/year)	0.5 Hg	1,500 V	U[b]	S[b]
Radioactivity releases (Ci/year)	0.02	0.0005	S	$250–500 \times 10^3$
Thermal discharge, power plant stack (billion $kWhr_t$/year)	1.64	1.71	2.2	0
Water				
Cooling water use (billion gal/year)	263	263	263	424
Process water use (billion gal/year)	1.46	1.75	1.42	0.095
Radioactivity releases (Ci/year)	0	0	0	500–1000
Other impacts (billion gal/year)	16.8	7.9	0	S
Thermal discharge, power plant (billion $kWhr_t$)	9	9	9	14

[a] From Comparative Risk-Cost-Benefit Study of Alternative Sources of Electrical Energy, WASH-1224 (December 1974).
[b] U = unevaluated, S = small.

These estimates are for coal plants in the heavily populated, and already polluted, northeastern portion of the United States, and are highly dependent upon siting variables.

The last category of risks in Table 16-1 involves public safety. Because of the large masses and volumes of fuel involved, coal transportation imposes a much more severe public safety hazard than transportation of nuclear fuels. Public injury rates for coal transport are, in fact, comparable to occupational injury rates in coal mining. The transportation of coal for a 1000 MW_e plant results in a statistical public death about every two years, which is almost entirely due to accidents at railroad grade crossings.

The environmental degradation by each of the four types of plants for generating electricity is shown in Table 16-2. For purposes of comparison, the electrical energy required for the gaseous diffusion enrichment of LWR fuel is assumed to be provided by a coal-fired plant, and prorated quantities of coal cycle pollutants are assigned to the LWR fuel cycle (although any source of electricity could be used). It is interesting to note that, on this basis, the LWR cycle "emissions" of SO_2 and particulates are greater than those of the natural gas fuel cycle. (The quantities of radioactive materials routinely released from nuclear facilities to the atmosphere are of primary concern to public health and have already been considered in Table 16-1 as a part of the public health and safety aspects.)

The possible long-term effects of another by-product of combustion, CO_2, upon the earth's climate were not directly tabulated. Likewise, damage from acid rain (i.e., moisture in the atmosphere combined with oxides of sulfur and nitrogen from combustion of fossil fuels) away from the plant site was not considered. These two problems ultimately should become a greater environmental burden for future generations than the wastes from nuclear power plants, although the latter receive much more public attention at present.

16-2 CONVENTIONAL AND NONCONVENTIONAL ENERGY SOURCES

16-2-1 Methods of Analysis

The risks from producing energy have been examined for eleven different sources: coal, oil, nuclear, natural gas, hydroelectric, wind, methanol, three forms of solar (space heating, thermal, and photovoltaic), and ocean thermal. The solar thermal system utilizes heliostats that focus the incident solar energy on a central receiver containing a fluid that is superheated and drives a turbine; the solar photovoltaic generates electricity directly from photocells.

The results from the Inhaber study [4–6] may be termed a "competing risk comparison," since nine of the eleven different techniques are for directly creating the same product, namely, electrical energy. For solar space heating, the thermal energy produced was taken to correspond to the electrical energy that would have been required to heat a building, while for methanol it was assumed that the energy would be used to power a gas turbine that produces electricity. It should be noted that for some of the low-temperature energy sources, this assumption leads to an inefficient end-use of the energy produced.

For a competing risk analysis, it is good to remember that the calculations must include damages from all aspects of the fuel cycle. For example, while a solar collector on a roof appears entirely safe as it absorbs sunlight, some risks are incurred in constructing and installing the device. Another consideration is the amount of construction materials required. For example, a solar collector requires a significantly larger input of construction material, per unit of energy output, than higher-energy-density systems such as nuclear; hence a solar collector could be termed *materials intensive*. (The emphasis upon the risks from acquiring construction materials is one of the features that distinguishes the Inhaber report from WASH 1224.) Finally, in a competing risk analysis, if both occupational and public health risks are considered, then the two should be carefully distinguished.

A major decision to make in a competing risk analysis is whether each energy source must be noninterruptible. Inhaber included risks from energy backup sources for three of the nonconventional energy systems (solar thermal, solar photovoltaic, and windpower sources), even though many people would argue that a nonconventional source should be considered to be auxiliary energy, and hence be interruptible. The backup system Inhaber selected was prorated according to the fraction of electricity generated in Canada–from hydro (58%), nuclear (5%), and coal/oil/gas (36%). (The backup energy source contributed major portions of the risks to those nonconventional energy systems, as we shall see later.) In addition to energy back-up, five other components in the risk analysis were considered [4–6].

1. material acquisition and construction,
2. emissions caused by material production,
3. operation and maintenance,
4. energy storage,
5. transportation.

For raw materials and component fabrication, the amount of materials required to produce a component was first determined. The number of hours required to produce this material was found, and accident statistics were used to estimate the number of deaths, injuries, and the amount of time lost due to disease per million hours worked. The number of man-hours required per unit of production was then multiplied by the damages per man-hour to produce the occupational risks. (As an example, if mining X tons of coal requires Y days and if the number of hours lost per day of work is Z, then the number of hours lost per ton of coal is YZ/X.)

The final materials required for system construction often require intermediate and raw materials; for example, steel requires iron ore, coal, and other basic materials. A summary of the materials and construction data that Inhaber assumed for each of the eleven different types of plants is shown in Table 16-3.

For transportation, estimates were available for risk incurred for coal; all other systems were prorated to that for coal on the basis of the risk per unit weight of material transported. The risk for operation and maintenance was based on estimates of the hours per year required and the risk per hour calculated from occupational statistics.

Public health risk falls into two categories. The first, and by far the largest, is risk due to air pollution. In contrast to WASH-1224, in which no relationship between health and concentration of oxides of sulfur and nitrogen in air was attempted, Inhaber chose to make this correlation, although sulfur oxides were the only air pollutant considered. The sulfur pollutants are produced from the burning of coal and oil, so these two technologies have a public health risk. Because nonconventional technologies often require coal and oil to produce their fabricated materials (for example, solar photovoltaic systems require steel, which, in turn, requires coal), nonconventional systems also have public health risks.

A second aspect of public health risk lies in potential catastrophic accidents. Nuclear power is acknowledged to have this potential so results from the Reactor Safety Study in Chapter 12 were used. Also, oil fires, natural gas explosions, and dam failures have occurred, and these risks were estimated using statistical data.

The risk of waste disposition was calculated for nuclear power. Other energy sources were assumed to have little or no risk from wastes [4, 5].

16-2-2 Risk Results

Figure 16-1 shows the occupational deaths, times 1000, per megawatt-year of energy, for each of the eleven energy systems [4]. The risk in-

Table 16-3 *Summary of material acquisition and construction data*[a]

	Materials*						
Energy system	Steel	Con-crete	Alu-mi-num	Glass	Cop-per	Total	CT**
Coal[b]	4.3[e]	6.8				11[i]	505
Oil[b]	3.2[e]	3.1				6.3	415
Natural gas[b]	1.5	2.4				3.9	302
Nuclear[b]	2.3[e]	12.7				15	633
Solar space heating	95		4.0	15	2.7	120[j]	12000
Wind	23	5		3.4[h]	0.6	36[k]	2400[g]
Methanol[b]	89[e,f,g]	23[f]				112	7300[g]
Solar thermal[c,d]	65	290	3.7	11		370	10000
Solar photovoltaic[d]	30	10	91	4.8		140[l]	1350[o]
Ocean thermal	12	80	2.8		0.5	96[m]	6800[g]
Hydroelectricity	0.7	92				630[n]	[p]

* In metric tons per MW-year net output over system lifetime.
** CT is construction time in manhours per MW-year over system lifetime.
[a] From H. Inhaber, AECB 1119/REV-3, 4th ed. (1980).
[b] Does not include weight of fuel.
[c] Not including 0.2 metric tons of silver.
[d] Multiplied by appropriate factor for climatic conditions.
[e] Includes all metals.
[f] Obtained by multiplying oil data by suitable factor.
[g] This is an average of two values.
[h] Fiberglass and plastics.
[i] Excluding lime for scrubbers.
[j] Including 3.7 metric tons of fiberglass insulation.
[k] Including 3.5 metric tons of fiberglass insulation.
[l] Including 4.8 metric tons of silicon semi-conductors.
[m] Including 1.1 metric tons of ammonia.
[n] Including rock and earth.
[o] Most labor is fabrication, not construction.
[p] Not available.

curred in activities related to gathering and handling fuels, acquiring material and equipment, and operating and maintaining power plants was included; risk incurred by the public was not included. The numerical values refer to a net output of 1 MW-yr of energy over the lifetime of the system; for example, coal would have an upper value of 0.0067 deaths per megawatt-yr output over the 30-year system life. The top of each bar indicates the upper end of the range of values and the dotted line within a bar is the lower; where no dotted line is shown, the upper and lower ends

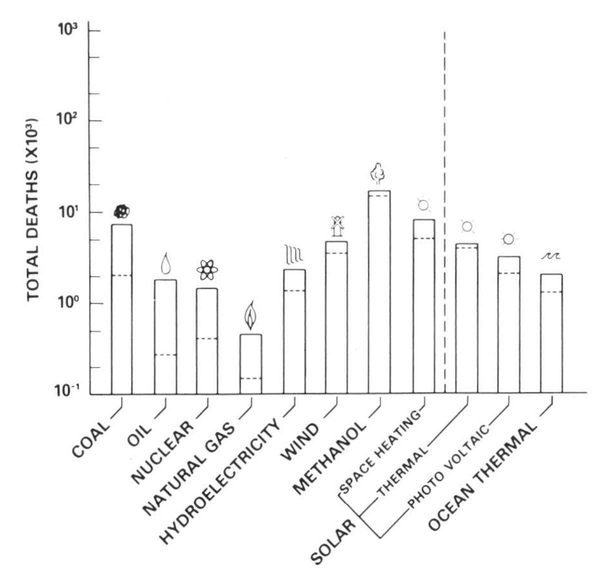

Fig. 16-1 *Occupational deaths per megawatt-year, as a function of energy system. (From H. Inhaber, Rep. AECB 1119/REV-3, 4th ed., 1980).*

of the range are similar. The bars to the right of the vertical dashed line indicate values for technologies that Inhaber felt were less likely to become a major source of energy in Canada.

The maximum number of occupational deaths in Fig. 16-1 results from methanol, followed by coal, solar space heating, and wind. The lowest value is for natural gas, and the next lowest is for nuclear. For most of the nonconventional systems, the cause of large risk values is high material acquisition and construction risk [4, 5].

Figure 16-2 shows the total occupational man-days lost (MDL) per unit energy. Deaths are incorporated in the total at 6000 MDL per fatality. The nonconventional technologies still have the highest values, and natural gas and nuclear have the lowest.

Figures 16-3 and 16-4, which show the risks to the public, are considerably different from Figs. 16-1 and 16-2. Much of the public risk is produced by emissions created after fuel is gathered (for the case of conventional technologies) or in the course of producing the materials for the power system (for nonconventional technologies). Natural gas has the

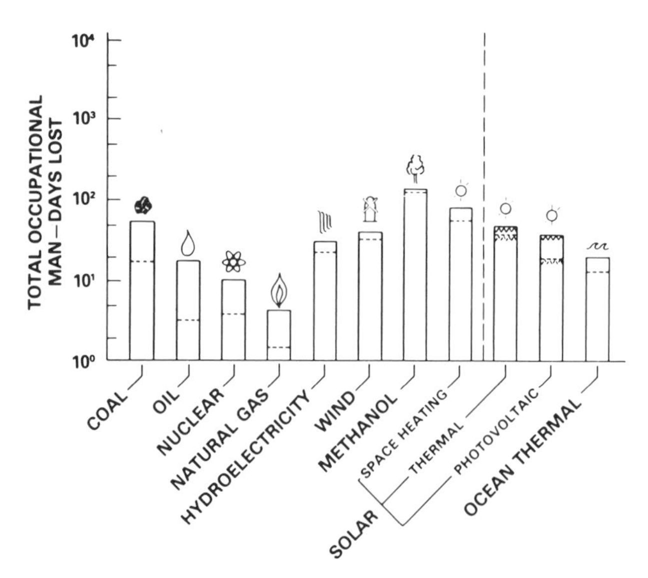

Fig. 16-2 *Occupational man-days lost per megawatt-year net output over lifetime of system. (From H. Inhaber, Rep. AECB 1119/REV-3, 4th ed., 1980).*

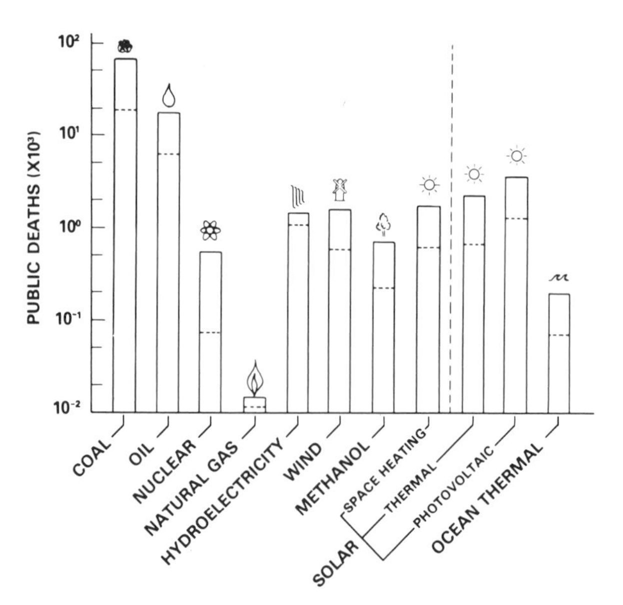

Fig. 16-3 *Public deaths per megawatt-year, as a function of energy system. (From H. Inhaber, Rep. AECB 1119/REV-3, 4th ed., 1980).*

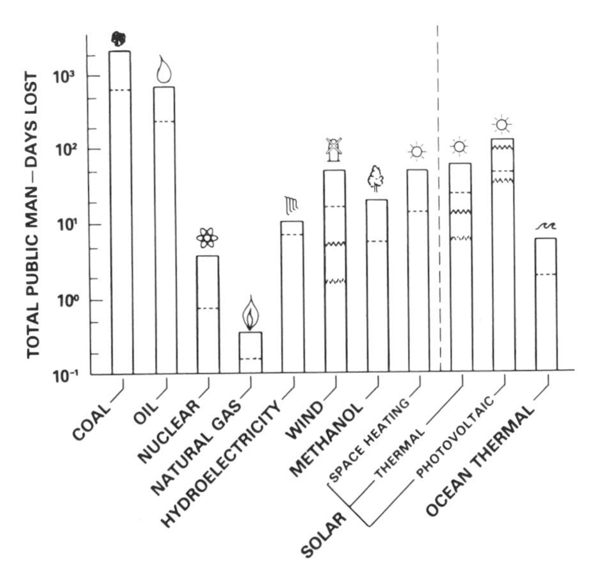

Fig. 16-4 *Public man-days lost per megawatt-year net output over lifetime of system. (From H. Inhaber, Rep. AECB 1119/REV-3, 4th ed., 1980).*

lowest public risk, followed by ocean thermal and nuclear. For nuclear power both the risks of waste management and possible reactor catastrophes were included.

An interesting breakdown is the sources of the risks, as shown in Fig. 16-5; the ordinate of each source is normalized to the sum of the maximum value calculated for the occupational and public man-days lost for that source. We can see that most of the risk for coal and oil is due to electricity production (i.e., air pollution), whereas most of the risk for natural gas, nuclear sources, and ocean thermal sources is due to material acquisition and construction. Although wind, solar thermal electric, and solar photovoltaic sources were calculated to have a large proportion of risk from energy backup, Inhaber concluded that the relative rankings of the 11 systems was not strongly influenced by whether the energy backup was included in the calculations [4].

16-2-3 Conclusions

The most important conclusion of the Inhaber study is that the risk from nonconventional energy sources can be as high as, or even higher than, that of some conventional sources [4]. This conclusion, however, has not gone unchallenged. The most extensive criticism (of an earlier version of

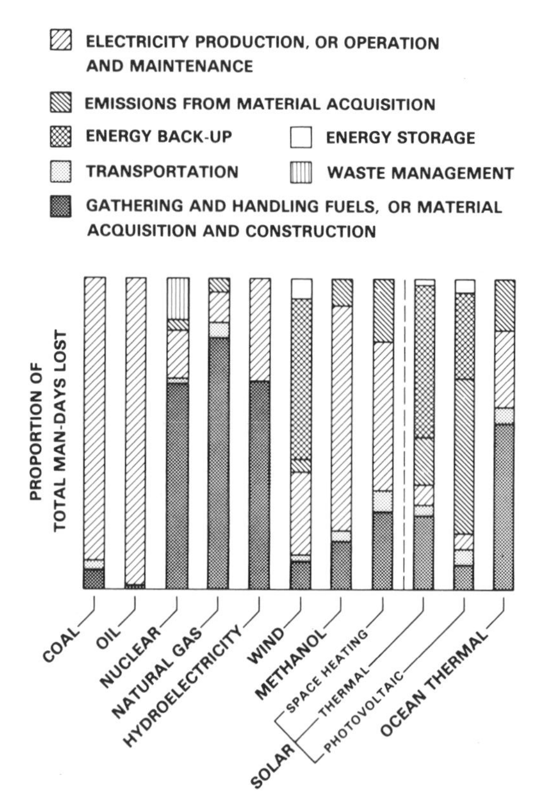

Fig. 16-5 *Proportions of risk by source–normalized to the sum of the occupational and public risks for each source. (From H. Inhaber, Rep.* AECB *1119/*REV-*3, 4th ed., 1980).*

the report) by Holdren *et al.* [7, 8], is a report nearly the size of Inhaber's. Particular features of the criticisms also have been published, along with counter-criticisms [9].

If nothing else, three things can be learned from the Inhaber study and its criticisms:

1. It is important to account for the risks from producing the materials used to construct an energy production system.
2. It is difficult to perform an accurate comparative risk assessment between technologies with ill-defined and incomplete data; for example, Inhaber has refined the magnitudes of some of his published results for nonconventional sources by factors as large as 7 or 8, while Holdren *et al.* have argued that some risk values should change by two orders of magnitude or more.

3. More work still needs to be done to assess accurately the comparative risks from producing energy from competing sources.

16-3 CANVEY ISLAND

This section presents results from a "complementary risk comparison" for which different technologies are used to create products with complementary end uses. The Canvey study was a site-specific investigation of the risks to people living in and around Canvey Island and the neighboring part of Thurrock [10]. The area measures about nine by two-and-a-half miles and is in the River Thames near the English Channel. The purpose of the study was to identify the risks from the different industrial installations on the island, both existing and proposed. Although there are no risks from nuclear power or other electricity production systems at Canvey Island, the results of the study illustrate what can be learned by assessing risks from more than one hazard in a specific locality.

The principal existing plants in the area at the time of the study were [10]:

1. Methane terminal for the importation of liquefied natural gas by sea and its storage (British Gas Corp.).
2. Tank farm for the importation by sea and storage of petroleum products (Texaco Ltd.).
3. Facility for the bulk storage in tanks of a wide variety of substances, including petroleum products and others that are flammable or toxic (London and Coastal Oil Wharves Ltd.).
4. Oil refinery and bulk distribution depot (Mobil Oil Co. Ltd.).
5. Liquefied petroleum gases cylinder-filling plant (Calor Gas Ltd.).
6. Distillation plant for the separation of imported crude oil into hydrocarbons that are stored; also a storage tank for the importation by sea of liquefied ammonia (Shell UK Oil).
7. Ammonium nitrate production plant (Fisons Ltd.).

Other industrial activity in the area includes ships anchored off one end of the island and carrying explosives and ship traffic near the jetties of some of the installations. Attention in the study was focused on the toxic liquefied gases, such as ammonia and hydrogen fluoride, and flammable substances like liquefied natural gas and liquefied petroleum gases, such as propane and butane.

One type of risk considered was the probability per year of death for the average individual. The analysis was broken down into different geographical areas on Canvey Island. The predictions ranged from a frequency of 1.3×10^{-3}/yr to a low of 1×10^{-4}/yr for the existing installations, which could be reduced to values from 6×10^{-4}/yr to 3×10^{-5}/yr with improved safety features. Also considered were the frequencies of death for proposed new facilities and extensions; it was shown that with the additional safety features on the existing plants and newly proposed facilities, the risks could be made less than those from the existing installations without the improved safety features (i.e., the status quo).

The second risk considered was societal risk, and consisted of the predicted frequency versus the number of deaths. Results for the different installations are illustrated in Table 16-4. The risk curves for the predicted frequencies of death from existing installations, without and with the safety improvements, are given in Fig. 16-6, while Fig. 16-7 shows the analogous curves for the proposed new developments. Results such as the latter enable the local licensing agencies to assess the additional risk burden incurred by inhabitants of the area if construction of additional installations were to be permitted.

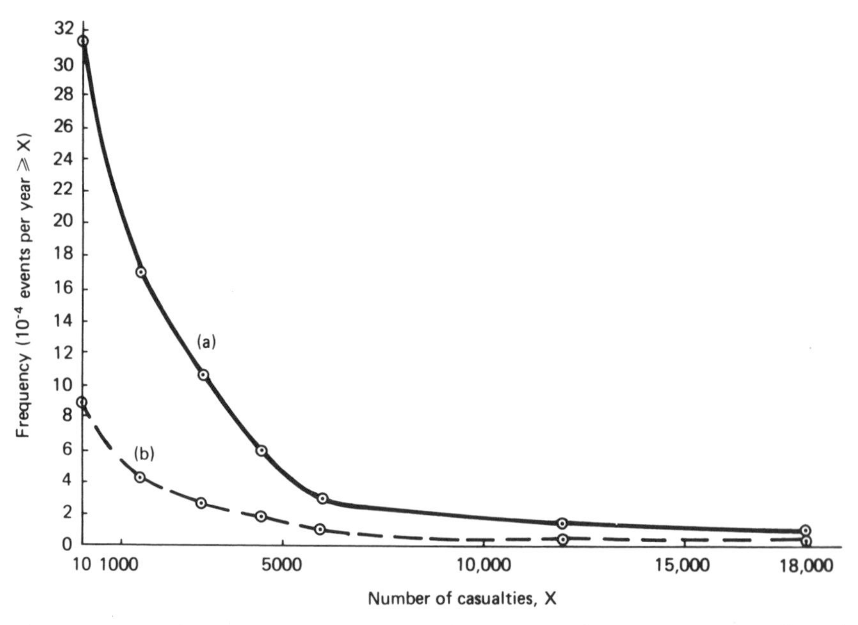

Fig. 16-6 *Societal risks from the existing installations* (*a*) *without improvements,* (*b*) *with improvements suggested.* (*From* CANVEY: An Analysis of Potential Hazards from Operations in the Canvey Island/Thurrock Area, *Her Majesty's Stationery Office, 1978.*)

Table 16-4 *Probability of Deaths per Year (times 10^4) from the Different Existing Installations*[a]

	Estimated upper limit of probabilities for numbers of casualties exceeding:[b]													
	10		1,500		3,000		4,500		6,000		12,000		18,000	
Location	a	b	a	b	a	b	a	b	a	b	a	b	a	b
Shell UK Oil	7.8	3.3	5	1.9	3.6	1.3	2.7	0.9	1.8	0.5	1.2	0.3	0.8	0.2
Mobil Oil Co. Ltd.	1.4	0.3	1.1	0.2	0.7	0.1	0.3	0.1						
British Gas Corporation	14.7	4	5.4	1.5	3	1	1.6	0.7	0.4	0.4	0.2	0.2	0.1	0.1
Fisons Ltd.	3.6	1	2.4	0.6	2	0.5	1.1	0.3	0.8	0.1	0.3	0.1	0.1	
Texaco Ltd. and London & Coastal Oil Wharves Ltd.	3.9		3.1		1.5		0.4							
Totals	31.4	8.6	17	4.2	10.8	2.9	6.1	2	3	1	1.7	0.6	1	0.3
Proposed developments														
Mobil extension	2.8	0.4	1.7	0.2	1.1	0.1	0.7	0.1	0.3		0.2		0.1	
Occidental Refinery	2.9	0.9	2.2	0.7	1.7	0.5	1.4	0.4	1.1	0.3	0.8	0.2	0.7	0.2
United Refinery	1.4	0.9	0.9	0.6	0.5	0.3	0.2	0.1						
Texaco and London & Coastal Oil Wharves Ltd.[c]	9		7.3		3.6		0.9							
Totals	16.1	2.2	12.1	1.5	6.9	0.9	3.2	0.6	1.4	0.3	1	0.2	0.8	0.2

[a] From CANVEY: An Investigation of Potential Hazards from Operations in the Canvey Island/Thurrock Area, Her Majesty's Stationary Office (1978).

[b] (1) Columns (a) show the assessed risks before improvements are made or, in the case of new developments, on the basis of the proposals as they were understood to be prior to January 1, 1978. (2) Columns (b) show the assessed risks if all the suggested action is taken and improvements are made at the existing installations and proposed developments. (3) Where a number is not shown the risk has been assessed as negligible.

[c] Relates to risks from proposed activities at the jetty of Occidental Refineries Ltd.

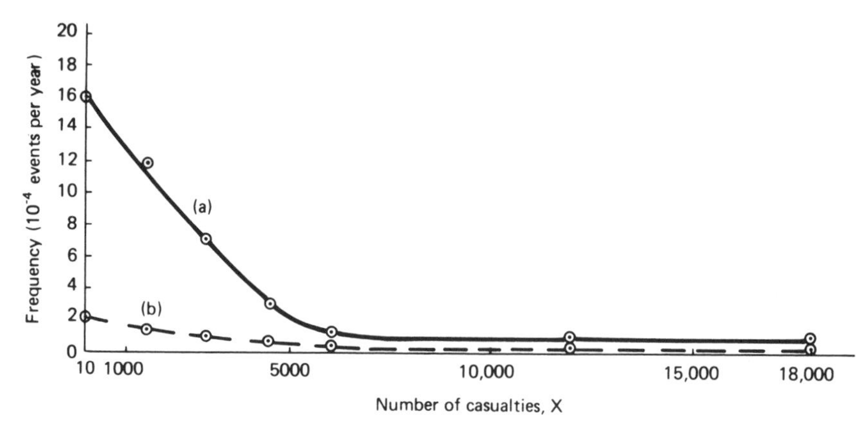

Fig. 16-7 *Societal risks from the proposed installations (a) without improvements, and (b) with improvements suggested. (From* CANVEY: An Analysis of Potential Hazards from Operations in the Canvey Island/Thurrock Area, *Her Majesty's Stationery Office, 1978.)*

Table 16-5 *Estimated Effects of Total and Instantaneous Failure of California Dams Filled to Capacity*[a]

Name of dam	Fatalities for failure at[b] Day	Night	Damage assessed in U.S. dollars
Van Norman	72,000	123,000	3×10^8
San Andreas / Lower Crystal Springs	21,000	33,000	1.1×10^8
Stone Canyon	125,000	207,000	5.3×10^8
Encino	11,000	18,000	5×10^7
San Pablo	24,000	36,000	7.7×10^7
Folsom	260,000	260,000	6.7×10^8
Chatsworth	14,000	22,000	6×10^7
Mulholland	180,000	180,000	7.2×10^8
Upper San Leandro / Lake Chabot	36,000	55,000	1.5×10^8
Shasta	34,000	34,000	1.4×10^8

[a] From P. Ayyaswamy *et al.*, Univ. of California at Los Angeles Rep. UCLA-ENG-7423 (1974).

[b] No allowance for evacuation.

16-4 DAMS

In the previous section, we briefly examined the risks from storage and transportation of various toxic and flammable liquefied gases. Dams provide another form of energy storage which can be considered.

Dam failures, which are not uncommon events, pose perhaps the greatest potential for large damages from man-made structures. For example, 33 dams failed in the United States between 1918 and 1958 and 5 of these were major disasters involving the loss of 1680 lives. With an average number of dams over the 40-year time interval of about 1000, these data suggest a failure rate approximately 8×10^{-4} per dam-year and a major disaster rate of approximately 1.3×10^{-4} per dam-year [11]. Between 1959 and 1965, nine major dams of the world failed in some manner, giving a world-wide failure rate of about 2×10^{-4} per dam-year for that period. These values agree with other people's conclusions that the frequency of a dam disaster is of the order of magnitude of 10^{-4} per dam-year [12, 13].

The causes of dam failure can generally be attributed to design, construction, or site inadequacies, or to natural phenomena, primarily floods or earthquakes, in excess of design criteria. The probability of failure of a particular dam is difficult to estimate, but it is strongly dependent upon the geologic site, surface faulting and seismicity, the type and size of dam, specific engineering and construction, and flood frequency.

A study of the probability of failure of 12 dams in California because of a severe earthquake has been performed and estimated fatalities and damages are presented in Table 16-5 [14]. This study is one of the few published reports that gives estimates of the maximum hazard and a crude failure probability for specific dams. The fatality and damage estimates are based on the assumption of total and instantaneous failure of dams filled to capacity.

Of particular concern to the State of California is the San Pablo Dam that supplies water to the city of Oakland, 15 incorporated cities, and 14 unincorporated areas along the eastern shores of San Francisco Bay. It was built in 1921 when the area downstream on San Pablo Creek was only sparsely settled. Today the potential flood area contains 26,000 residents, 10 schools, a hospital, shopping centers, and public buildings immediately below the dam.

The San Pablo Dam is almost a twin of the Lower San Fernando Dam, both in size, with a crest length of 381 m, and construction method. In February 1971, an earthquake caused the crest of the Lower San Fernando Dam to drop away on the upstream side, leaving only a meter-

and-a-half of dam above the water in its reservoir. If the dam would have breached and water would have cascaded down on 80,000 people, the loss of life and property almost certainly would have been so great as to stagger the imagination [14].

After the San Fernando Dam's near-mishap, an analysis showed that an earthquake of magnitude 8 on the Richter scale (equivalent to the one that struck San Francisco in 1906) would generate more than 10^3 times the

Table 16-6 *Annual Cancer Risks from a Variety of Nonnuclear Causes*[a]

Type of risk	Individual Risk/Year
Cosmic ray risks[b]	
One transcontinental flight/year	1 in 2,000,000
Airline pilot 50 hr/mo at 35,000 ft	1 in 20,000
Frequent airline passenger	1 in 65,000
Living in Denver compared to N.Y.	1 in 100,000
One summer (four months) camping at 15,000 ft	1 in 100,000
Other radiation risks[b]	
Average U.S. diagnostic medical x rays	1 in 100,000
Increase in risk from living in a brick building	1 in 200,000
Natural background at sea level	1 in 65,000
Eating and drinking	
One diet soda/day (saccharin)[c]	1 in 100,000
Average U.S. saccharin consumption[c]	1 in 500,000
Four tablespoons peanut butter/day (aflatoxin)[b,c]	1 in 25,000
One pint milk per day (aflatoxin)[b,c]	1 in 100,000
Miami or New Orleans drinking water[d]	1 in 800,000
1/2 lb charcoal broiled steak once a week (cancer risk only; heart attack, etc., additional)[e]	1 in 2,500,000
Alcohol—averaged over smokers and non-smokers[f]	1 in 20,000
Tobacco	
Smoker, cancer only[f]	1 in 800
Smoker, all effects (including heart disease)[f]	1 in 300
Person in room with smoker[f]	1 in 100,000
Miscellaneous	
Taking contraceptive pills regularly[f]	1 in 50,000
Skin cancer (curable) from sunbathing, rock climbing, and other outdoor activities.[f]	1 in 200

[a] Reprinted from P. B. Hutt, *Food Drug Cosmetic Law J.* 33, 558 (1978). © 1978 by Commerce Clearing House, Inc., Chicago, Illinois. All rights reserved.
[b] Linear extrapolation from human epidemiological data.
[c] Linear extrapolation from animal data.
[d] Multi-stage extrapolation from animal data.
[e] Based on equivalent concentration of benzo(a)pyrene in cigarette smoke.
[f] Human epidemiological data, no extrapolation.

energy of the San Fernando quake at the San Pablo site, so the San Pablo Dam has been strengthened [15].

16-5 OTHER RISKS

The incidence of cancer is a potential latent consequence of exposure to radiation, so it is especially useful to examine the individual risks for cancer from nonnuclear causes, as shown in Table 16-6 [16]. The table has been derived from data developed by Wilson [17].

Statistics for various natural, man-made, and occupational hazards may be obtained from various governmental agencies; a useful compilation of

Table 16-7 *Average Loss of Life Expentancy in Days Among Public Due to Energy Generation*[a]

Source	Fatalities per year	Average years lost	Days reduced life expectancy
Coal			
Air pollution	11,000	10	11.5
Transport accidents	300	35	1.0
			$\Sigma = 12.5$
Oil			
Air pollution	2,000	10	2.2
Fires	500	35	2.0
			$\Sigma = 4.2$
Gas			
Air pollution	200	10	0.2
Explosions	100	35	0.4
Fires	100	35	0.4
Asphyxiation	500	25	1.5
			$\Sigma = 2.5$
Hydroelectric			
Dam failures	50	35	0.2
			$\Sigma = 0.2$
Nuclear (400 GW)			
Routine emissions	8	20	0.018
Accidents	8	20	0.018
Tranposrt	<0.01	20	—
Waste	0.4	20	0.01
Plutonium toxicity	<0.01	20	$\Sigma = 0.037$
Electrocution	1,200	35	5.0
Total			24

[a] From B. L. Cohen and I-S. Lee, *Health Phys.* **36**, 707 (1979). Reproduced by permission of the Health Physics Society.

Table 16-8 *Average Loss of Life Expectancy in Days due to Various Causes*[a]

Cause	Days
Being unmarried—male	3500
Cigarette smoking—male	2250
Heart disease	2100
Being unmarried—female	1600
Being 30% overweight	1300
Being a coal miner	1100
Cancer	980
20% overweight	900
Cigarette smoking—female	800
Stroke	520
Dangerous job—accidents	300
Pipe smoking	220
Increasing food intake 100 cal/day	210
Motor vehicle accidents	207
Alcohol (U. S. average)	130
Accidents in home	95
Suicide	95
Being murdered (homicide)	90
Legal drug misuse	90
Average job—accidents	74
Drowning	41
Job with radiation exposure	40
Falls	39
Accidents to pedestrians	37
Safest jobs—accidents	30
Fire—burns	27
Generation of energy	24
Illicit drugs (U. S. average)	18
Firearms accidents	11
Natural radiation (BEIR)	8
Medical x rays	6
Coffee	6
Oral contraceptives	5
All catastrophes combined	3.5
Diet drinks	2
Reactor accidents—USC	2[b]
Reactor accidents—Rasmussen Report	0.02[b]
Radiation from nuclear industry	0.02[b]

[a] Extracted from B. L. Cohen and I-S. Lee, *Health Phys.* **36,** 707 (1979). Reproduced by permission of the Health Physics Society.

[b] These items assume that all United States power is nuclear. UCS is Union of Concerned Scientists, a prominent group of nuclear critics.

Table 16-9 *Actual and Postulated Worst-Consequence Events*[a]

Event and/or accident scenario	Actual, historical worst-consequence event			Postulated, worst-consequence event		
	Date, location, description	Number of fatalities (actual number)	Estimated liability if occurred today ($ billion)	Estimated probability (per year)	Estimated number of fatalities or complete disabilities	Estimated liability ($ billion)
Aircraft crash	3/29/77; Canary Islands (Spain); two 747's in ground collision	577	0.65	1×10^{-5} to 1×10^{-7} for a selected site adjacent to a selected airport	6,000 to 20,000	6.3 to 20.3
Chemical explosion and fire	12/6/17; Halifax, Canada	1,600	1.8	10^{-5} to 10^{-7} per site	12,000 to 25,000	12.5 to 25.5
Dam rupture	9/9/63; Vaiont, Italy	1,800	2.0	1×10^{-3} to 10^{-4} per site for high risk sites	20,000 to 750,000	20 to 800
Drug toxicity	Thalidomide Post-World War II	Hundreds of badly deformed babies	~0.5	10^{-4} to 10^{-6}	(?) 10,000 to 100,000	(?) 10 to 100
Earthquake	12/24/1556; Shensi, China	830,000	830	1×10^{-3} to 1×10^{-4}	100,000 to 1,000,000	100 to 1,000
Fire	9/9/1871; Peshtigo, Wisc.; forest fire	1,182	1.5	1×10^{-6} per high risk site	1,000 to 3,000	1.1 to 3.2
Flood	1897 Hwang-Lo River, China	900,000	1,000	1×10^{-2}	200,000 to 1,000,000	200 to 1,000
Meteorite impact		No known mortalities	No large liabilities	1×10^{-7}	10,000 on up	10 to 12

(Continued)

Table 16-9 (*Continued*)

Event and/or accident scenario	Actual, historical worst-consequence event			Postulated, worst-consequence event		
	Date, location, description	Number of fatalities (actual number)	Estimated liability if occurred today ($ billion)	Estimated probability (per year)	Estimated number of fatalities or complete disabilities	Estimated liability ($ billion)
Mine disaster	12/6/07; Monongah West Virginia	361	0.36	1×10^{-2} to 1×10^{-3}	500 to 1,000	0.5 to 1.0
Nuclear reactor accident		No known mortalities in civil program		1×10^{-7} for high risk site	10,000 to 20,000	15 to 25
Ship disaster	4/5/12; North Atlantic Ocean, Titanic hit iceberg	1,517	1.7	1×10^{-6} to 5×10^{-7}	2,500 to 5,000	2.8 to 5.5
Tidal wave/ hurricane	Several during 1963–65 in windstorm, E. Pakistan	79,000	80	1×10^{-2} to 1×10^{-3}	50,000 to 500,000	55 to 520
Tornado	3/18/25; in states of Mo., Ill., and Ind.	689	0.75	1×10^{-2} to 1×10^{-3}	1,000 to 10,000	1.2 to 10.3
Train crash	12/12/17; France	543	0.60	1×10^{-1} to 1×10^{-2}	200 to 2,000	0.22 to 1.1
Industrial pollutant	Pre-1970; mercury poisoning			10^{-4}	~10,000	10

[a] From K. A. Solomon, C. Whipple, and D. Okrent, *in* "Probabilistic Analysis of Nuclear Reactor Safety," Vol. 2, p. IV.7-1. American Nuclear Society, 1978.

these sources of data is available [18]. Cohen has used such data to calculate the "loss of life expectancy," which is the expected period of time the average individual in society loses as a result of performing a dangerous activity, compared to not performing that activity [19].

Table 16-7 shows a comparison of the loss of life expectancy to the public for the various sources of energy generation; the value for nuclear power is based on the assumption that all electricity in the United States is generated by nuclear reactors. The figure of 24 days for all sources in Table 16-7 seems large until it is compared with the average loss of life expectancies in Table 16-8, which were calculated from statistics for many other causes. More detailed comparisons as a function of age of the population were also calculated by Cohen for some of these causes [19].

Another interesting comparison of risks, in Table 16-9, gives the actual versus postulated consequences from worst-possible events [20]. Such a table shows that there are many ways in which more than 10,000 people could undergo death or disability, but a meteorite impact and a nuclear reactor accident are the only two that are not known to have killed members of the public.

EXERCISES

16-1 Compare the hazards and risks from the nuclear fuel cycle with those from coal, oil, and natural gas, as reported by Atomic Energy Commission Report WASH-1224. Your exposition should be aimed at the layperson who has little knowledge of how electricity is generated. A length of about 800 words, exclusive of tables, is preferred.

16-2 Compare the risks from the 11 energy sources Inhaber considered, and discuss some of the assumptions made in performing the study. Your exposition should be aimed at the layperson who has little knowledge of risks from energy systems. A length of about 800 words, exclusive of figures and tables, is preferred.

16-3 Compare the hazards and risks to the residents of Canvey Island, before and after safety improvements are made. Your exposition should be aimed at the layperson who has little knowledge of risks from industrial systems. A length of about 600 words, exclusive of figures and tables, is preferred.

16-4 Discuss the risks from dam failures. A length of about 200 words, exclusive of tables, is preferred.

16-5 Compare the statistical risks encountered by members of society from a wide variety of nonnuclear causes. A length of about 200 words, exclusive of tables, is preferred.

REFERENCES

1. Comparative Risk-Cost-Benefit Study of Alternative Sources of Electrical Energy, U.S. Atomic Energy Commission Rep. WASH-1224 (December 1974).
2. Comparative risk-cost-benefit study of alternative sources of electrical energy, *Nucl. Safety* **17,** 171 (1976).
3. Air Quality and Stationary Sound Emission Controls. National Academy of Sciences, Washington, D.C. (1975).
4. H. Inhaber, Risk of Energy Production. Atomic Energy Control Board Rep. AECB 1119/REV-3, 4th ed., Ottawa, Canada (1980).
5. H. Inhaber, Risk with energy from conventional and nonconventional sources, *Science* **203,** 718 (1979).
6. H. Inhaber, Is solar power riskier than nuclear? *Trans. Am. Nucl. Soc.* **30,** 11 (1978).
7. J. P. Holdren, K. Anderson, P. H. Gleick, I. Mintzer, and G. Morris, Risk of Renewable Energy Sources: A Critique of the Inhaber Report. University of California at Berkeley Energy and Resources Group Rep. ERG 79-3 (June 1979).
8. Critique of Inhaber report, *Nucl. News* 42 (September 1979).
9. J. P. Holdren, *Nucl. News,* 25 (March 1979); H. Inhaber, *ibid.* 25 (March 1979); J. P. Holdren, *ibid.* 32 (April 1979); H. Inhaber, *ibid.* 26 (May 1979).
10. Health and Safety Executive, CANVEY: An Investigation of Potential Hazards from Operations in the Canvey Island/Thurrock Area." Her Majesty's Stationery Office, London, 1978.
11. D. Okrent, Risk-benefit evaluation for large technological systems, *Nucl. Safety* **20,** 148 (1979).
12. P. F. Gast, Divergent public attitudes toward nuclear and hydroelectric plant safety, *Trans. Am. Nucl. Soc.* **16,** 40 (1973); U.S. Atomic Energy Commission Rep. CONF-730611-18 (June 1973).
13. R. K. Mark and D. E. Stuart-Alexander, Disasters as a necessary part of benefit-cost analyses, *Science* **197,** 1160 (1977).
14. P. Ayyaswamy *et al.,* Estimates of the Risks Associated with Dam Failure. Univ. of California at Los Angeles Rep. UCLA-ENG-7423 (1974).
15. J. Hazelwood, Quakes force new look at dam safety, *Ind. Res. Dev.* 95 (January 1980).
16. P. B. Hutt, Unresolved issues in the conflict between individual freedom and government control of food safety, *Food Drug Cosmetic Law J.* **33,** 558 (1978).
17. R. Wilson, Direct Testimony before the Occupational Health and Safety Administration, OSHA Docket No. H-090 (February 1978).
18. W. Baldewicz, G. Haddock, Y. Lee, Prajoto, R. Whitley, and V. Denny, Historical Perspectives on Risk for Large-Scale Technological Systems. Univ. of California at Los Angeles Rep. UCLA-ENG-7485 (1974).
19. B. L. Cohen and I-S. Lee, A catalog of risks, *Health Phys.* **36,** 707 (1979).
20. K. A. Solomon, C. Whipple, and D. Okrent, Insurance and catastrophic events: Can we expect de facto limits on liability recoveries? *in* "Probabilistic Analysis of Nuclear Reactor Safety," Vol. 2, p. IV.7-1. American Nuclear Society, LaGrange Park, Illinois, 1978.

CHAPTER 17

Risk-Benefit Assessments

17-1 ECONOMIC CONSIDERATIONS

The law of diminishing returns generally applies to risk reduction, as illustrated in Fig. 17-1. From a classic economics principle, safety expenditures should be made until the marginal (or "last dollar") exchange between social cost and risk control cost is equal; then the expenditure of one dollar for safety would reduce the social cost by one dollar [1].

As an example of the use of the marginal cost of engineered safety features, consider Fig. 17-2 in which risk is plotted versus the costs of risk reduction [2]. From this curve it follows that at some point the cost effectiveness (value received per unit expenditure) goes below some level of "acceptability." A practicable range of cost expenditures for risk reduction is also shown in the figure.

A problem arising in a *traditional* risk-benefit analysis is the need to identify both risks and benefits in the same units. When the benefits are measured in monetary units such as dollars, it is necessary to account for risks from disabilities and death (i.e., the value of human life) in the same way. One model that has been proposed accounts for the social costs of disabilities with the equation [1]

$$\begin{aligned}\text{Total social cost} &= NC(1+i)^t, \qquad t < 6000,\\ &= NC(1+i)^{6000}, \qquad t \geq 6000,\end{aligned} \tag{17-1}$$

where N is the number of individuals involved, C is the cost of one disability day, i is the so-called daily interest rate, and t is the time in consecutive days of disability. The National Safety Council value of 6000 days typically is selected to be equivalent to a fatality.

Several approaches have been suggested for placing a value on a human life [3]. Certainly the use of insurance premiums and court-decided com-

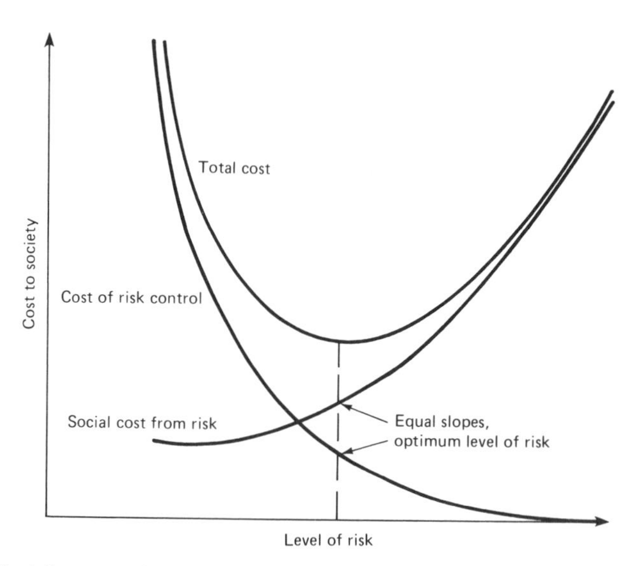

Fig. 17-1 *Social costs and control costs versus level of risk. (Reproduced with permission from the* Annual Review of Energy, ***1***, *629. Copyright © 1976 by Annual Reviews Inc.)*

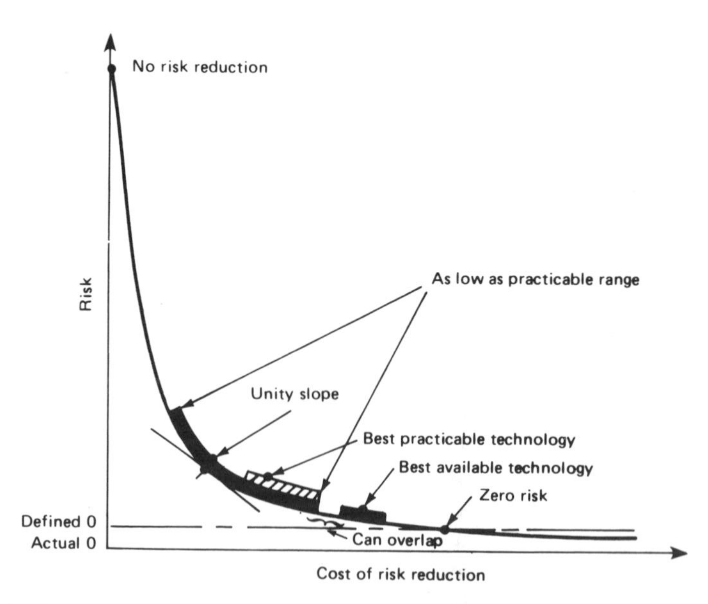

Fig. 17-2 *Some criteria for acceptance levels of cost effectiveness of risk reduction. (From W. D. Rowe, "An Anatomy of Risk," Wiley, 1977.)*

pensation is not a good way: insurance does not reduce the probability of death—it compensates the survivors. Furthermore, court cases vary widely in their dollar awards for loss of life. Three other methods have received considerable attention: the human-capital, the personal-risk, and the implicit societal approaches [3, 4].

In the human-capital method of evaluation, the discounted value of a person's future earnings are taken as the value of his or her life. Results of such analyses have varied from a low of $10,000 to a high of $400,000, with a typical value of about $300,000 (in late-1970s dollars) [3]. The value of $300,000 also has been obtained by using a flat rate of $50 per day of disability times the 6000 disability days equivalent to a death [5].

For the personal risk method, the necessary financial inducement required for an individual to accept an increased probability of accident is evaluated. The premiums paid to workers in hazardous occupations are compared with the increased risk, and the marginal risk-benefit trade-off for an individual is scaled upward to a life value. Implicit in such a calculation is the assumption that the individual knows the level of risk and that employment in a similar but safer occupation is available at a slightly lower wage; in addition, the increase in risk must be small, or nonlinear effects (risk aversion) will occur. Such an approach has given a value of life in the neighborhood of $200,000 (1973 dollars) [4].

In the implicit societal method for placing a value on human life, one examines records of what has been spent to avoid accidents in the past [4]. One example involves a military application in which the incremental costs spent on ejection seats for aircraft can be compared against the number of pilot lives saved. (Of course, credit must also be taken for improved morale of pilots flying aircraft equipped with such devices.) Another example arises with the costs spent on highway safety barriers and guard rails and the number of motorists estimated to have survived crashes only because of such facilities. This approach suffers from the limitation that past decisions dictated which safety systems should be built, and these decisions were not necessarily optimal.

Besides the costs for disability-days and deaths, there are also costs for radiation dose received. In the late 1970s, the Nuclear Regulatory Commission (NRC) used a guideline of $1000 per man-rem for this dose, and this value has been adopted for use in some nuclear power risk-benefit studies. Recently, a progressively increasing compensation has been proposed for exposure to equal increments of dose above a certain threshold (10^{-2} rem); this nonlinear relationship has the advantage of concentrating resources on the reduction of radiation exposure to people subjected to higher levels of risk [6].

17-2 DIFFERENT APPROACHES

There is no unique approach to performing a *value-impact analysis* or risk-benefit evaluation. It has been pointed out that the economist, the decision analyst, the politician, the engineer, the risk assessor, and the regulator each may have a different view [7]. To some people, a risk-benefit analysis is a special case of cost-benefit analysis [8]; to others it may mean the optimal use of insurance (as broadly defined) with which one protects an investment such as a human life [9]. A legal viewpoint is that a legislative decision imposing regulation reflects a societal decision that there are benefits, but that the risks can be excessive [10]; with this interpretation, a decision to support a technology that involves risk should be made a part of the political process.

A broad interpretation of value impact analyses can be obtained by considering the *cost of saving lives* (CSX), which is defined as the cost of preventing or avoiding an extra death before age 65 from any cause or the cost of saving an extra statistical life. Using this definition and estimates of the costs incurred because of *ratchetting* (the proliferation of nuclear reactor safety requirements and practice), for example, Siddall [11] has estimated that ratchetting from 1968 to 1979 may have saved a few statistical lives with a CSX value of about $\$190 \times 10^6$ (in 1979 dollars). As will be seen in Section 17-4, there are indications that many lives can be saved at a CSX value of less than $300,000 per life.

Perhaps the narrowest application of risk-benefit analysis (and one of the most useful) is to perform a comparative risk analysis on two slightly different designs of the same plant. In this way, the risks and costs from each of the two plants can be compared in order to determine the additional cost required to reduce the risk. For nuclear power applications, the methodology reduces to calculating the benefit of a particular design feature in terms of its ability to reduce the annual population radiation exposures due to normal and/or abnormal plant operation. This incremental benefit of reduced risk (Δ man-rem per year) is then balanced against the annualized incremental cost of the design feature (dollars per year) to obtain the cost-benefit ratio (dollars per man-rem) of the feature. Should this ratio be less than the currently accepted guideline of $1000 per man-rem, then the feature should be incorporated into the design. Such a quantitative risk-benefit methodology can be used to assess the radiological impact of normal plant operation on the environment and is required, to some degree, by NRC regulations [12].

A generalized expression for the cost-benefit ratio, which takes into consideration both the predicted frequency and consequences of events, is [13]:

$$\text{Cost-benefit ratio} = \frac{C}{\sum_{i=1}^{n} \mathscr{F}_i R_i - \sum_{i=1}^{n} \mathscr{F}_i' R_i'}, \tag{17-2}$$

where C = annualized cost of safety features, dollars per year; $\mathscr{F}_i$ = frequency (events per year) of the ith accident sequence without the safety feature installed; R_i = radiological consequences in man-rem of the ith accident sequence without the safety feature installed; and n = number of accident sequences upon which the proposed safety feature would have an effect in reducing frequency and/or consequences. The symbols $\mathscr{F}_i'$ and R_i' denote the corresponding frequency and consequence with the safety feature installed. From Eq. (17-2) it follows that the lower the cost-benefit ratio, the more cost-effective is the safety feature.

17-3 COST-BENEFIT ANALYSIS OF PWR ENGINEERED SAFETY FEATURES

The cost-benefit ratio in Eq. (17-2) has been calculated by O'Donnell and Mauro [13] for different key engineered safety features (ESFs) of the pressurized water reactor (PWR) described in the Reactor Safety Study (RSS) and Chapter 12. The base case was considered to be a plant built without an emergency core cooling system (ECCS) or containment. For such a plant, it may be assumed that any loss-of-coolant accident (LOCA) would result in core melting and rapid atmospheric dispersion of the fission product inventory. Thus the consequences of the event would be the same as for those RSS event sequences in which the ECCS and containment fail. In this case, however, the frequency of such severe consequences occurring is the same as the frequency of the initiating LOCA since the probability of failure of the ECCS and the containment is set equal to unity. The risks in man-rem per year for the base case and for the reactor equipped with an ECCS system are evaluated in Table 17-1. Using these values and the ECCS system price of $\$1.5 \times 10^6$ per yr, over the plant life, the ECCS cost-benefit ratio is found to be

$$\begin{aligned}\text{ECCS cost-benefit ratio} &= \frac{C}{\Sigma\mathscr{F}_i R_i - \Sigma\mathscr{F}_i' R_i'} \\ &= \frac{\$1.5 \times 10^6/\text{year}}{1.4 \times 10^5 - 2.5 \times 10^4} \\ &= \$14/\text{man-rem},\end{aligned} \tag{17-3}$$

which is much less than the NRC guideline value of $1000 per man-rem.

Table 17-1 *Risk Calculations for Determining the Cost-Benefit Ratio of the* ECCS[a]

ESF case	Accident event sequence[b]	Frequency, $year^{-1}$	Equivalent RSS consequence sequence[b]	Release category	Radiological consequences, man-rem	Risk, man-rem/year
		$\mathcal{F}_i$			R_i	$\mathcal{F}_i R_i$
No ESFs	A	1×10^{-4}	AB–α	1	8.0×10^{7}	8.0×10^{3}
	S_1	3×10^{-4}	S_1 B–α	1	8.0×10^{7}	2.4×10^{4}
	S_2	1×10^{-3}	S_2 B–α	1	8.0×10^{7}	8.0×10^{4}
	TMLB	3×10^{-4}	TMLB–α	1	8.0×10^{7}	2.4×10^{4}
						$\Sigma \mathcal{F}_i R_i = 1.4 \times 10^{5}$
		$\mathcal{F}_i'$			R_i'	$\mathcal{F}_i' R_i'$
ECCS only	A	1×10^{-4}	A–β	8	4.0×10^{4}	4.0×10^{0}
	S_1	3×10^{-4}	S_1–β	8	4.0×10^{4}	1.2×10^{1}
	S_2	1×10^{-3}	S_2–β	8	4.0×10^{4}	4.0×10^{1}
	AB	1×10^{-7}	AB–α	1	8.0×10^{7}	8.0×10^{0}
	AD	2×10^{-6}	ADC–α	1	8.0×10^{7}	1.6×10^{2}
	AH	1×10^{-6}	AH–α	3	4.4×10^{7}	4.4×10^{1}
	S_1 B	2×10^{-7}	S_1 B–α	1	8.0×10^{7}	1.6×10^{1}
	S_1 D	3×10^{-6}	S_1 DC–α	1	8.0×10^{7}	2.4×10^{2}
	S_1 H	3×10^{-6}	S_2 H–α	3	4.4×10^{7}	1.3×10^{2}
	S_2 B	8×10^{-7}	S_2 B–α	1	8.0×10^{7}	6.4×10^{1}
	S_2 D	9×10^{-6}	S_2 DC–α	1	8.0×10^{7}	7.2×10^{2}
	S_2 H	6×10^{-6}	S_2 H–α	3	4.4×10^{7}	2.6×10^{2}
	TMLB	3×10^{-4}	TMLB–α	1	8.0×10^{7}	2.4×10^{4}
						$\Sigma \mathcal{F}_i' R_i' = 2.5 \times 10^{4}$

[a] From E. P. O'Donnell and J. J. Mauro, *Nucl. Safety* **20,** 525 (1979).
[b] Key to PWR accident sequence symbols is in Table 12-6.

The frequencies $\mathscr{F}_i$ in Table 17-1 of the various accident sequences of interest were obtained from the RSS by using median estimates for accident sequence frequencies. The fractions of core fission products released for each accident were classified, in the manner of RSS, into nine release categories. The categories range from category 1, corresponding to a core melt condition with rapid, direct atmospheric dispersion (i.e., without effective ECCS or containment) to category 9, corresponding to no core melt with effective containment (i.e., ECCS and containment function as designed).

The radiological consequences (R_i) of each accident sequence of interest in Table 17-1 were calculated in terms of total integrated whole-body dose to an exposed population (man-rem), assuming a uniform population density of 400 persons per square mile surrounding the site [13]. This value, consistent with NRC guidelines with respect to population density [14], is typical of many existing nuclear plant sites, but differs from the population model considered in the RSS.

For the ECCS, it was assumed that a residual heat removal system would be provided for normal plant shutdown. Thus the ECCS annual costs of 1.5×10^6 are those associated with the additional equipment (high-pressure safety injection system and accumulators) required to perform the ECCS function. The annual costs were based on estimates for typical PWR plants in 1978 dollars with 8% interest over 40 years.

Besides the ECCS system, a cost-benefit ratio has been obtained for the containment system (including associated heat and fission product removal systems), and the emergency on-site power system, consisting of diesel–generator (DG) sets [13]. In each case, the costs include only the incremental cost of providing the ESF function with respect to equipment or structures that would be expected to be provided for normal plant operation.

The additional annual cost of a full-pressure-retaining containment structure and associated systems over the cost of a conventional-type power-plant structure housing the reactor coolant system was estimated to be $\$3.0 \times 10^6$ [13]. Included in this figure were expenses for a hydrogen recombiner system; although the PWR analyzed in the RSS was not equipped with a hydrogen recombiner system, NRC regulations now require such a system. (It was assumed that the hydrogen recombiner system would be capable of eliminating entirely the risk of those accident sequences in which the containment failed because of hydrogen-related overpressure.) The annual emergency diesel-generator system costs of $\$2.0 \times 10^6$ were based on replacing a small diesel generator used for plant equipment protection with two redundant full-capacity diesel generators capable of supplying the ESF loads and housed in a separate building [13].

Table 17-2 *Cost-Benefit Ratios for Engineered Safety Features Applied in Various Sequences*[a]

ESF sequence	Sequence of ESF application 1	2	3
1	DG	ECCS	Containment
Risk reduction, Δ man-rem/year	2.4×10^4	1.1×10^5	1.4×10^3
Cost–benefit ratio, $/man-rem	83	14	2083
2	DG	Containment	ECCS
Risk reduction, Δ man-rem/year	2.4×10^4	1.1×10^5	4.8×10^2
Cost–benefit ratio, $/man-rem	83	27	3125
3	ECCS	DG	Containment
Risk reduction, Δ man-rem/year	1.1×10^5	2.4×10^4	1.4×10^3
Cost–benefit ratio, $/man-rem	14	85	2083
4	ECCS	Containment	DG
Risk reduction, Δ man-rem/year	1.1×10^5	6.8×10^3	1.8×10^4
Cost–benefit ratio, $/man-rem	14	441	111
5	Containment	DG	ECCS
Risk reduction, Δ man-rem/year	1.2×10^5	1.8×10^4	4.8×10^2
Cost–benefit ratio, $/man-rem	25	111	3125
6	Containment	ECCS	DG
Risk reduction, Δ man-rem/year	1.2×10^5	4.0×10^2	1.8×10^4
Cost–benefit ratio, $/man-rem	25	3750	111

[a] From E. P. O'Donnell and J. J. Mauro, *Nucl. Safety* **20,** 525 (1979).

Table 17-2 presents the cost-benefit ratios for the ECCS, containment, and DG sets applied in various sequences. For example, the addition of the ECCS system to the basic reactor with no ECCS nor containment is illustrated as step 1 in either sequence 3 or 4 in Table 17-2. The risk reduction factors shown are the values of $(\Sigma\mathscr{F}_iR_i - \Sigma\mathscr{F}'_iR'_i)$ and correspond to the risk before and after the next ESF is installed. Thus, if the diesel generator sets are installed after the ECCS, as in ESF sequence 3, the incremental risk reduction over the plant with only the ECCS is 2.4×10^4 man-rem/yr, which corresponds to a cost-benefit ratio of $85/man-rem.

Table 17-2 shows that the cost-benefit ratio for any particular ESF is highly dependent on the sequence in which it is applied in the risk assessment. When considered first, the cost-benefit ratio for each of the three ESFs is well below $1000 per man-rem, indicating that, individually, the cost of these features would be well justified in the absence of any other ESFs. The ESF sequence 4 shows that the sequential additions of the ECCS, containment, and emergency power systems all lead to cost-benefit ratios less than $1000 per man-rem. Hence one can conclude that the regulations requiring such systems are not inconsistent with the guideline value of $1000 per man-rem.

It is interesting to note that if the hydrogen recombiner system is added after the three systems in Table 17-2, the incremental risk reduction is calculated to be less than 0.13 man-rem/yr for an annual cost of $\$4 \times 10^4$, giving a cost-benefit ratio for the recombiner system of more than $\$3 \times 10^5$/man-rem [13]. This conclusion may seen unwarranted in view of the role played by hydrogen recombiners in the Three Mile Island accident; however, a number of factors indicate that the actual risk of serious population exposures due to hydrogen-related containment failure in that event would not have been great even if recombiners had not been installed prior to the accident, provided portable ones could be obtained on short notice [13].

Finally, caution should be exercised when drawing conclusions from a methodology like that leading to Table 17-2. For example, if any ESF sequence other than no. 4 had been selected, it might have been concluded that one or more of the safety systems might not be justified because of an excessively high cost-benefit ratio. The methodology, however, does provide insight into the relative cost effectiveness of existing ESFs and the manner in which they contribute to reducing accident risk. Of course, it must be remembered that the assumed population distribution could differ significantly from actual site conditions; also, there may be other monetary costs or benefits, such as risk of property damage or plant outage, that also should have been included [13].

17-4 COST-BENEFIT ANALYSES OF VARIOUS HEALTH AND SAFETY MEASURES

Table 17-3 presents a comparison of cost-benefit values, expressed in terms of investment costs necessary to achieve a reduction in mortality risk, for nuclear and coal power plant design features, and various nonnuclear health and safety protective measures [13]; an extensive listing also has been published by Cohen [15]. The table shows that nuclear plant regulatory policy generally results in a considerably higher investment to achieve reductions in public mortality risk than for other activities. With the exception of the ECCS, all other nuclear plant design features have cost-benefit ratios of $1 million or more per life saved, with the hydrogen recombiners having a ratio in excess of $3 billion per life saved.

Table 17-3 is useful because it demonstrates the lack of consistency in health and safety policy among different agencies of the Federal Govern-

Table 17-3 *Cost-Benefit Ratios for Various Health and Safety Measures*[a]

	Cost–benefit ratio, \$10^6/life saved
Nuclear power-plant design features	
Radwaste effluent treatment systems	10
ECCS	0.1
Containment	4
DG sets	1
Hydrogen recombiners	>3000
Coal-fired power plant design features	
High-sulfur coal with SO_2 scrubbers, 85% removal	0.1–1.4
Low-sulfur coal with SO_2 scrubbers, 85% removal	0.7–10
Occupational health and safety	
OSHA[b] coke fume regulations	4.5
OSHA benzene regulations	300
Environmental protection	
EPA[c] vinyl chloride regulations	4
Proposed EPA drinking water regulations	2.5
Fire protection	
Proposed CPSC[d] upholstered furniture flammability standards	0.5
Smoke detectors	0.05–0.08
Automotive and highway safety	
Highway safety programs	0.14
Auto safety improvements, 1966–1970	0.13
Air bags	0.32
Seat belts	0.08
Medical and health programs	
Kidney dialysis treatment units	0.2
Mobile cardiac emergency treatment units	0.03
Cancer screening programs	0.01–0.08

[a] From E. P. O'Donnell and J. J. Mauro, *Nucl. Safety* **20,** 525 (1979).
[b] OSHA, Occupational Safety and Health Administration.
[c] EPA, Environmental Protection Agency.
[d] CPSC, Consumer Product Safety Commission.

ment. It appears, for example, that the Nuclear Regulatory Commission, Occupational Safety and Health Administration, and the Environmental Protection Agency set much higher standards for safety than do other agencies. This suggests that the regulatory emphasis on further reducing nuclear plant risks may not be justified in view of the availability of more cost-effective means of reducing risks that are not being fully pursued in other areas of society. Indeed, it would seem that money could be invested more effectively in emergency cardiac treatment units or cancer screening programs.

EXERCISES

17-1 Describe a technique for making a cost-benefit comparison for two similar plant designs, and illustrate the results for various Engineered Safety Feature systems of a PWR. A length of about 300 words, exclusive of tables, is preferred.

17-2 Compare the results of a cost-benefit analysis of various health and safety measures, as obtained by B. L. Cohen, *Health Phys.* **38,** 33 (1980). Your exposition should be aimed at the layperson who has little knowledge of risks in society. A length of about 300 words, exclusive of tables, is preferred.

REFERENCES

1. C. Starr, R. Rudman, and C. Whipple, Philosophical basis for risk analysis, *Ann. Rev. Energy* **1,** 629 (1976).
2. W. D. Rowe, "An Anatomy of Risk." Wiley, New York, 1977.
3. H. J. Otway, J. Linnerooth, and F. Niehaus, On the social aspects of risk assessment, *in* "Nuclear Systems Reliability Engineering and Risk Assessment" (J. B. Fussell and G. R. Burdick, eds.), p. 56. Soc. Industrial Applied Math., Philadelphia, Pennsylvania, 1977.
4. D. Okrent, Risk-benefit evaluation for large technological systems, *Nucl. Safety* **20,** 148 (1979).
5. L. A. Sagan, Human costs of nuclear power, *Science* **177,** 487 (1972).
6. National Radiological Protection Board, "The Application of Cost Benefit Analysis to the Radiological Protection of the Public." Her Majesty's Stationery Office, London, 1980.
7. D. Okrent, A General Evaluation Approach to Risk-Benefit for Large Technological Systems and its Application to Nuclear Power. Univ. of California at Los Angeles Rep. UCLA-ENG-7777 (December 1977).
8. J. Hirshleifer, The economic approach to risk-benefit analysis, in Risk-Benefit Methodology and Applications: Some Papers Presented at the Engineering Foundation Workshop, September 22–26, 1975, Asilomar, California (D. Okrent, ed.), UCLA-ENG-7598, p. 141 (December 1975).
9. J. M. Marshall, Optimum safety and production of information when risks are insured, in Risk-Benefit Methodology and Applications: Some Papers Presented at the Engineering Foundation Workshop, September 22–26, 1975, Asilomar, California (D. Okrent, ed.), UCLA-ENG-7598, p. 243, (December 1975).

10. H. P. Green, The risk-benefit calculus in safety determinations, *George Washington Law Rev.* **43,** 791 (1975); P. Handler, A rebuttal: The need for a sufficient scientific base for government regulation, *ibid.* **43,** 808 (1975).
11. E. Siddall, Control of spending on nuclear safety, *Nucl. Safety* **21,** 451 (1980).
12. Code of Federal Regulations Title 10, Energy, Part 50, Licensing of Production and Utilization Facilities, Appendix I, Numerical Guidelines for Design Objectives and Limiting Conditions for Operation to Meet the Criterion "As Low As Practicable" for Radioactive Materials in Light-Water-Cooled Nuclear Power Reactor Effluents, Revised (January 1, 1978).
13. E. P. O'Donnell and J. J. Mauro, A cost-benefit comparison of nuclear and nonnuclear health and safety protective measures and regulations, *Nucl. Safety* **20,** 525 (1979).
14. The Site Population Factor: A Technique for Consideration of Population in Site Comparison. U.S. Atomic Energy Commission Rep. WASH-1235 (October 1974).
15. B. L. Cohen, Society's valuation of life saving in radiation protection and other contexts, *Health Phys.* **38,** 33 (1980).

CHAPTER 18

Risk Acceptance

18-1 FACTORS AFFECTING RISK ACCEPTANCE

The acceptance of risk by individuals, and collectively by society, is affected by many parameters [1]. Some of the factors influencing acceptance are presented in Table 18-1. The distinction between voluntary and involuntary activities was the first to be studied extensively [2], and is probably the single most important factor.

Nuclear power has the disadvantage of many factors working against its acceptance. Although the risks may be willingly assumed by some people, many others consider a nearby nuclear reactor an involuntary risk. Cancer induced as a result of low-level radiation exposure is a delayed effect, and the possibility of genetic damages to future generations from such exposure is certainly an irreversible consequence; these factors are also highly correlated with "dread" risk. Likewise, because there are other conventional technologies available for generating electricity, exposure to risks from nuclear power is considered unnecessary by some individuals.

Another of the factors influencing the general public's acceptance of nuclear power and other advanced technologies is that the risks are not known or understood, although the studies cited in Chapters 12 through 16 are available at least to scientists and engineers. It is certainly true, however, that a probabilistic risk analysis can never yield results for "risk known with certainty."

A more elaborate breakdown of factors by which individual, group, and societal value judgments are made is given in Table 18-2 [3]. Voluntary versus involuntary risks are broken down in terms of factors already considered in Table 18-1; discounting in time involves the latency of consequences and the effects of time on risk values. The geographic distribution of risk requires consideration of whether the risk is distributed to individuals or groups, and it is a factor that cannot be overlooked.

Table 18-1 *Factors Affecting Risk Acceptance*[a]

Effect	Opposite Effect
Risk assumed voluntarily	Risk incurred involuntarily
Effect immediate	Effect delayed
No alternatives available	Many alternatives available
Risk known with certainty	Risk not known
Exposure is an essential	Exposure is a luxury
Encountered occupationally	Encountered nonoccupationally
Common hazard	"Dread" hazard
Affects average people	Affects especially sensitive people
Will be used as intended	Likely to be misused
Consequences reversible	Consequences irreversible

[a] From W. W. Lowrance, "Of Acceptable Risk." Kaufman, Los Altos, California, 1976. Copyright © 1976; all rights reserved.

Factors involving the nature of consequences in Table 18-2 include such diverse considerations as the values attached to premature death and aesthetics. Finally, there are other considerations, such as the fact that different individuals have different propensities for risk taking.

18-2 STATISTICAL RISK ACCEPTANCE ANALYSES

The normal United States death rate from disease (about one death per year per 100 people) serves as a reference for the highest level of involuntary accepted risk; the lowest level for involuntary risks is set by the risk of death from natural events—lightning, flood, earthquakes, insect and snake bites, etc. (about 1 death per year per 1 million people). Between these two extremes, the public is apparently willing to accept involuntary exposure in relation to the benefits derived [2, 4].

The survey by Otway and Erdmann [5], which covers risks in this general range and below, has produced the results shown in Table 18-3. Hazards with a death probability on the order of 10^{-6} per person per year were seen to be of little concern to most people. Considering this and other factors, Otway and Erdmann suggested the use of 10^{-7} per person per year as an upper bound for acceptable risk of death to an individual from nuclear plant accidents.

Starr [2, 6] used a *revealed preference* approach to study voluntary risk acceptance. The basis of this approach is that statistical data for various risks reveal society's degree of acceptance of the risks for each activity;

Table 18-2 *Factors Affecting the Value of Consequences*[a]

- I. Factors involving types of consequence
  - A. Voluntary and involuntary risks
    1. Equity and inequity
    2. Degree of knowledge
    3. Avoidability and alternatives
    4. Imposition—exogenous and endogenous
  - B. Discounting in time
  - C. Spatial distribution and discounting of risks
    1. Geographic distribution of risk
    2. Identification of risk agents
    3. Spreading of risk
  - D. Controllability of risk
    1. Perceived degree of control
    2. Systemic control of risk
    3. Crisis management
- II. Factors involving the nature of consequences
  - A. Hierarchy of need fulfillment
  - B. Variation in cultural values
  - C. Common versus catastrophic risks
  - D. National defense
  - E. Natural versus man-originated events
  - F. Knowledge as a risk
- III. Other factors
  - A. Factors involving the magnitude of frequency of occurrence of a consequence
    1. Low-frequency levels and thresholds
    2. Spatial distribution of risks and high-frequency levels
  - B. Situational factors
    1. Surprise and dissonant behavior
    2. Lifesaving systems
  - C. Propensity for risk taking
    1. Individual
    2. Group
    3. Conflict avoidance

[a] From W. D. Rowe, "An Anatomy of Risk." Wiley, New York, 1977.

that is, society is assumed to arrive by trial and error at a nearly optimal balance between risks and benefits for any activity. To the extent the revealed preference model is valid (i.e., equilibrium has been achieved), then the factors affecting risks are the same as those in Tables 18-1 and 18-2 influencing risk acceptance.

Starr's approach has the advantage of dealing with public behavior rather than with attitudes, and thus would seem to provide an appealing solution to the problem of determining acceptable levels of risk. One of the interesting results Starr observed is that there is a general benefit-risk

Table 18-3 *Acceptance of Annual Fatality Risk Levels*[a]

Annual fatality risk level, yr^{-1}	Conclusion
10^{-3}	This level is unacceptable to everyone
	Accidents providing hazard at this level are difficult to find
	When risk approaches this level, immediate action is taken to reduce the hazard
10^{-4}	People are willing to spend public money to control a hazard (traffic signs/control and fire departments)
	Safety slogans popularized for accidents in this category show an element of fear, i.e., "the life you save may be your own"
10^{-5}	People still recognize
	People warn children about these hazards (drowning, firearms, poisoning)
	People accept inconvenience to avoid, such as avoiding air travel
	Safety slogans have precautionary ring: "never swim alone," "never point a gun," "never leave medicine within a child's reach"
10^{-6}	Not of great concern to average person
	People aware of these accidents but feel that they can't happen to them
	Phrases associated with these hazards have element of resignation: "lightning never strikes twice," " an act of God"

[a] Extracted from H. J. Otway and R. C. Erdmann, *Nucl. Eng. Design* **13**, 365 (1970).

pattern for involuntary risks, as depicted in Fig. 18-1, in which the per capita benefits are compared to fatalities per unit time. Starr also argued that there is a risk multiplier of three orders of magnitude difference between voluntary and involuntary exposure for equivalent benefits. This multiplier is shown in Fig. 18-2, where the values for benefits B are the dollars the average involved individual spent on the activity, for the case of voluntary exposure, and the contribution of the activity to the individual's annual income, for involuntary exposure. The shape of the curve, which relates the risk R to benefit B, $R \sim B^3$, was obtained by examining the special case of miners exposed to high occupational risks.

The preliminary results of Starr have been examined in more detail by Otway and Cohen [7], who found that the methodology and results are "excessively sensitive to the assumptions made and the handling of data and that the existence of simple mathematical relationships, based upon the revealed preferences method, is unlikely." After performing a regression analysis on the same data base, they concluded that for voluntary risks in society, risk is proportional to the 1.8 power of benefit, as opposed to Starr's third power; for involuntary risks in society, the relationship between risk and benefit follows nearly a sixth power of proportionality (see Fig. 18-3). Rowe [3] then reassessed the whole subject of a relation-

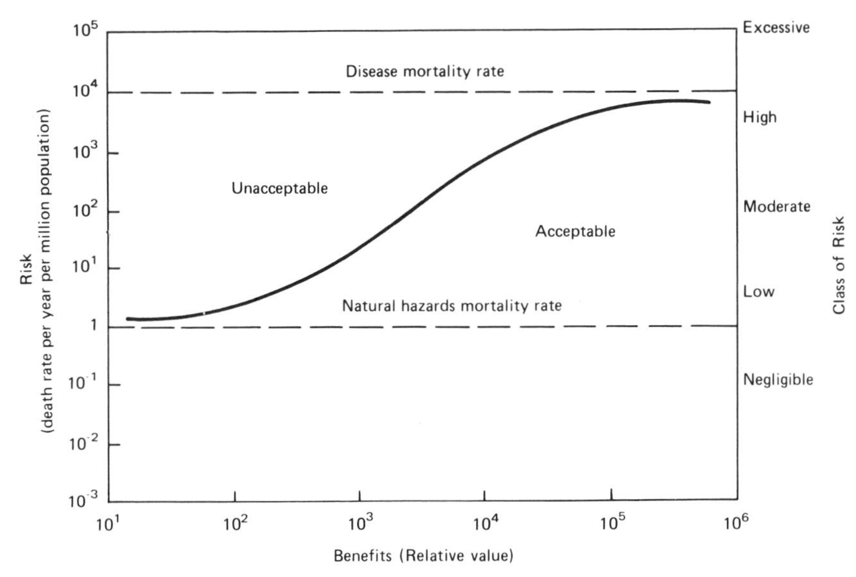

Fig. 18-1 *Benefit-risk pattern of involuntary exposure. (Reproduced with permission from the Annual Review of Energy,* **1**, *629. Copyright © 1976 by Annual Reviews, Inc.)*

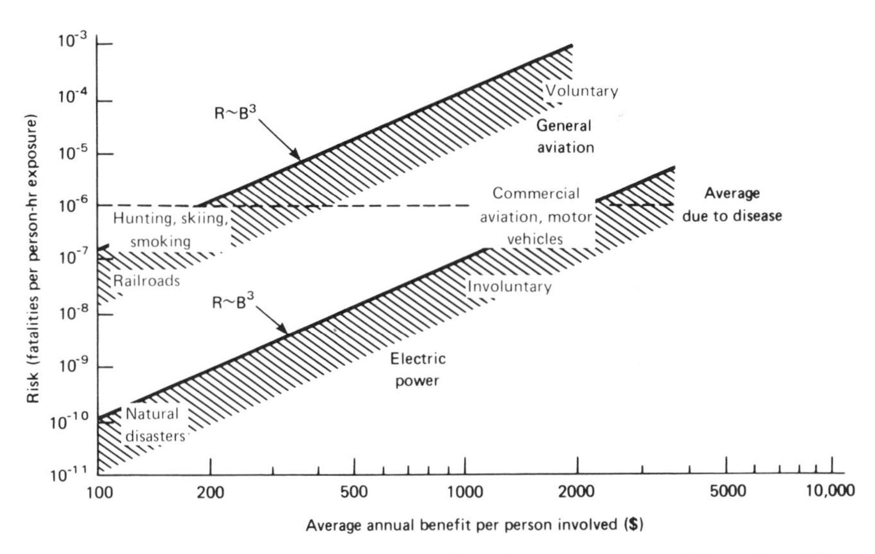

Fig. 18-2 *Risk versus benefit: voluntary and involuntary exposure. (Reproduced from C. Starr in* Perspectives on Benefit-Risk Decision Making, *1972, with the permission of the National Academy of Sciences, Washington, D.C.)*

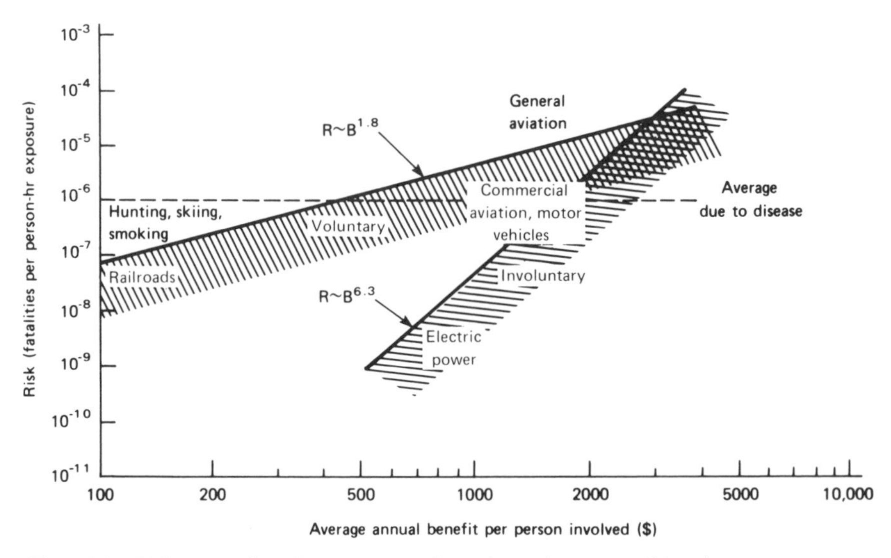

Fig. 18-3 *Risk versus benefit: regression lines for voluntary and involuntary exposure. (From H. J. Otway and J. J. Cohen, Rep. II* ASA-RM-*75-5, March 1975.)*

ship between risks and benefits and used Starr's data to develop a set of interrelationships among statistical risk factors based on differences in voluntary and involuntary risks. From this, Rowe ultimately concluded that $R \sim B$ for involuntary risks and $R \sim B^{0.5}$ for (regulated) voluntary risks.

There are other drawbacks to the revealed preference approach, however, besides the fact that different investigators can use the same data to reach different conclusions. First, the approach assumes that past behavior is a valid indicator of present preferences [8]. Second, the approach "does not serve to distinguish what is 'best' for society from what is 'traditionally acceptable' " [2]; furthermore, what is traditionally acceptable in the marketplace may not accurately reflect the public's safety preferences. A revealed preference approach also assumes that people not only have full information, but also can use that information optimally, an assumption that seems quite doubtful in the light of recent research into the psychology of decision making [9].

Another interesting approach to the analysis of risk acceptance is to examine historical records to see how risks from new technologies were accepted in the past. For example, societal attitudes in the nineteenth century toward the risks presented by the new technology of railway travel in the United Kingdom were similar to present-day attitudes toward nuclear power [10]. On the basis of this study, it was surmised that:

1. Arguments based on "comparability of risk" are unlikely to provide an effective means of gaining acceptance of nuclear power in the United Kingdom.
2. Some positive evidence that safety improvements have been made would improve the prospects of gaining acceptance.

18-3 PSYCHOMETRIC RISK ACCEPTANCE ANALYSES

An alternative approach to the revealed preference method employs questionnaires and psychometric surveys to measure the public's attitude toward the risks and benefits from various activities [8]. This general method of *expressed preferences* has been proposed because attitudes expressed in surveys often correlate quite well with behavior; furthermore, such surveys have the advantage of eliciting present values rather than historical preferences.

Figure 18-4 shows a psychometric analysis of some of the factors in Table 18-1 that affect risk [8]. Here the various factors are compared for nuclear and nonnuclear electric power, as measured on an arbitrary scale from one to seven, and as obtained from an admittedly small and unstatis-

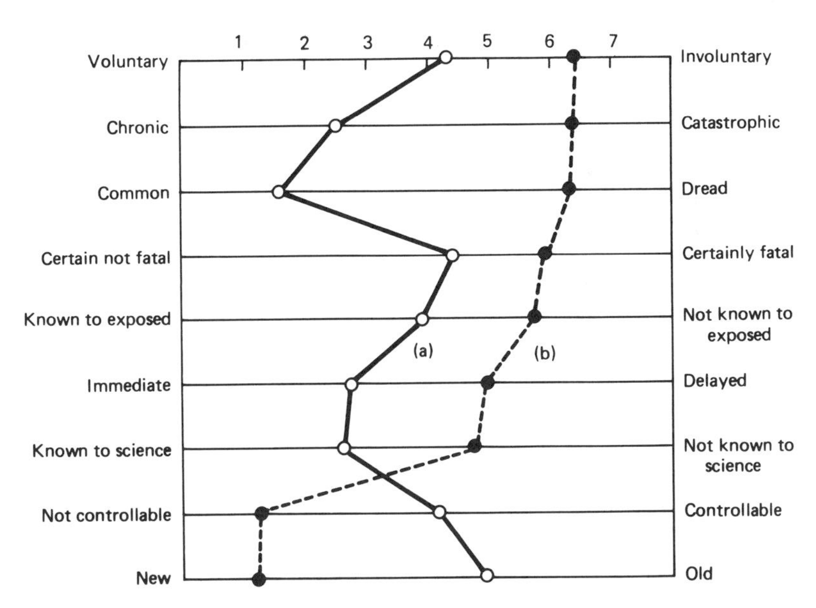

Fig. 18-4 *Psychometric analysis of factors affecting risk acceptance of* (*a*) *nonnuclear electric power and* (*b*) *nuclear power.* [*From B. Fischhoff, P. Slovic, S. Lichtenstein, S. Read, and B. Combs,* Policy Sciences *9, 127* (*1978*).]

tical sample of people. It may be noted that the "common-versus-dread" and "new-versus-old" factors are particularly important for the acceptance of nuclear power as compared to nonnuclear electric power.

An attitude model also has been applied for the psychometric analysis of risk acceptance [11]. Public responses to nuclear energy were characterized by negative responses to two of four attributes, the psychological attribute of dread or "anxiety" and a sociopolitical implications attribute; the degree of negativism differed dramatically for those who tended to favor nuclear power and those opposed to it, however. There was a positive response to the third attribute of economic benefit, with those in favor of nuclear power attaching much more importance than those against it. It was found that for the attribute of environmental and physical safety there was a very significant difference between the two groups, where the supporters of nuclear power had a positive attitude and those against nuclear power had the opposite attitude.

The answer to whether the revealed-preference approach of Section 18-2 or the expressed-preference approach is best depends upon one's conceptualization of the policy-making process. A definitive revealed-preference study would be an adequate guide to action only if one believes that rational decision making is best performed by experts who formalize past policies as prescriptions for future action; a definitive expressed-preference study would be an adequate guide only if one believes that people's present opinions should be society's final arbiter and that people act on their expressed preferences [8].

18-4 PERCEPTION OF RISKS

It has been pointed out the psychometric studies may help provide an indication of risk acceptance by an individual or, collectively, by society. A concern, however, is that those individuals who are queried about their risk aversion preferences may not be able to perceive the true nature of a postulated risk. Some of the factors affecting risk perception are [4, 12, 13]:

1. The manageability or controllability,
2. The probable size of an accident, which tends to be considered more than the frequency,
3. The probable severity of injury, if an accident were to occur,
4. The vividness with which the consequences can be pictured,
5. The publicity surrounding the risk,
6. The discounting of future risk,

7. The difficulty of updating risk judgments, once further valid data becomes available,
8. The impreciseness of intuitive estimates, where the tendency is to overestimate the reliability resulting from a small number of observations.

It also has been observed that factors influencing risk perception are broader than just the characteristics of the risk [14]. For example, during the 1960s and early 1970s, substantial negative changes occurred in the American public's attitudes toward societal groups like the military and the nuclear power industry. Some of these changes were an outgrowth of concern over the technologies that the groups were using; alarm over one technology of a given agency or group tended, in turn, to bring quick suspicion of other technologies, including the nuclear power industry [14].

In one survey, it has been concluded that perceived risk and perceived benefit cannot be independently evaluated [8]. This was shown by asking subjects to judge "the socially acceptable level of risk": those who first evaluated benefits consistently reported higher levels of risk acceptance than did subjects who first evaluated risk.

One acute problem of risk perception is the public's poor intuitive estimation of very small risks. A good illustration of this is shown in Fig. 18-5, where we can see that when people were asked to estimate the annual number of fatalities from specific causes, the perceived risk was found to vary by a factor of about 10^4, while the corresponding actual range of values is closer to a factor of 10^6 [15]. This observation suggests that an intuitive evaluation of risk is less sensitive to changes in probabilities than is a quantitative risk analysis.

The public's perception of very small probabilities has been postulated to be quite nonlinear [16]. The intuitive curve shown in Fig. 18-6 accounts for perception of a "very low" probability that is, indeed, "negligible," and also accounts for the perceived higher value for low-probability events seen in Fig. 18-5. If the curve in Fig. 18-6 is actually true, it helps to explain the controversy over catastrophic risks from nuclear power applications where high-consequence, low-frequency risks may be deemed more important because their frequencies are subjectively overestimated by some members of the public and deemed negligible by others.

The analysis of risk perception is a research area that is still undergoing rapid development. In one recent attempt to quantify perception, a mathematical formulation has been proposed to separate an individual's attitude toward a certain consequence, such as loss of life, into an adversion index under certainty and an attitude toward uncertainty [17]. For the latter factor, an individual decision maker who considers an almost certain loss a sure loss, but an almost certain gain not a sure gain, is classified

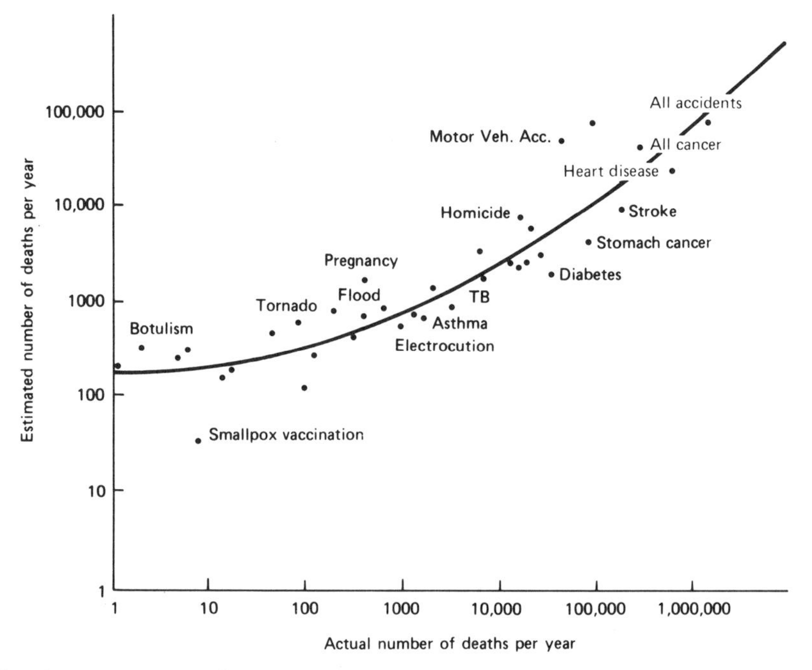

Fig. 18-5 *Comparison of perceived risk to actual risk. [From B. Fischhoff, C. Hohenemser, R. Kasperson, and R. Kates,* Environment *20, No. 7, p. 16 (September 1978).]*

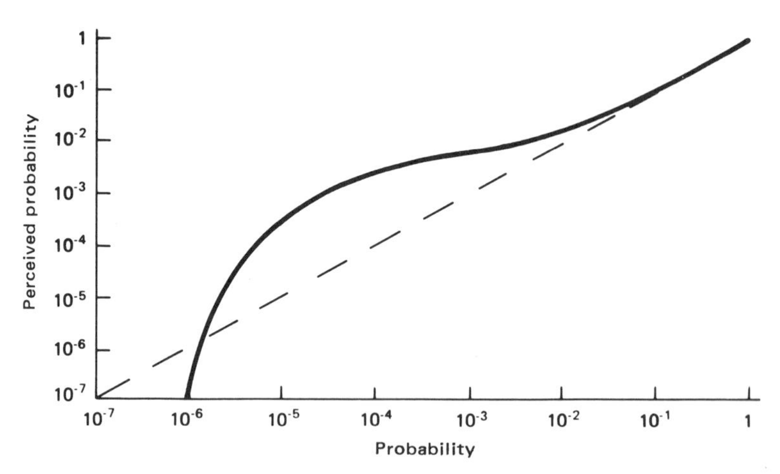

Fig. 18-6 *Postulated perception of probability. [From C. Starr and C. Whipple,* Science *208, 1114–1119 (6 June 1980). Copyright ©1980, by the American Association for the Advancement of Science.]*

as pessimistic; an optimistic decision maker considers an almost certain gain as a sure gain, but an almost certain loss not a sure loss. These attitudes are uncoupled from the adversion to the consequence itself, which has been analyzed using a probability distribution function for the damages from a particular consequence [17].

18-5 CRITERIA FOR RISK ACCEPTANCE

The management of risk is a part of risk acceptance: it is the stage in which a response is made to unacceptable risks that have been identified. To manage risk, it is appropriate to see if criteria for its acceptance can be developed. McDonald and Temme suggest that it is an insurmountable task to perform an objective mathematical convolution of risks, benefits, and values to determine "societal risk acceptance criteria," since the results always will be inconsistent with the values of some segment of society [18]. Hence they concluded that it is not productive for risk analysts to aim toward the definition of such criteria, but instead "maximum risk criteria" should be developed; the difference is that there is no implication of societal acceptance in the latter term.

There is a body of opinion among social scientists, with some supporting evidence, that the only way to achieve societal acceptance of risk is by participation of divergent elements in an open and decentralized decision-making process and that this can be done without criteria for acceptable levels of risk; from the point of view of the nuclear power industry, however, there are strong incentives to develop quantitative risk criteria since, in some cases, without such criteria a goal of zero risk has been established [19]. Zero risk ignores the trade-off between risk and benefit, and ignores the difficulty or impossibility of reaching zero risk [20].

The need for realistic guidelines for acceptable levels of risk perhaps is greatest for federal regulatory agencies given a congressional mandate to protect the public from "unreasonable risk," but no guidance of how to judge reasonableness. Such mandates create the potential for criticisms of the regulators, both by those working in the regulated industry and by those "intervenors" who wish to ensure that the regulators are doing their job. One way to reduce the influence of bias and arbitrariness is to substitute for variable judgments of regulators a numerical definition of "reasonable risk."

A possible by-product of establishing risk criteria is that the nuclear industry is given an incentive to develop methods of controlling risk. This approach is preferable to one in which a regulatory authority specifies the technology for meeting risk targets rather than the targets themselves [16].

Several quantitative goals have been suggested for acceptance levels of risk. The "limit line" of Farmer [21] was the first to be introduced and consisted of a proposed release criterion for the equivalent ground-level release of ^{131}I. In this approach, the risk is acceptable provided the risk curve lies entirely to the left of the limit line in Fig. 18-7 for frequency versus damage. The line was developed by selecting an interval of 1000 reactor-years for a release of a few thousand curies of ^{131}I, and by the slope of -1.5 on the log–log scale; the transition in the slope in the range of 10 to 1000 curies served to minimize the frequency of small releases, largely on the basis of their nuisance value [21]. If one were to modify the Farmer line approach to take into account risks *and benefits,* then one might consider a family of lines in the frequency-versus-damage plane, each corresponding to a different benefit [22].

In another effort to take credit for the benefits as well as the risks, Okrent and Whipple categorized major facilities or technologies as essential, beneficial, or peripheral to society, and proposed a decreasing level of acceptance-risk to the most exposed individual [23]. The values selected for the individual risks of death were 2×10^{-4} per year, 10^{-5} per year and 2×10^{-6} per year, respectively, at a 90% confidence level. The value of 10^{-5} per year for the probability of dying from any hazard that carries no benefit or is easily avoided also was suggested by Comar [24].

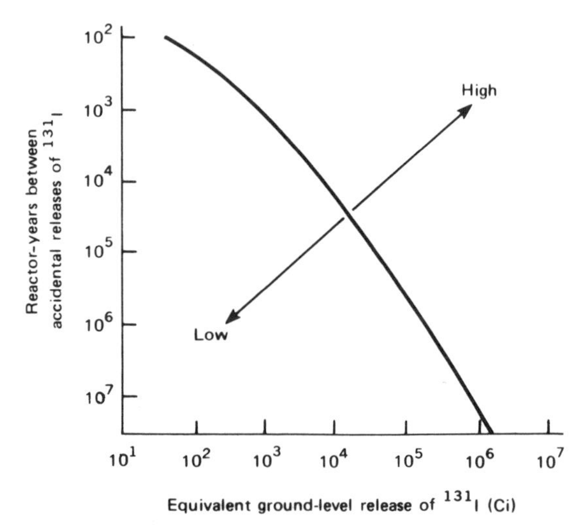

Fig. 18-7 *Proposed release criterion of Farmer. [From F. R. Farmer,* Nucl. Safety *8, 539 (1967).]*

It is important to note that any quantitative risk criterion appears quite attractive when there is disagreement over the *level* of risk rather than the *value assigned* to that level. In cases where values are assigned to the benefits of technology or human lives (as discussed in Section 17-1), however, there must be a social consensus on the relative benefits and costs; otherwise the quantitative approach to risk analysis will break down, leaving only the intuitive risk assessment approach as a viable method [16].

18-6 PATHWAYS TOWARD RISK ACCEPTANCE

Public response to x-rays provides some clue for the acceptance of risks from nuclear power. Widespread acceptance of x-rays shows that a radiation technology can be tolerated once its use becomes familiar, its benefits clear, and its practitioners trusted; the slow path by which nuclear power might gain better acceptance requires "an incontrovertible long-term safety record, a responsible agency that is rejected and trusted, and a clear appreciation of benefit" [25].

A major problem for the acceptance of risk is associated with communication. Communication is required from nuclear power proponents to the general public and its individuals, and vice versa. In the communication to the public, three difficulties have seemed to exist [26]: (1) the problem of communicating scientific and technical knowledge, (2) the problem of transparency of the process of decision making, and (3) the problem of convincing the public of the necessity of nuclear power.

Probably the most important variables for communicating to the public are credibility of the communicator and the discrepancy between the message and an individual's initial attitude [27]. The first characteristic obviously is a necessary attribute, but the second effect is less well known. People have a "tolerance band," about their initial attitude, in which they are willing to accept new input information; furthermore, extreme messages falling outside the tolerance band of a neutrally oriented person tend to have an effect opposite to that intended [27].

One of the objectives for improved communication in the next few years should be to increase the public's appreciation of the risks from hazards other than nuclear radiation. These hazards include (see Section 16-5) cigarette smoke, NO_x from automobile exhausts, low-level concentrations of chemicals in water and foods, and other environmental effects such as electromagnetic radiation and noise. Once the general public becomes better informed about these other hazards, perhaps the disproportionate concerns over the risks from nuclear power will diminish.

For all such hazards, correlations of low-level exposure with consequence are difficult to measure and can only be composites of individuals' average responses. At very low exposures, effects may be only marginally detectable and the presence of the offending agent may be difficult to gauge. Frequently, an impossibly large number of people would have to be examined to obtain results with statistical significance; such uncertainties become a part of *trans-science* [28], that part of science that cannot be quantified without an unattainably large expenditure of capital and/or effort. On the other hand, at the opposite extreme where responses to high exposure are sought, the correlation may remain in doubt because accidental exposures are rare and ethical considerations prohibit deliberately exposing human beings to large doses of the hazard; also, extrapolating animal test findings to estimate human experience is unavoidably imprecise [1].

The problem of improving the correlation of consequences with very low doses is especially important for nuclear radiation since not only can the uncertainties in nuclear risk analyses be reduced, but the communication of the meaning of these risks to the public can be improved. Figure

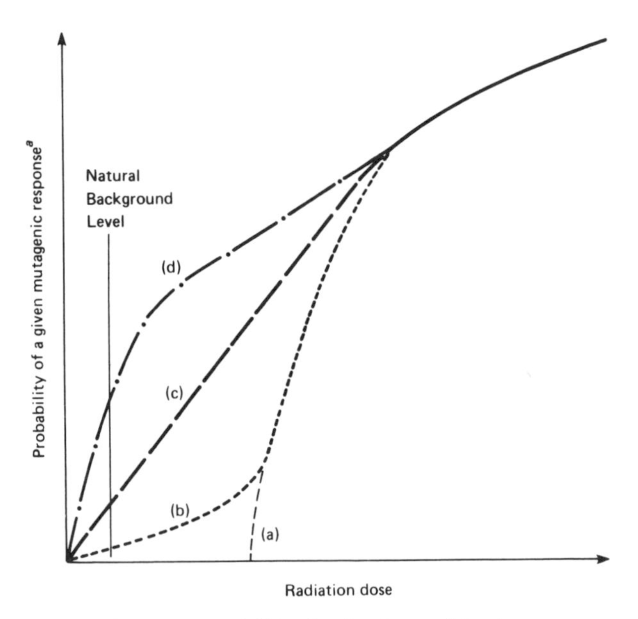

Fig. 18-8 *Schematic of various possibilities for the shape of the dose-response curve. (Reproduced from* Risks Associated with Nuclear Power: A Critical Review of the Literature, Summary and Synthesis Chapter, 1979, *with permission of the National Academy of Sciences, Washington, D.C.)*

[a] For example, the increase, produced by irradiation, in the probability of initiation of a cancer.

18-8 shows four possible extrapolations below the region of the dose-response curve where the response (solid line) is roughly known from observations [29]. The threshold theory [curve (a) in the figure] has gradually become less accepted; curve (b) generally conforms to the conclusions of the International Commission on Radiological Protection. The linear dose-response curve (c) is that now selected by the Biological Effects of Ionizing Radiation (BEIR) Committee, and curve (d) is approximately that of several highly disputed studies that attribute high rates of cancer incidence to very low levels of radiation, including background.

EXERCISES

18-1 Discuss the factors affecting risk acceptance as they pertain to the following sources of energy: (a) nuclear, (b) coal, and (c) solar thermal.

18-2 Describe the differences between statistical and psychometric risk acceptance analyses, and cite example results of each.

18-3 Provide arguments as to why the intuitive perception of deaths per year, illustrated in Fig. 18-5, differs from the statistical records.

18-4 Give arguments in favor of and against the use of a quantitative risk criterion for the nuclear power industry.

18-5 Explain the difficulties in obtaining a dose–response curve for exposures to air pollution (SO_2 and NO_x emissions).

REFERENCES

1. W. W. Lowrance, "Of Acceptable Risk." Kaufman, Los Altos, California, 1976.
2. C. Starr, Social benefit versus technological risk, *Science* **165,** 1232 (1969).
3. W. D. Rowe, "An Anatomy of Risk." Wiley, New York, 1977.
4. C. Starr, R. Rudman, and C. Whipple, Philosophical basis for risk analysis, *Ann. Rev. Energy,* **1,** 629 (1976).
5. H. J. Otway and R. C. Erdmann, Reactor safety and design from a risk viewpoint, *Nucl. Eng. Design* **13,** 365 (1970).
6. C. Starr, Benefit-cost studies in sociotechnical systems, *in* "Perspectives on Benefit-Risk Decision Making," p. 17, National Academy of Engineering, Washington, D.C., 1972.
7. H. J. Otway and J. J. Cohen, Revealed Preferences: Comments on the Starr Benefit Risk Relationships. International Institute of Applied Systems Analysis Rep. IIASA RM-75-5 (March 1975).
8. B. Fischhoff, P. Slovic, S. Lichtenstein, S. Read, and B. Combs, How safe is safe enough? A psychometric study of attitudes towards technological risks and benefits, *Policy Sci.* **9,** 127 (1978).

9. P. Slovic, B. Fischhoff, and S. Lichtenstein, Behavioral decision theory, *Ann. Rev. of Psychol.* **28,** 1 (1977).
10. L. Cave, R. E. Holmes, and P. J. Holmes, Public attitudes in relation to the risks presented by new technologies, *in* "Probabilistic Analysis of Nuclear Reactor Safety," Vol. II, p. IV.8-1. American Nuclear Society, La Grange Park, Illinois, 1978.
11. H. J. Otway and K. Thomas, The contribution of safety issues to public perceptions of energy systems, *in* "Probabilistic Analysis of Nuclear Reactor Safety," Vol. II, p. IV.1-1. American Nuclear Society, La Grange Park, Illinois, 1978.
12. A. Tversky and D. Kahneman, Judgment under uncertainty, *Science* **185,** 1124 (1974).
13. E. L. Zebroski, Attainment of balance in risk-benefit perceptions, in Risk-Benefit Methodology and Application: Some Papers Presented at the Engineering Foundation Workshop, September 22–26, 1975, Asilomar, California (D. Okrent, ed.). Univ. of California at Los Angeles Rep. UCLA-ENG-7598, p. 633 (December 1975).
14. E. W. Lawless, "Technology and Social Shock." Rutgers Univ. Press, New Jersey, 1977.
15. B. Fischhoff, C. Hohenemser, R. Kasperson, and R. Kates, Handling hazards, *Environment* **20,** 16 (1978).
16. C. Starr and C. Whipple, Risks of risk decisions, *Science* **208,** 1114 (1980).
17. H. A. Munera and G. Yadigaroglu, A new methodology to quantify risk perception, *Nucl. Sci. Eng.* **75,** 211 (1980).
18. A. McDonald and M. I. Temme, The role of risk criteria in nuclear plant decisions, *in* "Probabilistic Analysis of Nuclear Reactor Safety, Vol. II, p. VII.2-1. American Nuclear Society, La Grange Park, Illinois, 1978.
19. S. Levine, The role of risk assessment in the nuclear regulatory process, *Nucl. Safety* **19,** 556 (1978).
20. P. B. Hutt, Unresolved issues in the conflict between individual freedom and government control of safety, *Food Drug Cosmet. Law J.* **33,** 558 (1978).
21. F. R. Farmer, Reactor safety and siting: A proposed risk criterion, *Nucl. Safety* **8,** 539 (1967).
22. D. Okrent, A General Evaluation Approach to Risk-Benefit for Large Technological Systems and its Application to Nuclear Power. Univ. of California at Los Angeles Rep. UCLA-ENG-7777 (December 1977).
23. D. Okrent and C. Whipple, An Approach to Societal Risk Acceptance Criteria and Risk Management. Univ. of California at Los Angeles Rep. UCLA-ENG-7746 (June 1977).
24. C. L. Comar, Risk: A pragmatic de minimus approach, *Science* **203,** 319 (1979).
25. P. Slovic, S. Lichtenstein, and B. Fischhoff, Images of disaster: Perception and acceptance of risks from nuclear power, *in* "Fast Reactor Safety Technology," Vol. 1, p. 501. American Nuclear Society, La Grange Park, Illinois, 1979.
26. K. J. Wirtz, Acceptable levels of risk and why, *Proc. Fast Reactor Safety Meeting.* U.S. Atomic Energy Commission Rep. CONF-740401-P3, p. 1708 (1974).
27. H. J. Otway, The nuclear-societal interface, *Proc. At. Ind. Forum Conf. Nucl. Power and the Public, New Orleans, Louisiana* (March 27–30, 1977).
28. A. M. Weinberg, Science and trans-science, *Science* **177,** 211 (1972); *Minerva* **10,** 209 (1972).
29. Risks Associated With Nuclear Power: A Critical Review of the Literature, Summary and Synthesis Chapter. National Academy of Sciences, Washington, D.C., 1979.

CHAPTER 19

Epilogue

Commercial nuclear power is not risk free, nor has it been entirely without accidents. Indeed, no technology is entirely without risk.

One reason for performing a risk analysis is to determine the accident sequences that lead to system failure and to remove, wherever possible, the weakest links of the system. Such system changes usually lead to either improved plant availability or better protection of the capital investment in a plant.

Methods of risk analysis build upon those developed in Part I of this book. Substantial additional efforts are needed, however, to standardize the models, data, and quantitative approaches used in risk analyses [1]. For example, fault trees need to be codified for the kinds of failure and the detail of failures covered. Also, more work needs to be done in developing maintenance and repair models, and on human error and common cause failure modeling [1, 2]. Although we recognize that more data should be collected and processed for use in risk analyses, there is an increased need to ferret out low quality information and process the remaining data into a more useable form.

A second reason for performing risk assessments is to better help those who regulate the nuclear industry: "It is becoming more and more clear that risk-assessment techniques will have an increasingly significant role in the nuclear regulatory process" [3]. Many risk studies for nuclear power systems have already been done, as discussed in Part II, and many are currently in progress. For example, in an important new study by the Electric Power Research Institute, the risks for the entire nuclear fuel cycle are being assessed in an integrated fashion. The study considers routine risks incurred from mining and milling operations, and predicted accident risks from reprocessing, mixed-oxide (MOX) fuel fabrication, transportation of recovered material, and waste disposal. The risk curve for latent cancer health effects, obtained from a preliminary status report of the study, is shown in Fig. 19-1.

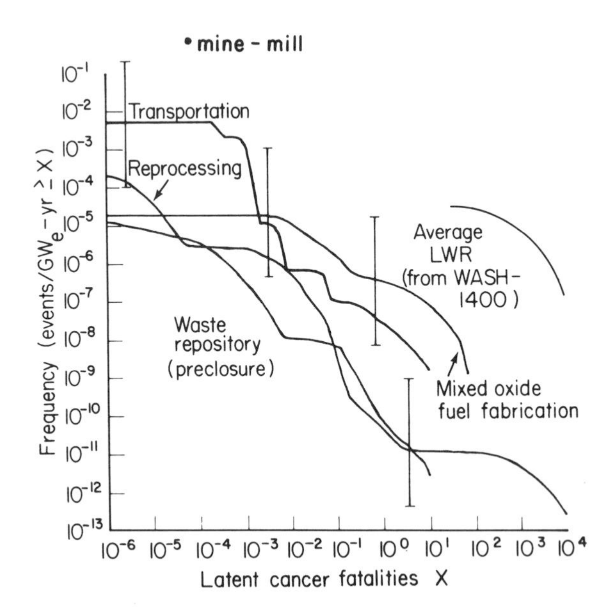

Fig. 19-1 *Risk of latent cancer fatality incidence per* GW_e *reactor year. [From R. C. Erdmann* et al., *Electric Power Research Institute Rep. EPRI NP-1128 (1979).*]

Two important conclusions from the fuel cycle risk assessment by Erdmann *et al.* are [4]:

1. The risk from a light water reactor (see Chapter 12) reasonably approximates the risk of nuclear power, since the supporting fuel cycle contributes only about 1% to the total risk.

2. The different steps in the supporting fuel cycle may be ranked, in decreasing order of risk, as: mining and milling, transportation, mixed-oxide fuel fabrication, reprocessing, waste disposal preclosure, and waste disposal postclosure. (See a graphic illustration of these results, Fig. 19-2, in which the volume of each cube is proportional to the radiological health effects of that fuel cycle step; the large block for nuclear power plant risk rests on a very large plateau of natural background.)

It is interesting and ironic that the ordering of the risk from the different steps in the fuel cycle is approximately the reverse of that which is often reported to the general public in the news.

The 1970s saw the passage of major environmental legislation, some of which impacted negatively on nuclear power development. This legislation was enacted perhaps because [5]:

> The richest, longest-lived, best-protected, most resourceful civilization, with the highest degree of insight into its own technology, is on its way to becoming the most frightened.

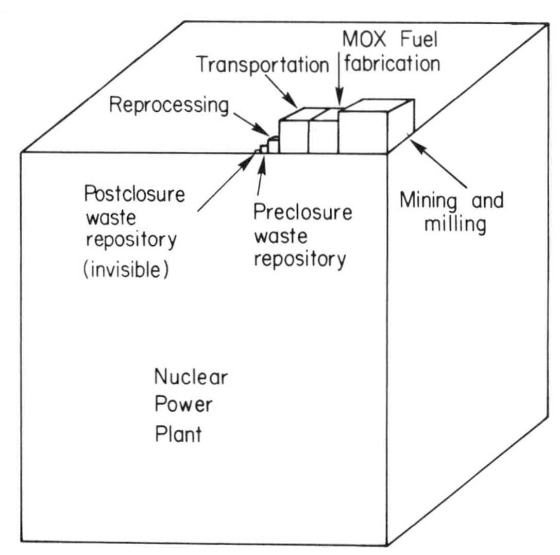

Fig. 19-2 *Relative risks of nuclear power according to the volume of each block.* [*From R. C. Erdmann* et al., *Electric Power Research Institute Rep. EPRI NP-1128 (1979).*]

Those charged with making decisions about public safety, including legislators and regulators, tend to strike a balance between those individuals who would like to shut down all future (and even existing) nuclear power plants and those who feel that "the real risk of nuclear power is not having enough of it" [6].

The attitude of many nuclear engineers toward risk-benefit analysis, discussed in Part III, is basically that of Zebroski [7], who in 1975 expressed concern that the hazards and risks from various energy sources were being evaluated unevenly. Thus, even though the risks from alternate (nonnuclear) energy sources appear to be larger than those from nuclear power (see Chapter 16), these considerations are not adequately being factored into current debates over nuclear power. Furthermore, there are risks to society from deprivation of energy that do not appear to have yet been included in risk studies; for example, people can be killed from lack of sufficient heat or food (obtained in large part because of energy), and people have been murdered by others who became irrational over deprivation of energy (as in the case of deaths that occurred in the queues for gasoline during the 1973 oil embargo). Perhaps someday the dream of Starr and Whipple can be achieved [8]:

> Assuming that the total funds allocated for risk reduction could be transferred freely between different risk reduction opportunities (which is certainly not always possible), the maximum number of lives that could be saved nationally is found when the marginal cost of saving life is uniform among the opportunities.

REFERENCES

1. S. Levine and W. E. Vesely, Prospects and problems in risk analyses: Some viewpoints, *in* "Nuclear Systems Reliability Engineering and Risk Assessment" (J. B. Fussell and G. R. Burdick, eds.), p. 5. Soc. Ind. Applied Math., Philadelphia, Pennsylvania, 1977.
2. J. B. Fussell and J. S. Arendt, System reliability engineering methodology: A discussion of the state of the art, *Nucl. Safety* **20,** 541 (1979).
3. S. Levine, The role of risk assessment in the nuclear regulatory process, *Ann. Nucl. Energy* **6,** 281 (1979).
4. R. C. Erdmann *et al.,* Status report on the EPRI fuel cycle accident risk assessment, Electric Power Research Institute Rep. EPRI NP-1128 (July 1979).
5. A. Wildavsky, No risk is the highest risk of all, *Am. Sci.* **67,** 32 (1979).
6. R. P. Hammond, The real risk of nuclear power, *in* "Energy and the Environment, Cost-Benefit Analysis in Energy" (R. A. Karam and K. Z. Morgan, eds.), Energy, Supplement 1, p. 290. Pergamon, Oxford, 1976.
7. E. L. Zebroski, Attainment of balance in risk-benefit perceptions, in Risk-Benefit Methodology and Applications: Some Papers Presented at the Engineering Foundation Workshop, September 22–26, Asilomar California (D. Okrent, ed.). Univ. of California at Los Angeles Rep. UCLA-ENG-7598, p. 633 (December 1975).
8. C. Starr and C. Whipple, Risks of risk decisions, *Science* **208,** 1114 (1980).

Some Useful Mathematical Functions

The *gamma* function is defined as

$$\Gamma(x) = \int_0^\infty y^{x-1} e^{-y} dy, \quad x \neq -n \qquad (n = 0, 1, 2, \cdots), \tag{A-1}$$

and is tabulated in standard references on mathematical functions [1]. The gamma function obeys a recursion relation of the form

$$\Gamma(x + 1) = x\Gamma(x). \tag{A-2}$$

For the special case of an integer r,

$$\Gamma(r) = (r - 1)!, \tag{A-3}$$

while another special result is

$$\Gamma(0.5) = \pi^{1/2}. \tag{A-4}$$

A plot of $1/\Gamma(x)$ is given in Fig. A-1.

The derivative of the natural logarithm of the gamma function is called the *digamma* function (or sometimes the *psi* function) and is defined as

$$\psi(x) = d[\ln \Gamma(x)]/dx. \tag{A-5}$$

A plot of this function is shown in Fig. A-2; one value of the function that is of particular interest to mathematicians is *Euler's constant*, γ, defined to be

$$\gamma = -\psi(1) = 0.5772. \tag{A-6}$$

The *incomplete gamma* function, $\gamma(x, z)$, is a two-parameter function defined by

$$\gamma(x, z) = \int_0^z y^{x-1} e^{-y} dy. \tag{A-7}$$

This function tends to the (complete) gamma function $\Gamma(x)$ as $z \to \infty$. A plot of some incomplete gamma functions is shown in Fig. A-3.

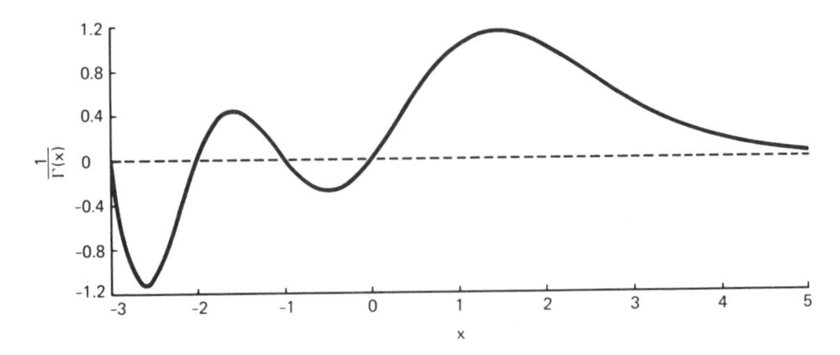

Fig. A-1 *Reciprocal of the gamma function. (From E. Jahnke and F. Emde,* Tables of Higher Functions, *Teubner (Leipzig), 1952; reprinted by Dover Publications).*

A special case of the incomplete gamma function is extensively tabulated and finds widespread use in probability theory [1]. The cumulative probability for the *chi-square* distribution, $P(\chi^2|r)$, for integer r, $r \geq 1$, is defined as

$$P(\chi^2|r) = [2^{r/2}\Gamma(r/2)]^{-1} \int_0^{\chi^2} t^{r/2-1}e^{-t/2}\,dt$$
$$= \gamma(r/2, \chi^2/2)/\Gamma(r/2). \qquad \text{(A-8)}$$

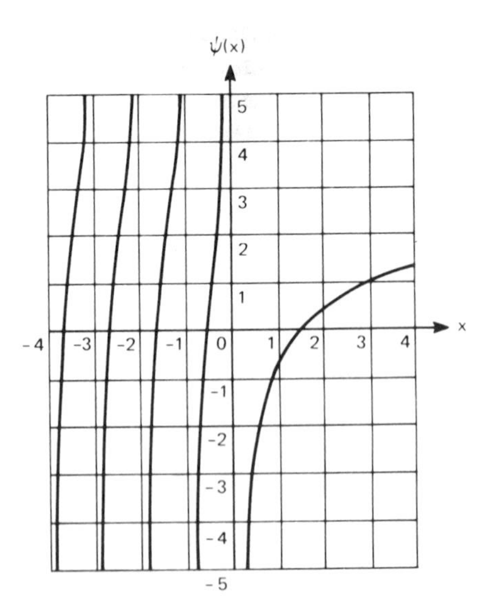

Fig. A-2 *Digamma function. (From M. Abramowitz and I. A. Stegun, eds.,* Handbook of Mathematical Functions, *Appl. Math. Series 55, National Bureau of Standards (1964) and Dover Publications).*

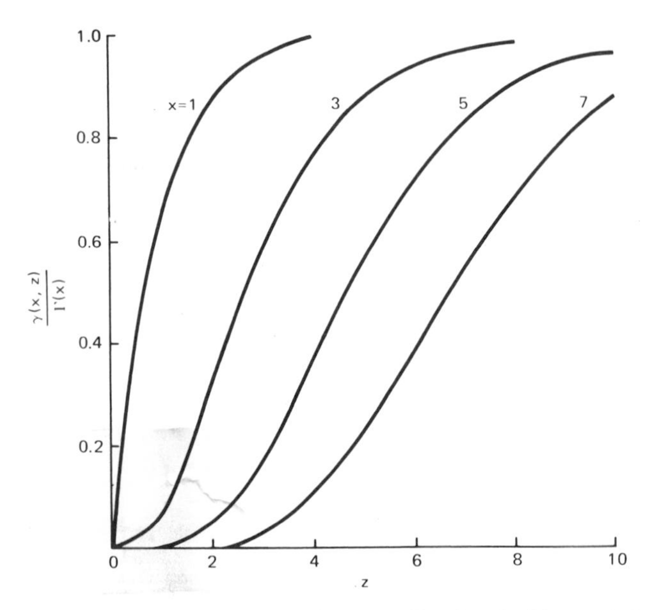

Fig. A-3 *Normalized incomplete gamma function. (Reprinted with permission from M. Tribus, "Rational Descriptions, Decisions and Designs," Pergamon, copyright © 1969.)*

The integer r in this distribution is called the "number of degrees of freedom." A special case of chi-square and incomplete gamma distributions is the *error* function

$$\text{erf } z = \frac{2}{\sqrt{\pi}} \int_0^z e^{-u^2}\, du$$

$$= \gamma(0.5, z^2)/\Gamma(0.5) = P(2z^2|1). \tag{A-9}$$

Another function related to the incomplete gamma function is the *exponential integral* function, which is defined as

$$E_n(z) = \int_1^\infty x^{-n} e^{-zx}\, dx = z^{n-1} \int_z^\infty y^{-n} e^{-y}\, dy, n = 1, 2, \ldots . \tag{A-10}$$

The exponential integral function can be written in terms of gamma functions as

$$E_n(z) = z^{n-1}[\Gamma(1 - n) - \gamma(1 - n, z)], \tag{A-11}$$

and likewise is tabulated [1]. Some important properties of $E_n(z)$ are

$$E_n(z) = \int_z^\infty E_{n-1}(y)dy \tag{A-12}$$

$$dE_n(z)/dz = -E_{n-1}(z) \tag{A-13}$$

$$E_n(0) = (n-1)^{-1},\ n > 1. \tag{A-14}$$

REFERENCE

1. M. Abramowitz and I. Stegun (eds.), "Handbook of Mathematical Functions." National Bureau of Standards, Washington, D.C., 1964, reprinted by Dover, New York.

APPENDIX B

Failure Data

Table B-1 *Hazard Rates λ and Demand Failure Probabilities Q_d for Mechanical Hardware*[a,b]

Components	Failure mode	Assessed range on probability of occurrence	Computational median	Error factor
1. Pumps (includes driver)	Failure to start on demand $Q_d{}^c$	3×10^{-4} –3×10^{-3}/d	1×10^{-3}/d	3
	Failure to run, given start λ_0 (normal environments)	3×10^{-6} –3×10^{-4}/hr	3×10^{-5}/hr	10
	Failure to run, given start λ_0 (extreme, post-accident environments inside containment)	1×10^{-4} –1×10^{-2}/hr	1×10^{-3}/hr	10
	Failure to run, given start λ_0 (post-accident, after environmental recovery)	3×10^{-5} –3×10^{-3}/hr	3×10^{-4}/hr	10
2. Valves				
a. Motor operated:	Failure to operate (includes driver) $Q_d{}^d$	3×10^{-4} –3×10^{-3}/d	1×10^{-3}/d	3
	Failure[e] to remain open (plug) Q_d	3×10^{-5} –3×10^{-4}/d	1×10^{-4}/d	3
	λ_s	1×10^{-7} –1×10^{-6}/hr	3×10^{-7}/hr	3
	Rupture λ_s	1×10^{-9} –1×10^{-7}/hr	1×10^{-8}/hr	10
b. Solenoid operated:	Failure to operate $Q_d{}^f$	3×10^{-4} –3×10^{-3}/d	1×10^{-3}/d	3
	Failure to remain open, Q_d (plug)	3×10^{-5} –3×10^{-4}/d	1×10^{-4}/d	3
	Rupture λ_s	1×10^{-9} –1×10^{-7}/hr	1×10^{-8}/hr	10
c. Air-fluid operated:	Failure to operate $Q_d{}^f$	1×10^{-4} –1×10^{-3}/d	3×10^{-4}/d	3
	Failure to remain open Q_d (plug)	3×10^{-5} –3×10^{-4}/d	1×10^{-4}/d	3
	λ_s	1×10^{-7} –1×10^{-6}/hr	3×10^{-7}/hr	3
	Rupture λ_s	1×10^{-9} –1×10^{-7}/hr	1×10^{-8}/hr	10
3. Check valves	Failure to open Q_d	3×10^{-5} –3×10^{-4}/d	1×10^{-4}/d	3
	Internal leak λ_o (severe)	1×10^{-7} –1×10^{-6}/hr	3×10^{-7}/hr	3
	Rupture λ_s	1×10^{-9} –1×10^{-7}/hr	1×10^{-8}/hr	10
4. Vacuum valve	Failure to operate Q_d	1×10^{-5} –1×10^{-4}/d	3×10^{-5}/d	3
5. Manual valve	Failure to remain open Q_d (plug)	3×10^{-5} –3×10^{-4}/d	1×10^{-4}/d	3
	Rupture λ_s	1×10^{-9} –1×10^{-7}/hr	1×10^{-8}/hr	10
6. Relief valves	Failure to open Q_d	3×10^{-6} –3×10^{-5}/d	1×10^{-5}/d	3
	Premature open λ_o	3×10^{-6} –3×10^{-5}/hr	1×10^{-5}/hr	3
7. Test valves, flow meters, orifices	Failure to remain open Q_d (plug)	1×10^{-4} –1×10^{-3}/d	3×10^{-4}/d	3
	Rupture λ_s	1×10^{-9} –1×10^{-7}/hr	1×10^{-8}/hr	10
8. *Pipes*				
a. Pipe $\leq$ 7.5 cm diam per section	Rupture/plug λ_s, λ_o	3×10^{-11}–3×10^{-8}/hr	1×10^{-9}/hr	30

(Continued)

Table B-1 (*Continued*)

Components	Failure mode	Assessed range on probability of occurrence	Computational median	Error factor
b. Pipe > 7.5 cm diam per section	Rupture λ_s, λ_o	3×10^{-12}–3×10^{-9}/hr	1×10^{-10}/hr	30
9. Clutch, mechanical	Failure to operate Q_d	1×10^{-4} –1×10^{-3}/d	3×10^{-4}/d	3
10. Scram rods (single)	Failure to insert	3×10^{-5} –3×10^{-4}/d	1×10^{-4}/d	3

[a] From *Reactor Safety Study,* Appendix III, Failure Data, WASH-1400, October 1975.
[b] See Section 5-3 for discussion of use.
[c] Demand probabilities are based on the presence of proper input control signals. For turbine driven pumps, the effect of failures of valves, sensors, and other auxiliary hardware may result in significantly higher overall failure rates for turbine driven pump systems.
[d] Demand probabilities are based on presence of proper input control signals.
[e] Plug probabilities are given in demand probability, and per hour rates, since phenomena are generally time dependent, but plugged condition may only be detected upon a demand of the system.

Table B-2 *Hazard Rates λ and Demand Failure Probabilities Q_d for Electrical Equipment*[a,b]

Component	Failure mode	Assessed range	Computational median	Error factor
1. Clutch, electrical	Failure to operate Q_d[c]	1×10^{-4}–1×10^{-3}/d	3×10^{-4}/d	3
	Premature disengagement λ_0	1×10^{-7}–1×10^{-5}/hr	1×10^{-6}/hr	10
2. Motors, electric	Failure to start Q_d	1×10^{-4}–1×10^{-3}/d	3×10^{-4}/d	3
	Failure to run, given start λ_0 (normal environment)	3×10^{-6}–3×10^{-5}/hr	1×10^{-5}/hr	3
	Failure to run, given start λ_0 (extreme environment)	1×10^{-4}–1×10^{-2}/hr	1×10^{-3}/hr	10
3. Relays	Failure to energize Q_d	3×10^{-5}–3×10^{-4}/d	1×10^{-4}/d	3
	Failure of NO contacts to close, given energized λ_0	1×10^{-7}–1×10^{-6}/hr	3×10^{-7}/hr	3
	Failure of NC contacts by opening, given not energized λ_0	3×10^{-8}–3×10^{-7}/hr	1×10^{-7}/hr	3
	Short across NO/NC contact λ_0	1×10^{-9}–1×10^{-7}/hr	1×10^{-8}/hr	10
	Coil open λ_0	1×10^{-8}–1×10^{-6}/hr	1×10^{-7}/hr	10
	Coil short to power λ_0	1×10^{-9}–1×10^{-7}/hr	1×10^{-8}/hr	10
4. Circuit breakers	Failure to transfer Q_d	3×10^{-4}–3×10^{-3}/d	1×10^{-3}/d	3
	Premature transfer λ_0	3×10^{-7}–3×10^{-6}/hr	1×10^{-6}/hr	3
5. Switches				
a. Limit	Failure to operate Q_d	1×10^{-4}–1×10^{-3}/d	3×10^{-4}/d	3
b. Torque	Failure to operate Q_d	3×10^{-5}–3×10^{-4}/d	1×10^{-4}/d	3
c. Pressure	Failure to operate Q_d	3×10^{-5}–3×10^{-4}/d	1×10^{-4}/d	3
d. Manual	Failure to transfer Q_d	3×10^{-6}–3×10^{-5}/d	1×10^{-5}/d	3

Table B-2 (*Continued*)

Component	Failure mode	Assessed range	Computational median	Error factor
6. Switch contacts	Failure of NO contacts to close given switch operation λ_0	1×10^{-8}–1×10^{-6}/hr	1×10^{-7}/hr	10
	Failure of NC by opening, given no switch operation λ_0	3×10^{-9}–3×10^{-7}/hr	3×10^{-8}/hr	10
	Short across NO/NC contact λ_0	1×10^{-9}–1×10^{-7}/hr	1×10^{-8}/hr	10
7. Battery power systems (wet cell)	Failure to provide proper output λ_s	1×10^{-6}–1×10^{-5}/hr	3×10^{-6}/hr	3
8. Transformers	Open circuit primary or secondary λ_0	3×10^{-7}–3×10^{-6}/hr	1×10^{-6}/hr	3
	Short primary to secondary λ_0	3×10^{-7}–3×10^{-6}/hr	1×10^{-6}/hr	3
9a. Solid state devices hi power applications (diodes, transistors, etc.)	Fails to function λ_0	3×10^{-7}–3×10^{-5}/hr	3×10^{-6}/hr	10
	Fails shorted λ_0	1×10^{-7}–1×10^{-5}/hr	1×10^{-6}/hr	10
b. Solid state devices, low power applications	Fails to function λ_0	1×10^{-7}–1×10^{-5}/hr	1×10^{-6}/hr	10
	Fails shorted	1×10^{-8}–1×10^{-6}/hr	1×10^{-7}/hr	10
10a. Diesels (complete plant)	Failure to start Q_d	1×10^{-2}–1×10^{-1}/d	3×10^{-2}/d	3
	Failure to run, emergency conditions, given start λ_0	3×10^{-4}–3×10^{-2}/hr	3×10^{-3}/hr	10
b. Diesels (engine only)	Failure to run, emergency conditions, given start λ_0	3×10^{-5}–3×10^{-3}/hr	3×10^{-4}/hr	10
11. Instrumentation—general (includes transmitter, amplifier, and output device)	Failure to operate λ_0	1×10^{-7}–1×10^{-5}/hr	1×10^{-6}/hr	10
	Shift in calibration λ_0	3×10^{-6}–3×10^{-4}/hr	3×10^{-5}/hr	10
12. Fuses	Failure to open Q_d	3×10^{-6}–3×10^{-5}/d	1×10^{-5}/d	3
	Premature open λ_0	3×10^{-7}–3×10^{-6}/hr	1×10^{-6}/hr	3
13. Wires (typical circuits, several joints)	Open circuit λ_0	1×10^{-6}–1×10^{-5}/hr	3×10^{-6}/hr	3
	Short, to ground λ_0	3×10^{-8}–3×10^{-6}/hr	3×10^{-7}/hr	10
	Short to power λ_0	1×10^{-9}–1×10^{-7}/hr	1×10^{-8}/hr	10
14. Terminal boards	Open connection λ_0	1×10^{-8}–1×10^{-6}/hr	1×10^{-7}/hr	10
	Short to adjacent circuit λ_0	1×10^{-9}–1×10^{-7}/hr	1×10^{-8}/hr	10

[a] From *Reactor Safety Study*, Appendix III, Failure Data, Washington-1400, October 1975.

[b] See Section 5-3 for discussion of use.

[c] Demand probabilities are based on presence of proper input control signals.

Table B-3 *Human Error Probabilities*[a,b]

Demand failure probability	Activity
10^{-4}	Selection of a key-operated switch rather than a nonkey switch. (This value does not include the error of decision where the operator misinterprets situation and believes key switch is correct choice.)
10^{-3}	Selection of a switch (or pair of switches) dissimilar in shape or location to the desired switch (or pair of switches), assuming no decision error. For example, operator actuates large handled switch rather than small switch.
3×10^{-3}	General human error of commission, e.g., misreading label and, therefore, selecting wrong switch.
10^{-2}	General human error of omission when there is no display in the control room of the status of the item omitted, e.g., failure to return manually operated test valve to proper configuration after maintenance.
3×10^{-3}	Errors of omission where the items being omitted are embedded in a procedure rather than at the end as above.
3×10^{-2}	Simple arithmetic errors with self-checking but without repeating the calculation by redoing it on another piece of paper.
1/x	Given that an operator is reaching for an incorrect switch (or pair of switches), he or she selects a particular similar appearing switch (or pair of switches), where $x =$ the number of incorrect switches (or pairs of switches) adjacent to the desired switch (or pair of switches). The $1/x$ applies up to 5 or 6 items. After that point the error rate would be lower because the operator would take more time to search. With up to 5 or 6 items, the operator doesn't expect to be wrong and therefore is more likely to do less deliberate searching.
10^{-1}	Given that an operator is reaching for a wrong motor operated valve MOV switch (or pair of switches), he or she fails to note from the indicator lamps that the MOV(s) is (are) already in the desired state and merely changes the status of the MOV(s) without recognizing that he or she had selected the wrong switch(es).
~1.0	Same as above, except that the state(s) of the incorrect switch(es) is (are) *not* the desired state.
~1.0	If an operator fails to operate correctly one of two closely coupled valves or switches in a procedural step, he or she also fails to correctly operate the other valve.
10^{-1}	Monitor or inspector fails to recognize initial error by operator. Note: With continuing feedback of the error on the annunciator panel, this high error rate would not apply.
10^{-1}	Personnel on different work shift fail to check condition of hardware unless required by checklist or written directive.

Table B-3 (*Continued*)

Demand failure probability	Activity
5×10^{-1}	Monitor fails to detect undesired position of valves, etc., during general walk-around inspections, assuming no check list is used.
0.2–0.3	General error rate, given very high stress levels, where dangerous activities are occurring rapidly
$2^{(n-1)}x$	Given severe time stress, as in trying to compensate for an error made in an emergency situation, the initial error rate x, for an activity doubles for each attempt, n, after a previous incorrect attempt, until the limiting condition of an error rate of 1.0 is reached or until time runs out. This limiting condition corresponds to an individual's becoming completely disorganized or ineffective.
~1.0	Operator fails to act correctly in the first 60 seconds after the onset of an extremely high stress condition, e.g., a large LOCA.
9×10^{-1}	Operator fails to act correctly after the first 5 minutes after the onset of an extremely high stress condition.
10^{-1}	Operator fails to act correctly after the first 30 minutes in an extreme stress condition.
10^{-2}	Operator fails to act correctly after the first several hours in a high stress condition.
x	After 7 days after a large LOCA, there is a complete recovery to the normal error rate x, for any task.

[a] Reactor Safety Study, Appendix III, Failure Data, WASH-1400 (October 1975).

[b] See Section 5-4 for discussion of use.

APPENDIX C

Some Matrix Mathematics

To evaluate the general equations in Chapter 7 for the time-dependent availability, the steady-state availability, or the mean-time-to-failure, it is necessary to understand a few basic rules about matrices. The $(N + 1) \times (N + 1)$ matrix $\mathbf{M}$ has elements M_{nm}, where n denotes the row and m the column location, and can be written as

$$\mathbf{M} = \begin{bmatrix} M_{00} & M_{01} & \cdots & M_{0N} \\ M_{10} & M_{11} & \cdots & M_{1N} \\ \vdots & \vdots & & \vdots \\ M_{N0} & M_{N1} & \cdots & M_{NN} \end{bmatrix}. \tag{C-1}$$

(The matrix $\mathbf{M}$ is the unit matrix $\mathbf{I}$ if $M_{nn} = 1$, $n = 0$ to N, and $M_{nm} = 0$, $n \neq m$.) When $\mathbf{M}$ acts on the vector $\mathbf{P}$, defined as

$$\mathbf{P} = \begin{bmatrix} P_0 \\ P_1 \\ \vdots \\ P_N \end{bmatrix}, \tag{C-2}$$

the result is the vector

$$\mathbf{MP} = \begin{bmatrix} M_{00}P_0 + M_{01}P_1 + \cdots + M_{0N}P_N \\ M_{10}P_0 + M_{11}P_1 + \cdots + M_{1N}P_N \\ \vdots \qquad\qquad \vdots \qquad\qquad\qquad \vdots \\ M_{N0}P_0 + M_{N1}P_1 + \cdots + M_{NN}P_N \end{bmatrix}. \tag{C-3}$$

In a similar way, the matrix product $\mathbf{M}^2$ is a matrix with elements

$$[\mathbf{M}^2]_{nm} = \sum_{j=0}^{N} M_{nj}M_{jm}. \tag{C-4}$$

The transpose of $\mathbf{M}$, denoted by $\mathbf{M}^T$, has the elements

$$M^T_{nm} = M_{mn}, \tag{C-5}$$

and may be considered to be an interchange of the matrix elements about the diagonal elements M_{nn}. The *determinant* of $\mathbf{M}$, $|\mathbf{M}|$, may be written in terms of its *cofactors* by expanding the determinant about any row or column,

$$\begin{aligned} |\mathbf{M}| &= M_{n0}(\mathrm{cof}\,\mathbf{M})_{n0} + M_{n1}(\mathrm{cof}\,\mathbf{M})_{n1} + \cdots \\ &\quad + M_{nN}(\mathrm{cof}\,\mathbf{M})_{nN}, \qquad n = 0 \text{ to } N, \\ |\mathbf{M}| &= M_{0m}(\mathrm{cof}\,\mathbf{M})_{0m} + M_{1m}(\mathrm{cof}\,\mathbf{M})_{1m} + \cdots \\ &\quad + M_{Nm}(\mathrm{cof}\,\mathbf{M})_{Nm}, \qquad m = 0 \text{ to } M. \end{aligned} \tag{C-6}$$

Here $(\mathrm{cof}\,\mathbf{M})_{ij}$ is the cofactor for the element M_{ij} and is obtained by deleting the ith row and jth column, calculating the determinant of the remaining array, and multiplying by $(-1)^{i+j}$. For example,

$$\begin{aligned} (\mathrm{cof}\,\mathbf{M})_{12} &= \mathrm{cof}\begin{bmatrix} M_{00} & M_{01} & M_{02} & M_{03} & \cdots \\ M_{10} & M_{11} & M_{12} & M_{13} & \cdots \\ M_{20} & M_{21} & M_{22} & M_{23} & \cdots \\ M_{30} & M_{31} & M_{32} & M_{33} & \cdots \\ \vdots & \vdots & \vdots & \vdots & \end{bmatrix} \\ &= (-1)^{1+2}\begin{vmatrix} M_{00} & M_{01} & M_{03} & \cdots \\ M_{20} & M_{21} & M_{23} & \cdots \\ M_{30} & M_{31} & M_{33} & \cdots \\ \vdots & \vdots & \vdots & \end{vmatrix}. \end{aligned} \tag{C-7}$$

Failure Modes and Effects Analysis

Failure modes and effects analysis (FMEA) is a tool to systematically analyze all contributing component failure modes and identify the resulting effects on the system. It is frequently used for nuclear power applications whenever a detailed analysis involving fault trees (see Chapter 8) and event trees (see Chapter 9) is not required. Many times an FMEA will be performed as a preliminary system reliability analysis to assist development of a more quantitative event tree/fault tree analysis.

Several slightly different formats are used for an FMEA, but all require that the failure or malfunction of each component of the system, including the mode of failure, be considered. Then the effects of the failure are traced through the system in order to assess the ultimate effect on the system performance.

Successful development of an FMEA requires that the analyst know all the significant failure modes, such as failure to start, stop, open, close, or continue to operate, and the scheduled periods between service and the design lifetime. It also requires the analyst to assess the effect of any failure mode on the overall performance of the system according to the following hazard classification [1]:

Criticality category	Effect
I. Safe	Negligible; no effect on system.
II. Marginal	Failure will degrade system to some extent but will not cause major system damage or injury to personnel.
III. Critical	Failure will degrade system's performance and/or cause personnel injury, and if immediate action is not taken, serious injuries or deaths to personnel and/or loss of system will occur.
IV. Catastrophic	Failure will produce severe system degradation causing loss of system and/or multiple deaths or injuries.

Example D-1 Construct an FMEA for the domestic hot water system as shown in Fig. D-1. The water system works as follows: the gas valve is operated by the controller, which, in turn, is operated by a temperature measuring and comparing device. The gas valve operates the main burner in full-on/full-off modes. The check valve in the water inlet line prevents reverse flow due to overpressure in the hot water system. The pressure relief valve opens when pressure in the system exceeds 100 psi.

When the temperature of water is below the desired range (140° to 180°F), the temperature measuring and comparing device signals the controller to open the gas valve and turn on the gas burner, which is lit by a pilot burner. When the water temperature reaches the desired level, the temperature measuring and comparing device signals the controller to turn off the gas valve and thus turn off the main gas burner.

The FMEA is shown in Table D-1. ◇

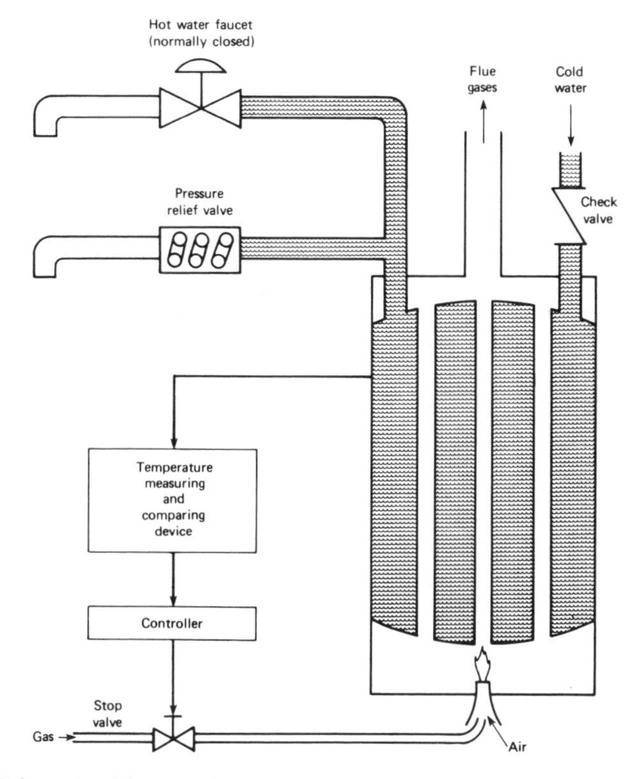

Fig. D-1 *Schematic of domestic hot water system. [From H. E. Lambert, Lawrence Livermore Laboratory Report* UCID-*16238 (1973).]*

Table D-1 *Failure Mode and Effects Analysis of a Domestic Hot Water System*[a]

Component	Failure or error mode	Effects on		Class				Failure frequency	Detection methods	Compensating provisions and remarks
		other components	whole system	I	II	III	IV			
Pressure relief valve	Jammed open	Increased operation of temperature sensing controller; gas flow due to hot water loss	Loss of hot water; greater cold water input; greater gas consumption	X				Reasonably probable	Observe at pressure relief valve	Shut off water supply, reseal or replace relief valve
	Jammed closed	None	None	X				Probable	Manual testing	Unless combined w/other component failure, this failure has no consequence
Gas valve	Jammed open	Burner continues to operate, pressure relief valve opens	Water temperature and pressure increase; water $\rightarrow$ steam			X		Reasonably probable	Water at faucet too hot; pressure relief valve open (observation)	Open hot water faucet to relieve pressure. Shut off gas supply. Pressure relief valve compensates.

	Jammed closed	Burner ceases to operate	System fails to produce hot water	X		Remote	Observe at output (water temperature too low)	
Temperature measuring and comparing device	Fails to react to temperature rise above preset level	Controller, gas valve, burner continue to function "on." Pressure relief valve opens	Water temperature too high; water → steam		X	Remote	Observe at output (faucet)	Pressure relief valve compensates. Open hot water faucet to relieve pressure. Shut off gas supply.
	Fails to react to temperature drop below preset level	Controller, gas valve, burner continue to function "off"	Water temperature too low	X		Remote	Observe at output (faucet)	

[a] From H. E. Lambert, Lawrence Livermore Laboratory Rep. UCID-16238 (1973).

An FMEA is one of several variations of the *inductive* approach. The *failure mode effect and criticality analysis* method is similar to FMEA except that the criticality of the failure is analyzed in greater detail [1, 2]; a *preliminary hazard analysis* is similar but emphasizes the potential hazards posed to plant personnel and other humans by the system. Other inductive systems analysis methods are discussed elsewhere [1, 3].

The main disadvantage of an FMEA is that it considers only one failure at a time and not multiple or common cause failures; the advantages are that it is simple to apply and it provides an orderly examination of the hazard conditions in a system.

REFERENCES

1. H. E. Lambert, System Safety Analysis and Fault Tree Analysis. Lawrence Livermore Laboratory Rep. UCID-16238 (1973).
2. W. E. Jordan, Failure modes, effects, and criticality analysis, *Proc. Ann. Reliability and Maintainability Symp., San Francisco, California* p. 30. Institute of Electrical and Electronics Engineers, 1972.
3. W. E. Vesely, F. F. Goldberg, N. H. Roberts and D. F. Haasl, Fault Tree Handbook, Nuclear Regulatory Commission Rep. NUREG-0492 (1981).

APPENDIX E

Light Water Reactor Safety Systems

The Reactor Safety Study (RSS) for United States commercial light water reactors [1] was directed at assessing the effectiveness of the Engineered Safety Features (ESFs) of a typical boiling water reactor (BWR) and a pressurized water reactor (PWR). Part of the study addressed the consequences of a loss of coolant accident (LOCA) after a pipe break initiating event. The various ESF functions, and the systems to accomplish those functions, are summarized in the glossary of definitions in Table E-1. Figures E-1 and E-2 give some explanation of the function of the various ESFs.

Regardless of the design details of a particular reactor, the ESFs perform a uniform set of functions that cover:

(a) Reactor shutdown or "trip" (RT) to stop the fission process during the LOCA. (The trip system was discussed in Section 8-5.)
(b) Emergency core cooling (ECC) to minimize any release of radioactivity from the fuel into the containment.
(c) Post-accident radioactivity removal (PARR) to remove from the containment atmosphere any radioactivity released from the core.
(d) Post-accident heat removal (PAHR) to remove the core decay heat from the containment in order to prevent its overpressure.
(e) Containment integrity (CI) to prevent the radioactivity not removed by PARR from being dispersed into the environment.

Because of the inherent complexity of a nuclear reactor safety system, it was necessary in the RSS to assign to each unique basic input event (diamond or circle) shown on the fault trees a unique eight-character code name. Then, even though a component failure event may appear several places on a fault tree and on more than one fault tree, if it is the same component with the same failure mode, it will have the same name. The RSS coding scheme identifies events by system, component type, compo-

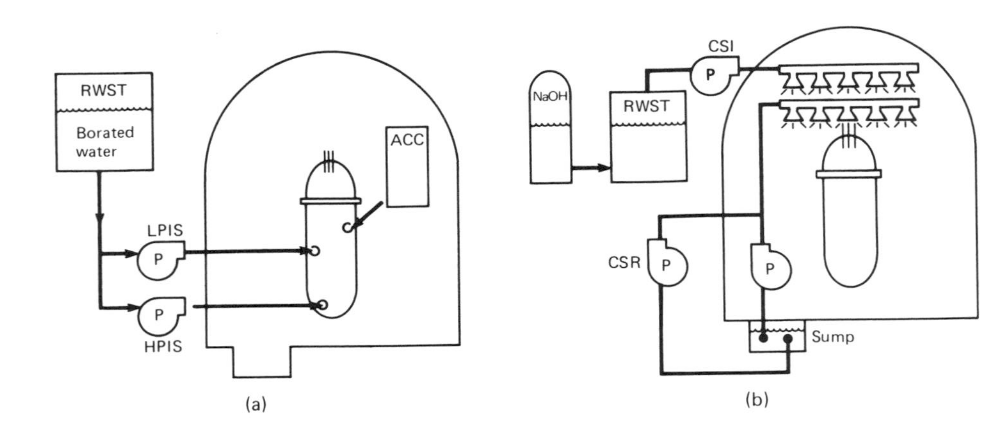

Fig. E-1 *Illustrations of* PWR *systems used to perform* ESF *functions.*

(a) *Emergency coolant injection* (ECI). *Borated water is furnished to cool the core by means of the accumulators, the low-pressure injection system* (LPIS), *and the high-pressure injection system* (HPIS).

(b) *Post-accident radioactivity removal* (PARR). *Radioactivity is collected from the containment atmosphere by the containment spray injection system* (CSIS), *the containment spray recirculation system* (CSRS), *and sodium hydroxide addition* (SHA) *to spray water.*

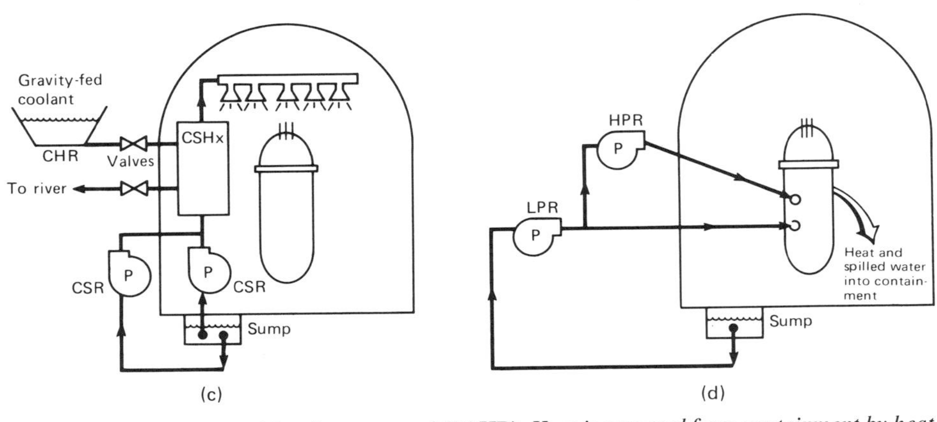

(c) *Post-accident heat removal* (PAHR). *Heat is removed from containment by heat exchangers that involve the containment spray recirculation system and the containment heat removal system* (CHRS).

(d) *Emergency cooling recirculation* (ECR). *The core is cooled by heat being transferred to containment by the low-pressure recirculation system* (LPRS) *and the high-pressure recirculation system* (HPRS). *Both systems, using injection pumps aligned to a recirculation mode, pump water from a containment sump into the core.* [*From Reactor Safety Study, Appendix I,* WASH-*1400 (October 1975*).]

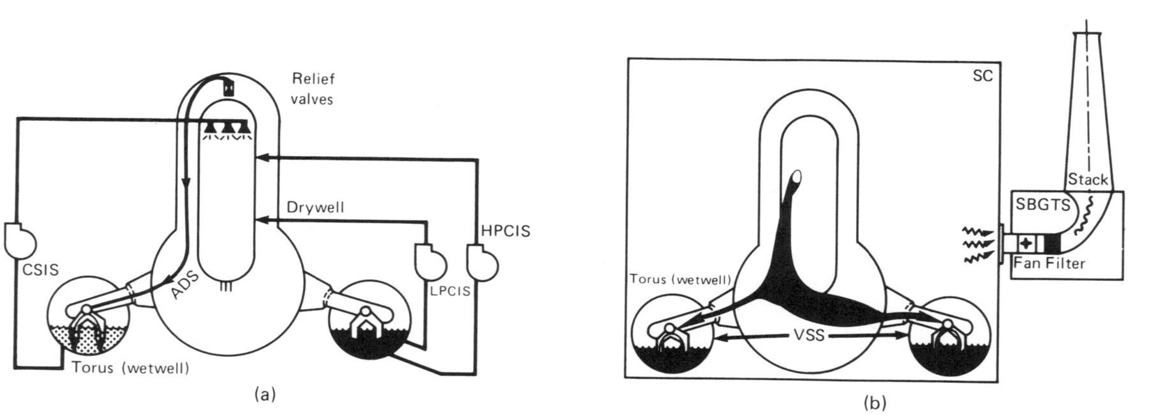

Fig. E-2 *Illustrations of* BWR *systems used to perform* ESF *functions.*

(a) *Emergency coolant injection* (ECI). *Water is pumped from the torus to cool the core by means of the low-pressure coolant injection system* (LPCIS), *the core spray injection system* (CSIS), *and the high-pressure coolant injection system* (HPCIS). *The automatic depressurization system* (ADS) *is used where necessary to reduce primary coolant system pressure.*

(b) *Post-accident radioactivity removal* (PARR). *Radioactivity is scrubbed from the containment atmosphere as it passes through the vapor suppression system* (VSS). *Radioactivity that leaks from the containment is also guided by the secondary containment* (SC) *to the standby gas treatment system* (SBGTS) *and an elevated stack release.*

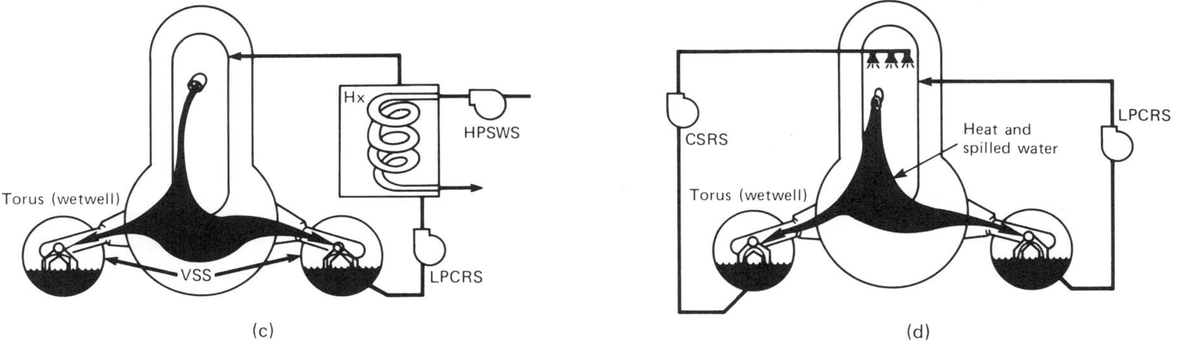

(c) *Post-accident heat removal* (PAHR). *Initially, blowdown heat is removed by the vapor suppression system* (**VSS**). *After blowdown, heat is removed from containment by heat exchangers that involve the low-pressure coolant recirculation system* (**LPCRS**), *and the high-pressure service water system* (**HPSWS**).

(d) *Emergency coolant recirculation* (ECR). *Water is pumped from the torus to the core by means of the core spray recirculation system* (CSRS) *and the low-pressure recirculation system* (LPCRS). [*From Reactor Safety Study, Appendix I,* WASH-*1400* (*October 1975*).

Table E-1 *Glossary of Terms for Reactor Safety*[a]

Category	Code	System name
General	BWR	Boiling water reactor
	DBA	Design basis accident
	ESF	Engineered safety feature
	EP	Electric power
	LOCA	Loss-of-coolant accident
	PB	Pipe break
	PWR	Pressurized water reactor
	RCS	Reactor coolant system
	RWST	Refueling water storage tank
	TE	Transient event
ESF systems	CI	Containment integrity
	ECC	Emergency core cooling
	ECI	Emergency coolant injection
	ECF	Emergency cooling function
	ECR	Emergency coolant recirculation
	PAHR	Post-accident heat removal
	PARR	Post-accident radioactivity removal
	RT	Reactor trip
ESF functions (PWR)	ACC	Accumulators
	AFWS	Auxiliary feedwater system
	CL	Containment leakage
	CLCS	Consequence limiting control system
	CHRS	Containment heat removal system
	CSIS	Containment spray injection system
	CSRS	Containment spray recirculation system
	CVCS	Chemical volume control system
	HPIS	High-pressure injection system
	HPRS	High-pressure recirculation system
	LPIS	Low-pressure injection system
	LPRS	Low-pressure recirculation system
	PCS	Power conversion system
	NaOH	Sodium hydroxide
	RPS	Reactor protection system
	SICS	Safety injection control system
ESF functions (BWR)	ADS	Automatic depressurization system
	CL	Containment leakage
	CSIS	Core spray injection system
	CSRS	Core spray recirculation system
	HPCIS	High-pressure coolant injection system
	HPSWS	High-pressure service water system
	LPCIS	Low-pressure coolant injection system
	LPCRS	Low-pressure coolant recirculation system
	RCICS	Reactor core isolation cooling system
	RHRS	Residual heat removal system
	RPS	Reactor protection system
	SBGTS	Standby gas treatment system
	SC	Secondary confinement
	SHA	Sodium hydroxide addition

[a] From Reactor Safety Study, Appendix I, WASH-1400 (October 1975).

Table E-2 *Coding Scheme for Naming Events in the Reactor Safety Study*

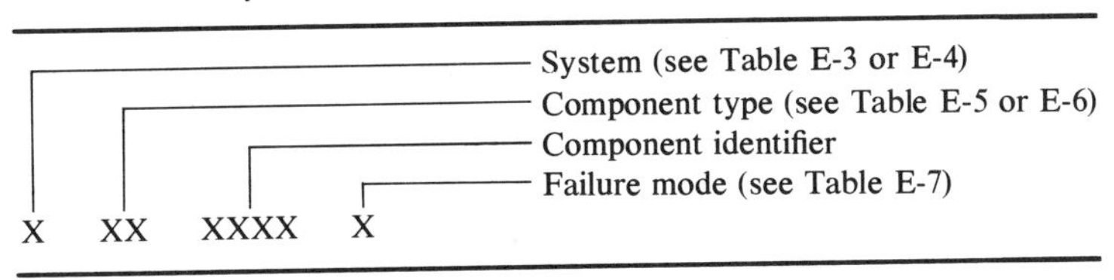

nent identification, and failure mode, as shown in Table E-2, according to the codes in Tables E-3 through E-6. Thus CCVBOO1C for a PWR denotes a check valve in the containment spray injection system, numbered B001, which fails closed.

Table E-3 PWR *System Identification Code Used in the Reactor Safety Study*[a]

Code	System name
A	Accumulator (ACC)
G	Containment leakage (CL)
N	Consequence limiting control system (CLCS)
K	Containment heat removal system (CHRS)
C	Containment spray injection system (CSIS)
D	Containment spray recirculation system (CSRS)
J	Electrical power (EPS)
F	High-pressure injection system (HPCIS)
H	High-pressure recirculation system (HPCRS)
B	Low-pressure injection system (LPIS)
E	Low-pressure recirculation system (LPRS)
L	Sodium hydroxide addition system (SHAS)
I	Reactor protection system (RPS)
M	Safety injection control system (SICS)
P	Auxiliary feedwater (AF)

[a] From Reactor Safety Study, Appendix II, WASH-1400 (October 1975).

Table E-4 BWR *System Identification Code Used in the Reactor Safety Study*[a]

Code	System name
1	Main stream
2	Reactor recirculation and automatic depressurization (ADS)
3	Reactor protective system (RPS)
6	Feedwater
8	Offgas system
9	Vapor suppression system (VSS)
A	Core spray injection system (CSIS)
B	Low-pressure coolant injection system (LPCIS)
C	Nuclear boiler vessel instrumentation
D	High-pressure service water system (HPCIS)
E	Service water
F	High-pressure coolant injection system (HPCIS)
L	Reactor core isolation cooling system (RSICS)
M	Emergency service water system (ESWS)
N	Radwaste
Q	Electrical power
R	Hydraulic control unit
S	Containment leakage (CL)
U	Chilled water
V	Standby gas treatment system (SBGTS)
X	Diesel generator
Y	Containment atmosphere dilution system
Z	Reactor building heating and ventilation
#	Fuel pool

[a] From Reactor Safety Study, Appendix II, WASH-1400 (October 1975).

Table E-5 *Code Used in the Reactor Safety Study for Mechanical Components*[a]

Code	Component	Code	Component
AC	Accumulator	SL	Sluice gate
BL	Blower	SP	Sump
CD	Control rod drive unit	ST	Subtree
FA	Cover plate	TK	Tank
DM	Damper	TG	Tubing
DL	Diesel	TB	Turbine
XJ	Expansion joint	CV	Valve, Check
FL	Filter or strainer	EV	Explosive operated
GB	Gas bottle	HV	Hydraulic operated
GK	Gasket	XV	Manual
HE	Heat exchanger	MV	Motor operated
NZ	Nozzle	AV	Pneumatic operated
OR	Orifice	RV	Relief
PP	Pipe	SV	Safety
CP	Pipe cap	KV	Solenoid operated
PV	Pressure vessel	DV	Stop Check
PM	Pump	VV	Vacuum relief
ED	Reactor control rod	VT	Vent
RF	Refrigeration unit	WL	Well

[a] From Reactor Safety Study, Appendix II, WASH-1400 (October 1975).

Table E-6 *Code Used in the Reactor Safety Study for Electrical Components*[a]

Code	Component	Code	Component
AM	Amplifier	GS	Ground switch
AN	Annunciator	RE	Relay
BY	Battery	CN	Relay or switch contact
BC	Battery charger	RS	Reset switch
BS	Bus	RT	Resistor, temp device
CA	Cable	AD	Signal Comparator
CB	Circuit breaker	PS	Switch, Pressure
CL	Clutch	QS	Torque
CS	Control switch	TS	Temperature
CO	Coil	TM	Terminal Board
DI	Detector	DE	Diode or Rectifier
DC	dc Power supply	FU	Fuse
FS	Flow switch	GE	Generator
HG	Heating element	HT	Heat tracing
IM	Input module	SB	Test pushbutton
IV	Inverter (solid state)	OL	Thermal overload
ES	Level switch	TI	Timer
LT	Light	CT	Transformer, Current
LS	Limit switch	OT	Potential (or control)
SW	Manual switch	TR	Power
MO	Motor	TF	Transmitter, Flow
MS	Motor starter	TL	Level
ND	Neutron detector	TP	Pressure
PT	Potentiometer	TT	Temperature
RC	Recorder	WR	Wire
LA	Lightning arrester	OO	Event (where no component involved)

[a] From Reactor Safety Study, Appendix II, WASH-1400 (October 1975).

Table E-7 *Failure Mode Code Used in the Reactor Safety Study*[a]

Failure mode	Code
Closed	C
Disengaged	G
Does not close	K
Does not open	D
Does not start	A
Engaged	E
Exceeds limit	M
Leakage	L
Loss of function	F
Maintenance fault	Y
No input	N
Open	O
Open circuit	B
Operational fault	X
Overload	H
Plugged	P
Rupture	R
Short circuit	Q
Short to ground	S
Fault transfer	T

[a] From Reactor Safety Study, Appendix II, WASH-1400 (October 1975).

REFERENCE

1. Reactor Safety Study—An Assessment of Accident Risks in U.S. Nuclear Power Plants. U.S. Nuclear Regulatory Commission Rep. WASH-1400, NUREG 75/014 (October 1975).

APPENDIX F

Additional Light Water Reactor Safety Study Fault Trees

The top of the fault tree for ''Failure to Trip for Small LOCA'' is given in Section 8-6. The subtrees in Figs. F-1 through F-5 for that tree illustrate the intricate linking of events in a fault tree for a complicated system.

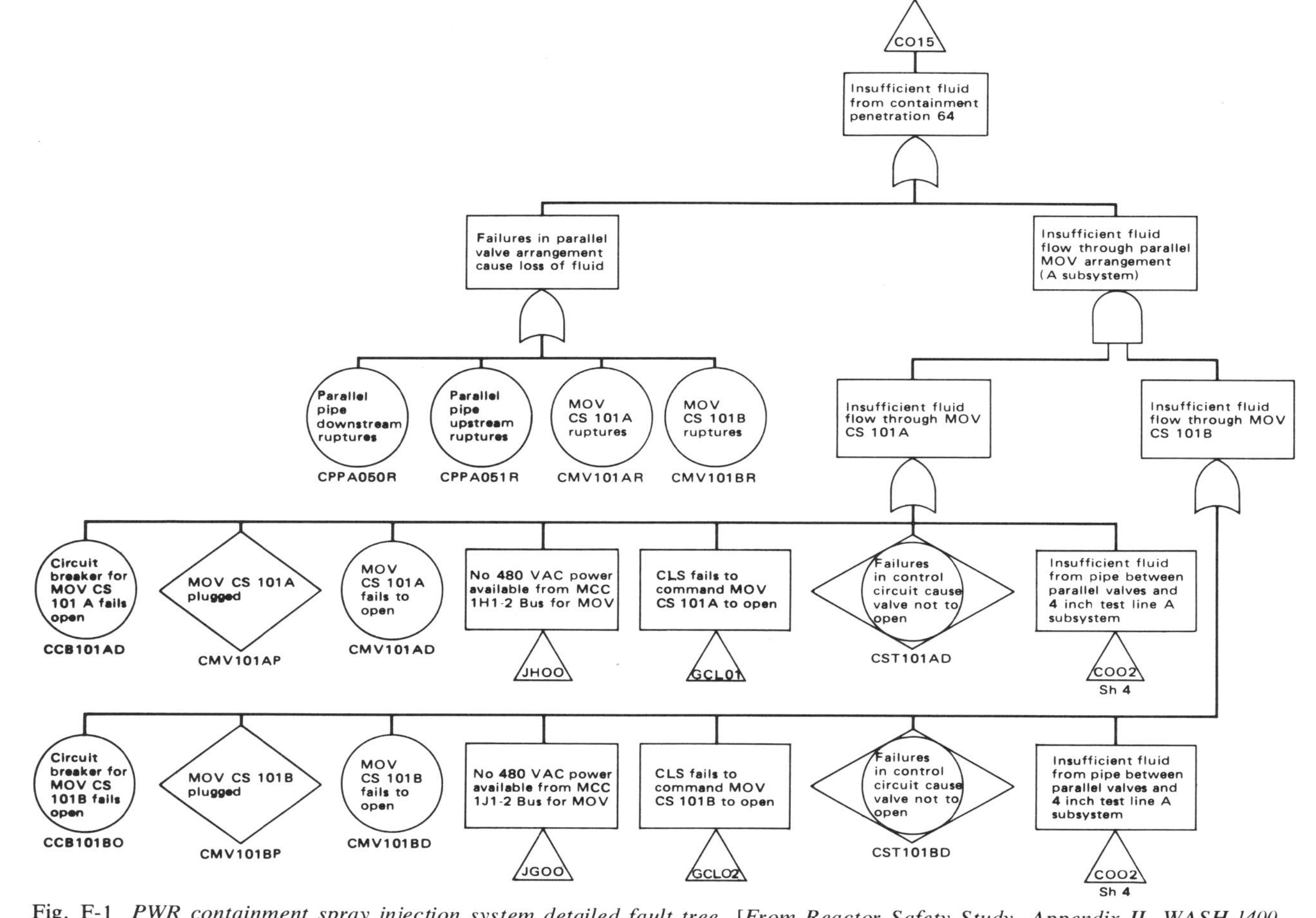

Fig. F-1 *PWR containment spray injection system detailed fault tree.* [*From Reactor Safety Study, Appendix II, WASH-1400 (October 1975).*]

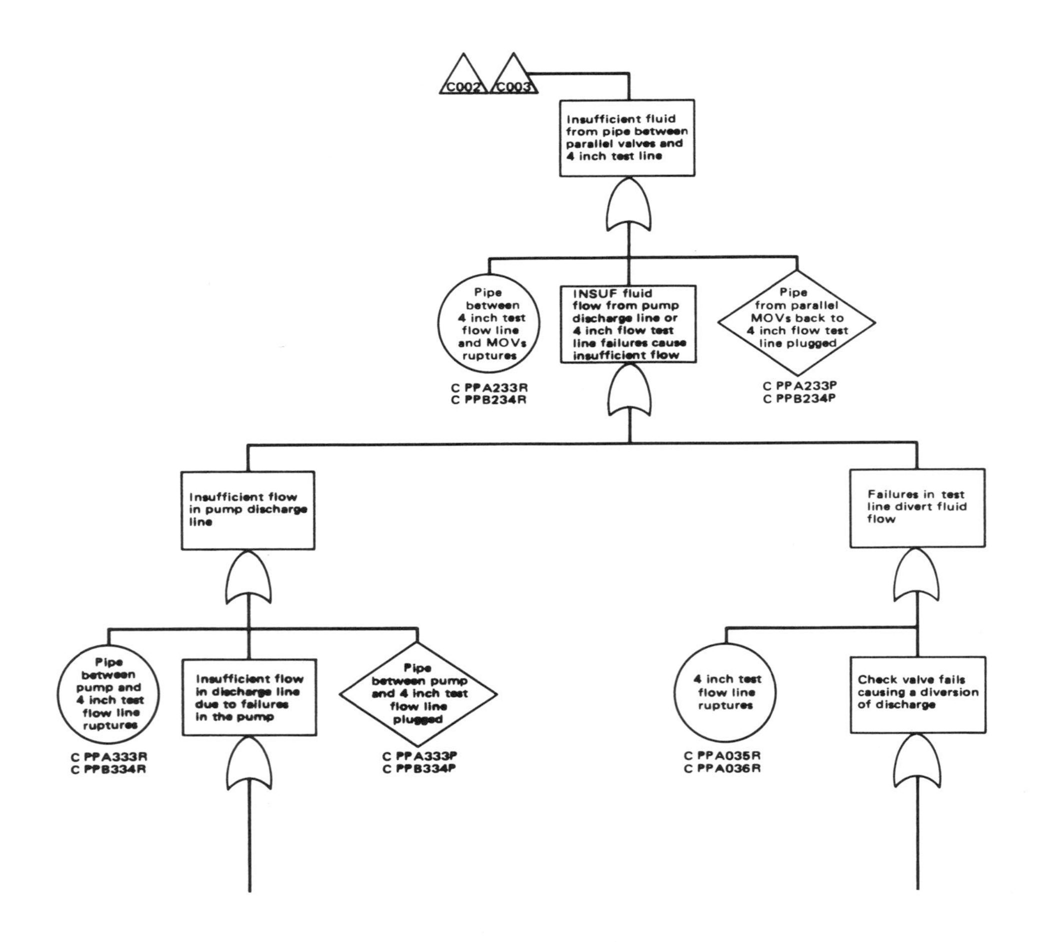
C002
C003
Insufficient fluid from pipe between parallel valves and 4 inch test line
Pipe between 4 inch test flow line and MOVs ruptures
C PPA233R
C PPB234R
INSUF fluid flow from pump discharge line or 4 inch flow test line failures cause insufficient flow
Pipe from parallel MOVs back to 4 inch flow test line plugged
C PPA233P
C PPB234P
Insufficient flow in pump discharge line
Failures in test line divert fluid flow
Pipe between pump and 4 inch test flow line ruptures
C PPA333R
C PPB334R
Insufficient flow in discharge line due to failures in the pump
Pipe between pump and 4 inch test flow line plugged
C PPA333P
C PPB334P
4 inch test flow line ruptures
C PPA035R
C PPA036R
Check valve fails causing a diversion of discharge

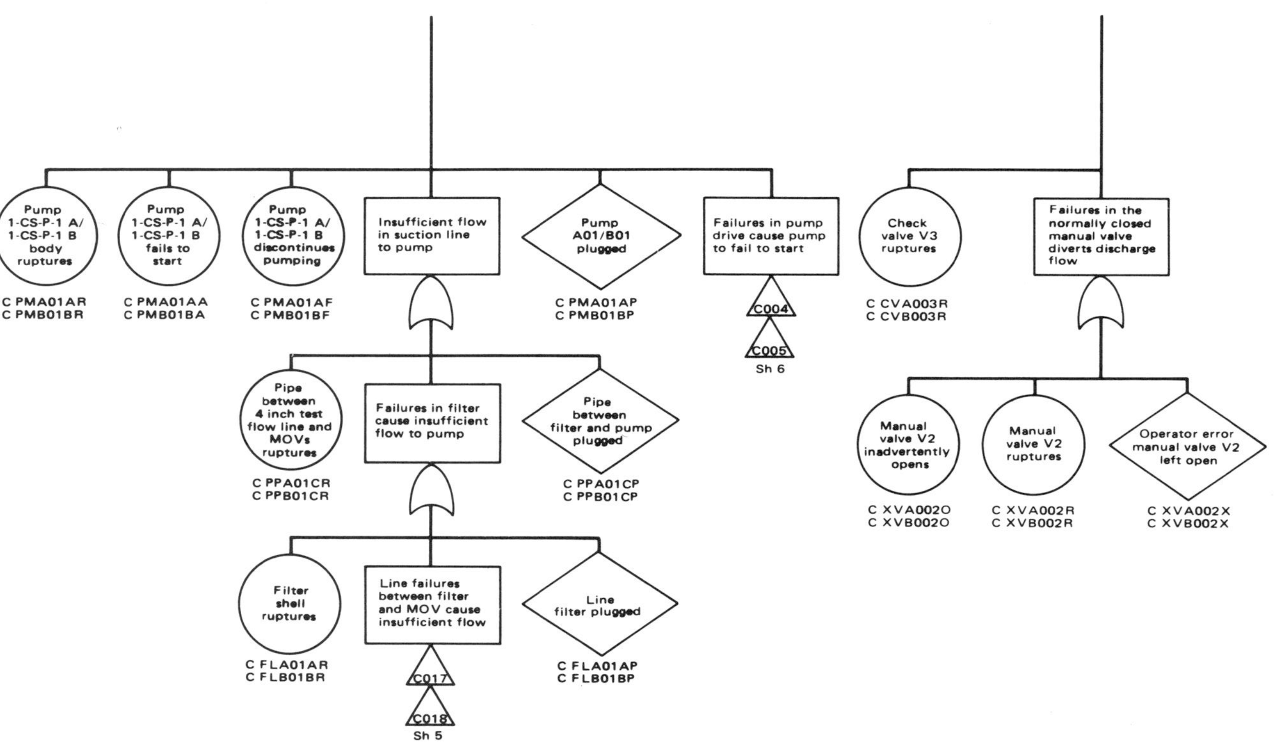

Fig. F-2 *PWR containment spray injection system detailed fault tree.* [*From Reactor Safety Study, Appendix II, WASH-1400* (*October 1975*).]

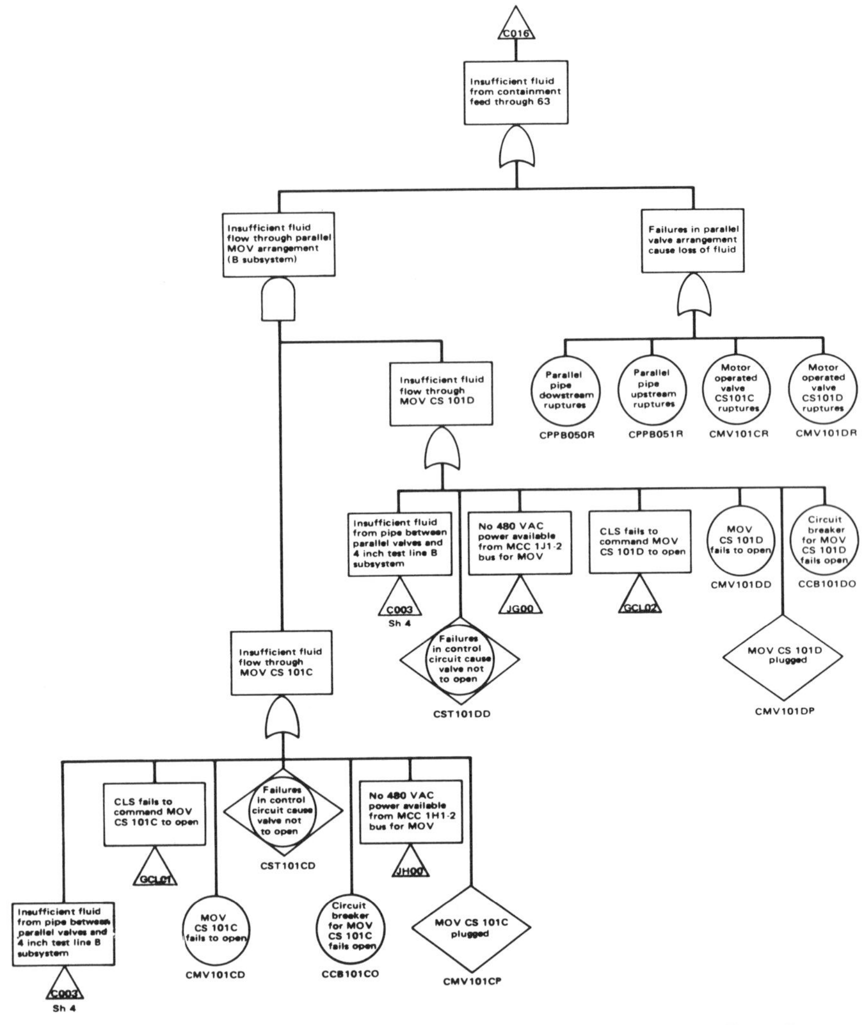

Fig. F-3 *PWR containment spray injection system detailed fault tree.* [*From Reactor Safety Study, Appendix II, WASH-1400 (October 1975).*]

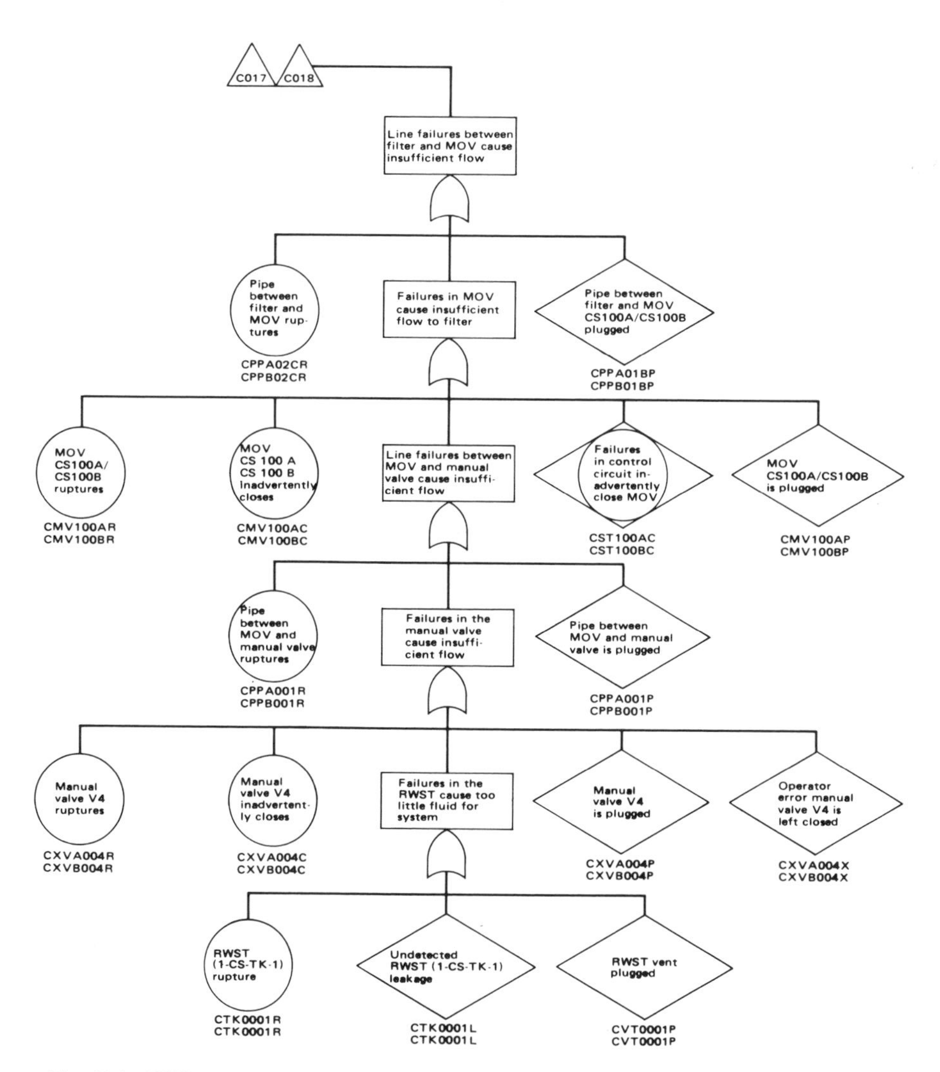

Fig. F-4 *PWR containment spray injection system detailed fault tree.* [*From Reactor Safety Study, Appendix II, WASH-1400 (October 1975).*]

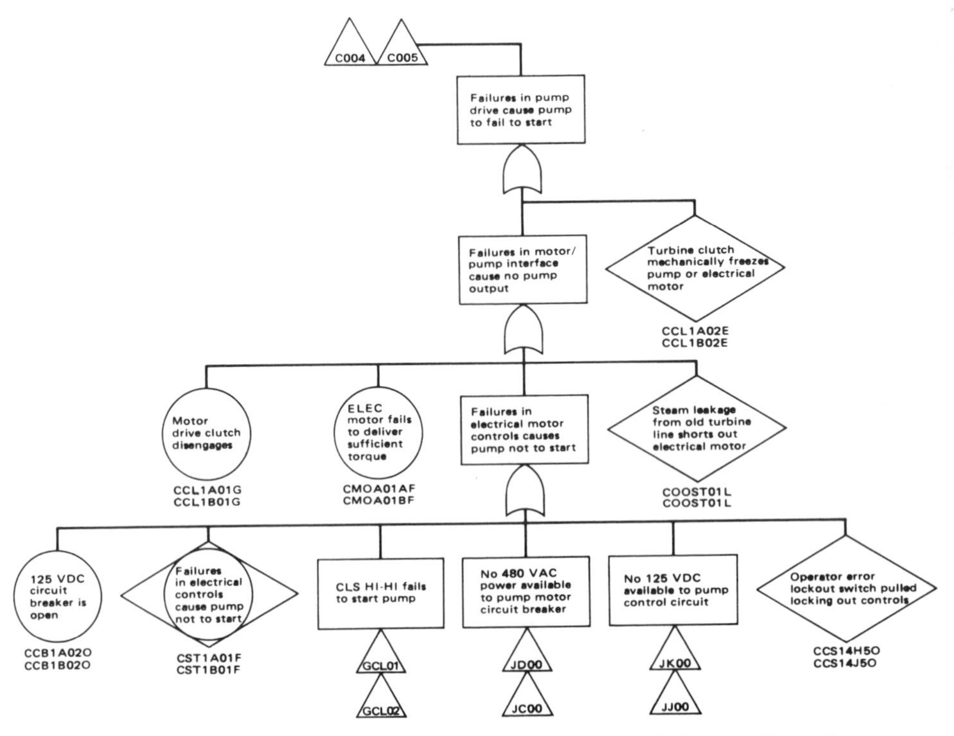

Fig. F-5 *PWR containment spray injection system detailed fault tree.* [*From Reactor Safety Study, Appendix II, WASH-1400* (*October 1975*).]

APPENDIX G

The GO Method

The GO method for evaluating the probability that a system is in one of several states was introduced in Section 10-1. Here a few more details of the construction of the GO chart will be presented, along with two examples [1–4].

The term *event* refers to the occurrence of an output from a system component, where the information content can be abstract (a logic flow) or physical (voltage, water in a pipe, pneumatic pressure, torque on a shift, mechanical force on a gear, etc.). Each event can occur in one or more *states* (numbered from 0 up to a maximum of 128), depending upon the input and the operating mode of the component. The states are frequently used to model operational sequences in which each state refers to a time of occurrence of the output event; then the first time period (state 0) represents premature operation, and the last represents failure to operate (output never occurs). Alternatively, the states can be used to model quantities of material; then the probability of various quantities of material flowing may be calculated.

A set of 16 standardized functional operators are used to model the physical components of a system. These functional operators combine the input event probability distribution with the component operating mode probabilities to produce the distribution of the output event in the various states. The set of functional operators (sometimes called *building blocks*) is shown in Fig. G-1, while their use to describe common system components is illustrated in Fig. G-2. Each functional operator is defined by a "type" number (1 through 17, with type 4 nonexistent) inside the operator symbol. The input signal numbers (S) and resulting output signal numbers (R) are used to keep track of logic flow.

The collection of specific parameters for a given operator is referred to as the "kind" data, and usually consists of probabilities. A brief description of some of the operators given in Fig. G-1 and their kind data follows.

Type 1 is a simple component with one input and two operational states (good or failed); it is used extensively to model anything from a resistor to

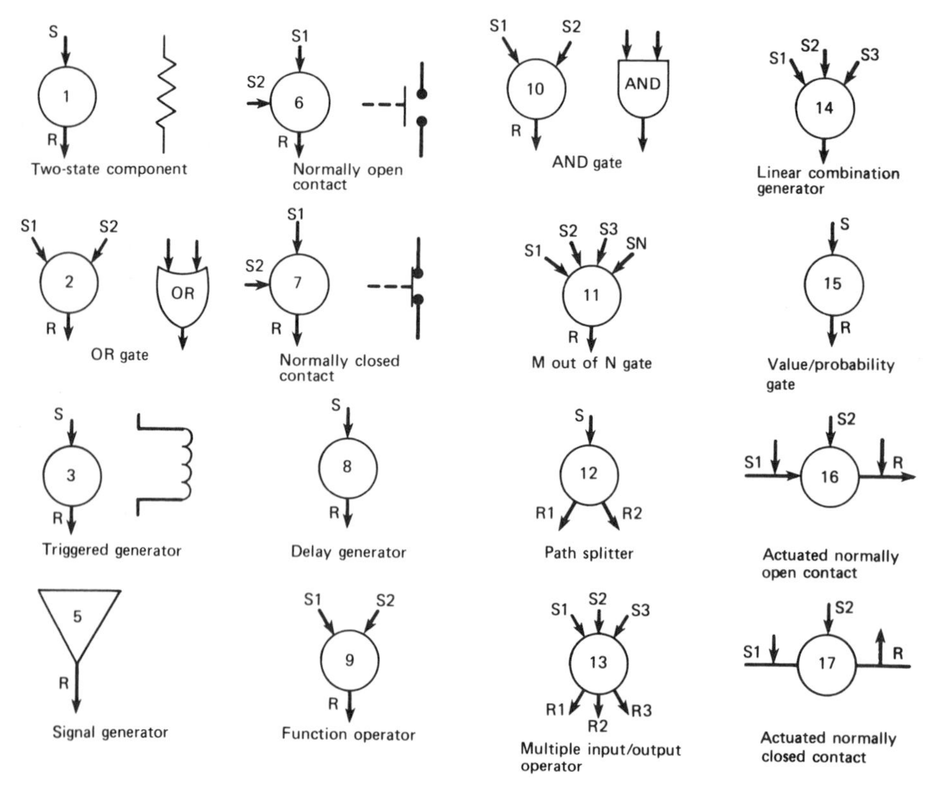

Fig. G-1 *The GO functional operators.* (*From R. L. Williams and W. Y. Gateley, GO Methodology–Overview, Rep. EPRI NP-765* (*May 1978*).]

a complete subsystem module; the kind data include P_1 (good) and P_2 (failed). The type 3 operator is a component with three operating modes: premature (output occurs with no input), good (output occurs when input occurs), and failed (no output); the kind data are P_1 (good), P_2 (failed), and P_3 (premature).

Type 2 is a logical OR gate with up to 10 inputs; the output event occurs as soon as any input occurs. This type is perfect and kind data are not used.

Type 5 is a problem initiator that has no input of its own; this type indicates the presence, absence, or distribution of inputs (electricity, water, etc.) at the start of the problem. The kind data are the probabilities of occurrence in the various time periods or the amounts of material flowing.

Type 6 is a component with a primary input S_1 and a secondary (or trigger) input S_2; output occurs when both inputs are present, but the

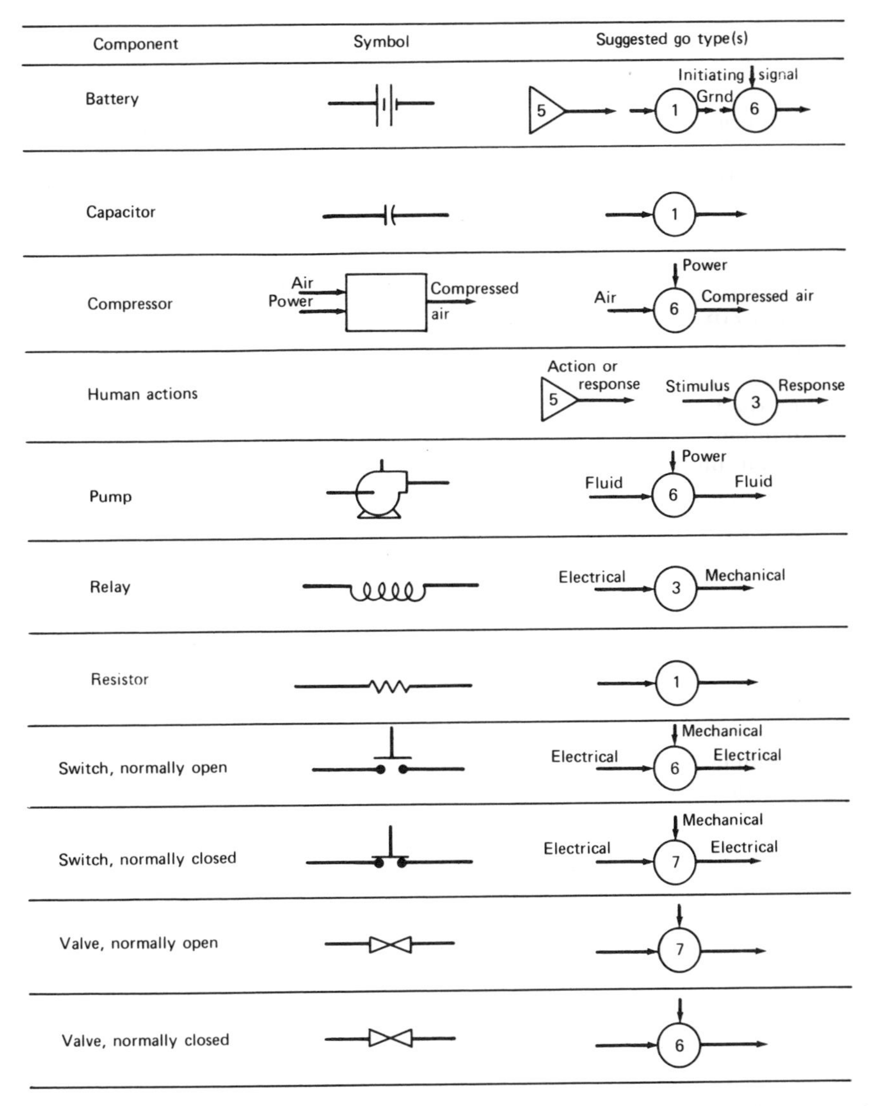

Fig. G-2 *Suggested GO representations for some common system components. [Extracted from R. L. Williams and W. Y. Gateley, GO Methodology–Overview, Rep. EPRI NP-765 (1978).]*

inputs are not interchangeable. Type 6 typically models a normally open switch or a normally closed valve; when used to model a switch, a premature mode of operation (when only S_1 is present) represents "contact shorted," while failure represents "stuck open." The kind data are P_1

(good), P_2 (failed), and P_3 (premature). The type 7 operator is similar to type 6, but is a normally closed switch or normally open valve.

Type 9 is a general-purpose, state-change operator that produces an output at a time determined by the difference between the times of S_1 and S_2. It can be used to model the more complex logical gates (ANDNOT, exclusive OR, etc.). The kind data define the operator logic.

Type 10 is a logical AND gate with up to 10 inputs; the output occurs only if all inputs are present. There are no kind data.

Once the GO chart has been constructed to model the system, and the kind numbers have been specified, the GO program is used to calculate the system unavailibility from the event probabilities.

Example G-1 Consider the simple electrical subsystem shown in Fig. G-3 with four external inputs, three relays, three normally open switch contacts, a resistor, and two final output signals. The GO chart in Fig. G-4 shows the operator type-kind representation and the sequential operator numbers that indicate the order of the analysis to be followed in the GO program. Details about how to complete the calculation are available [1]. ◇

Example G-2 Consider the alarm system of Fig. G-5 for the detection and annunciation of an event [1]. When the event occurs, it is detected by four sensors whose outputs are sent to two 2-out-of-3 logic detectors, each having its own alarm device. The system operates normally if at least one of the alarms operates when the alarm event occurs, it fails if neither

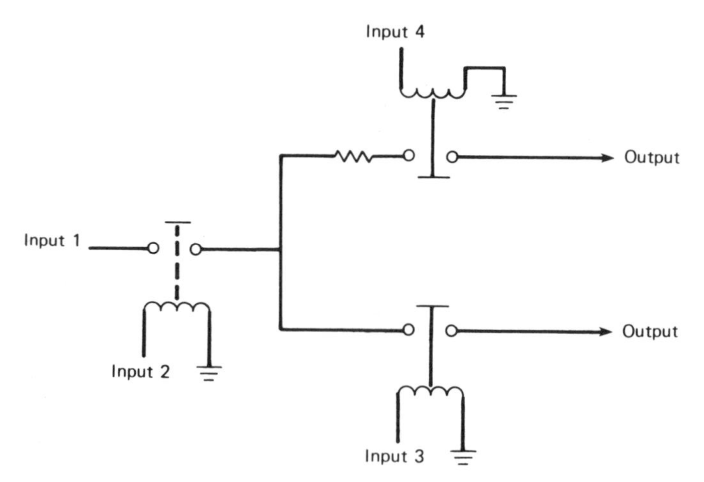

Fig. G-3 *Electrical subsystem of Example G-1. [From R. L. Williams and W. Y. Gateley, GO Methodology–Overview, Rep. EPRI NP-765 (1978).]*

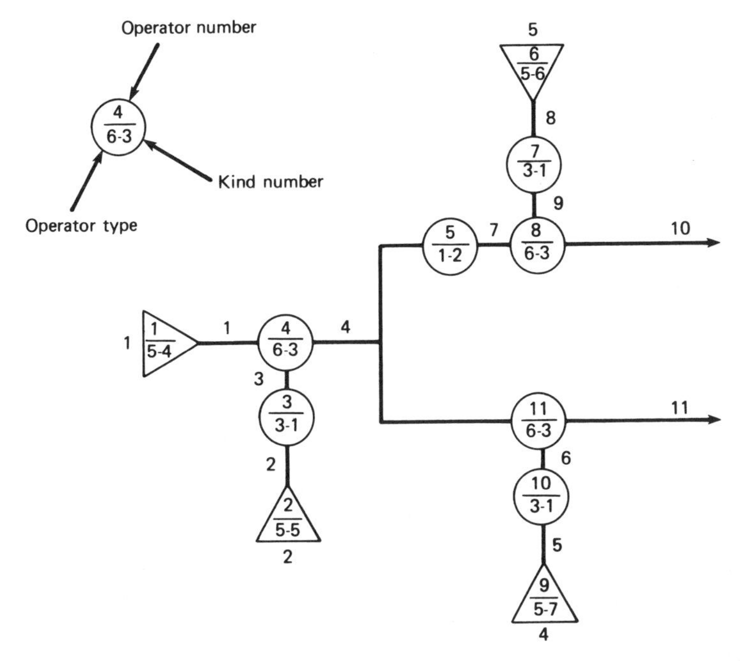

Fig. G-4 *GO chart for subsystem of Example G-1.* Operator number *assigned by processing sequence.* Operator type *determines manner of manipulating inputs and outputs.* Kind number *specifies probabilities of operation. Type 5 triangular symbols are annotated similar to circles for other types, with operator number always on top and type-kind numbers below. Numbers on line segments identify signals in order of analysis, while numbers outside triangles indicate most likely signal value (time). [From R. L. Williams and W. Y. Gateley, GO Methodology–Overview, Rep. EPRI NP-765 (1978).]*

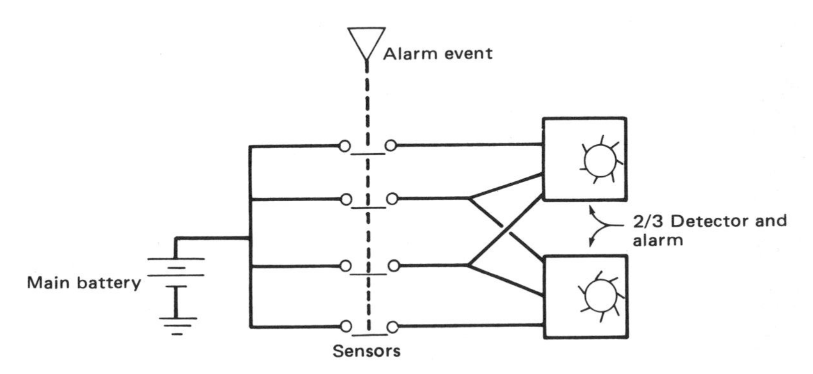

Fig. G-5 *Alarm system of Example G-2. [From R. L. Williams and W. Y. Gateley, GO Methodology–Overview, Rep. EPRI NP-765 (1978).]*

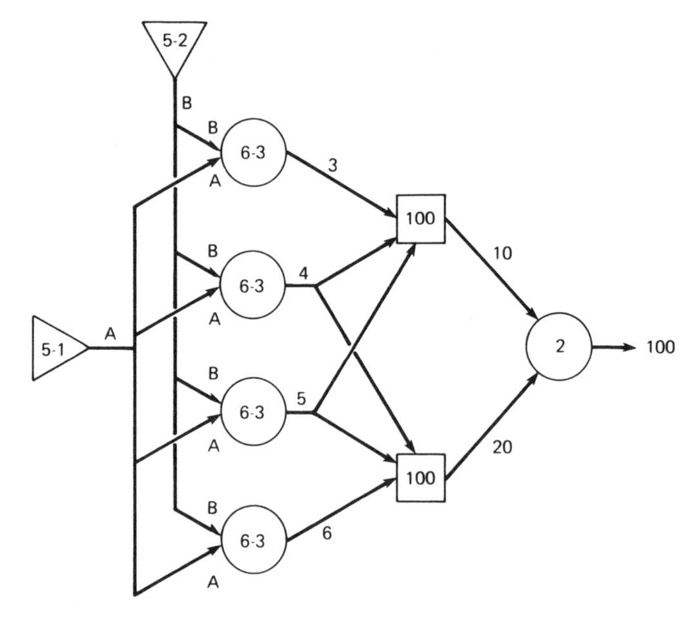

Fig. G-6 *GO chart for alarm system of Example G-2. A is primary, B is secondary.* [*From R. L. Williams and W. Y. Gateley, GO Methodology–overview, Rep. EPRI NP-765 (1978).*]

alarm operates when the alarm event occurs, and it gives a false alarm if at least one of the alarms operates before the event occurs.

The GO chart for the system is shown in Fig. G-6. The principal probability of interest in this problem is denoted in the figure as signal 100. ◇

REFERENCES

1. R. L. Williams and W. Y. Gateley, GO Methodology—Overview. Electric Power Research Institute Rep. EPRI NP-765 (May 1978).
2. W. Y. Gateley and R. L. Williams, GO Methodology—System Reliability Assessment and Computer Code Manual. Electric Power Research Institute Rep. EPRI NP-766 (May 1978).
3. Kaman Science Corporation, GO Methodology—Fault Sequence Identification and Computer Code Manual. Electric Power Research Institute Rep. EPRI NP-767 (May 1978).
4. R. L. Williams and W. Y. Gateley, Use of the GO Methodology to directly generate minimal cut sets, *in* "Nuclear Systems Reliability Engineering and Risk Assessment" (J. B. Fussell and G. R. Burdick, eds.), p. 825. Soc. Industrial Appl. Math., Philadelphia, Pennsylvania, 1977.

Answers to Selected Exercises

Chapter 2

2-2 (a) $\cdots, P(A_4|B) = 0.207, \cdots$ (b) $\cdots, P(A_4|B) = 0.170, \cdots$
2-3 (a) $\cdots, P(A_4|B) = 0.253, \cdots,$ (b) $\cdots, P(A_4|B) = 0.681, \cdots$
2-4 1753
2-5 (a) $t/(t + 2)$ (*b*) $2/(t + 2)^2$
2-6 (a) $\exp(-\lambda t)$, $t < t_1$, $\exp[-\lambda t - k(t - t_1)^2/2]$, $t \geq t_1$
2-7 (a) 177 days (b) 1770 days
2-8 (a) $2a$ (*b*) $1 - 2E_3(at)$ (*c*) $2E_3(at)$ (*d*) $aE_2(at)/E_3(at)$
(*e*) $2/3a$

Chapter 3

3-1 0.143
3-2 (a) 0.0846 (b) 0.016
3-3 (a) $\Theta = (\Theta' - \Theta_1)/(\Theta_2 - \Theta_1)$
3-4 (a) 0.09889 (b) 0.0947
3-5 0.159
3-6 (a) 0.1353 (b) 0.2706 (c) 0.2706 (d) 0.0000382
3-7 (a) 0.8187 (b) 0.8187 (c) 0.8187
3-8 (a) 0.031 (b) 0.99902
3-9 (a) 34.45 hr (b) 14.75 hr
3-10 (a) $1/\Gamma(a + 1)$ (b) $m = \sigma^2 = a + 1$
3-11 (a) 0.729 (b) 0.493 (c) 0.368 (d) 20
3-12 (a) 0.995 (b) 88,950 hr
3-13 (a) 0.81 (b) 0.039
3-14 (a) 0.0304 (b) 911.3 yr

Chapter 4

4-1 $\hat{\lambda} \approx 3.7 \times 10^{-4}$ hr^{-1}, $\hat{\tau} \approx 1000$ hr
4-2 $\hat{\alpha} \approx 0.75$, $\hat{\beta} \approx 1.6 \times 10^4$ hr
4-3 $\hat{\alpha} \approx 0.25$, $\hat{\beta} \approx 475$ hr

4-4 $\hat{\alpha} \approx 2$, $\hat{\beta} \approx 95$ days
4-5 $\hat{\alpha} \approx 1.5$, $\hat{\beta} \approx 5.7 \times 10^5$ cycles
4-6 (a) 0.7264 cm
4-7 (a) 0.824 ± 0.006 cm (b) 0.824 ± 0.008 cm
4-8 (a) 0.824 ± 0.0069 cm (b) 0.824 ± 0.0049 cm
(c) 0.824 ± 0.0089 cm
4-9 $\hat{\alpha} \approx 1.4$, $\hat{\beta} \approx 1.85 \times 10^5$ sec
4-10 $\hat{\alpha} \approx 2$, $\hat{\beta} \approx 1.5 \times 10^6$ cycles

Chapter 6

6-1 $R_1(R_2 + R_3 - R_2R_3)$
6-2 $R^n(2 - R)^n$
6-3 Show that $(2 - R)^n/(2 - R^n) > 1$
6-4 (a) $2R + 3R^2 - 12R^3 + 13R^4 - 6R^5 + R^6$
(b) $2R + R^2 - 4R^3 + R^4 + 2R^5 - R^6$
6-5 (a) 0.504 (b) 0.954
6-6 (b) 0.241 (c) 0.754
6-7 (a) 0.9995 (b) 0.9950
6-8 0.931
6-9 0.537
6-10 MTTF = 1.25×10^6 hr
6-11 (b) For (1), (2), and (3), MTTF = $11/6\lambda$, $5/6\lambda$, $3/\lambda$
6-12 (a) 0.99998 (b) 0.998
6-13 (b) $\lambda_1^{-1} + \lambda_2^{-1}$ (c) $\text{MTTF}_{\text{standby}} > \text{MTTF}_{\text{parallel}} > \text{MTTF}_{\text{series}}$
6-14 $$\frac{\lambda_2\lambda_3\, e^{-\lambda_1 t}}{(\lambda_2 - \lambda_1)(\lambda_3 - \lambda_1)} + \frac{\lambda_1\lambda_3\, e^{-\lambda_2 t}}{(\lambda_1 - \lambda_2)(\lambda_3 - \lambda_2)} + \frac{\lambda_1\lambda_2\, e^{-\lambda_3 t}}{(\lambda_1 - \lambda_3)(\lambda_2 - \lambda_3)}$$
6-18 $R_A(2R_C - R_C^2) + (1 - R_A)R_BR_C(2 - R_BR_C)$
6-21 $5R^3 - 5R^4 + R^5$
6-22 $4R^3 - 2R^4 - 2R^5 + R^6$
6-23 $R^2 + 3R^3 - 3R^4 - R^5 + R^6$
6-24 $2R^2 + 2R^3 - 5R^4 + 2R^5$
6-27 (a) *AD, BCD* (b) *AF, BEF, CE, D* (c) *AB, CD, EF*
AC is also a minimal cut set.

Chapter 7

7-1 (a) 5.15×10^4 hr (b) 1.5×10^3 hr
7-2 (d) $\sum_{n=1}^{4} (1/\lambda_n)$
7-3 (d) $\mu_1\mu_2/(\mu_1\mu_2 + \mu_1\lambda_2 + \mu_2\lambda_1)$ (e) $(\lambda_1 + \lambda_2)^{-1}$

7-4 (c) $R(t) = -(s_2e^{s_1t} - s_1e^{s_2t})/(s_1 - s_2)$
$s_1 = -[(5\lambda + \mu) + (\lambda^2 + 10\lambda\mu + \mu^2)^{1/2}]/2$
$s_2 = -[(5\lambda + \mu) - (\lambda^2 + 10\lambda\mu + \mu^2)^{1/2}]/2$
(d) $(5\lambda + \mu)/6\lambda^2$
7-5 (c) $3(\lambda t)^2$ (d) $n\mu(3\lambda + \mu)/[n\mu(3\lambda + \mu) + 6\lambda^2]$
7-6 (a) $n\mu(\lambda + \lambda^* + \lambda_c + \mu)/[n\mu(\lambda + \lambda^* + \lambda_c + \mu) + (\lambda + \lambda_c)(\lambda + \lambda^*) + \lambda_c(\lambda + \lambda_c + \mu)]$
(b) $(2\lambda + \lambda^* + \lambda_c + \mu)/[(\lambda + \lambda_c)(\lambda + \lambda^* + \lambda_c) + \mu\lambda_c]$
7-9 $1/\lambda + (\lambda + \mu)(\lambda + \mu^*)/[\lambda(\lambda + \lambda^*)(\lambda + \mu + \mu^*)]$

Chapter 8

8-3 (a) *AB, CD* (b) *AC, AD, BC, BD* (c) *A, BC*
(d) *AD, AEF, BCD, BCEF*
8-4 (c) $\approx 10^{-8}$ (f) 4×10^{-6}
8-5 (c) 2×10^{-9}
8-10 (b) 3.6×10^{-4} (c) 3.04×10^{-4}
8-11 (b) 3.22×10^{-2} (c) tank rupture

Chapter 9

9-1 (b) 1.22×10^{-4}
9-2 (c) 8.38×10^{-4}/yr
9-3 (c) 4.61×10^{-2}

Chapter 15

15-1 (b) 2.3×10^{-5}, 1.1×10^{-2}

Index

G

N

O

W

X

Z